D1230232

ADVANCES IN PROTEIN CHEMISTRY

Volume 29

CONTRIBUTORS TO THIS VOLUME

C. B. Anfinsen

Gunter Blankenhorn

Henry B. F. Dixon

Robert E. Feeney

N. Michael Green

H. A. Scheraga

Gregorio Weber

ADVANCES IN PROTEIN CHEMISTRY

EDITED BY

C. B. ANFINSEN

National Institute of Arthritis, Metabolism and Digestive Diseases
Bethesda, Maryland

JOHN T. EDSALL

Biological Laboratories
Harvard University
Cambridge, Massachusetts

FREDERIC M. RICHARDS

Department of Molecular Biophysics and Biochemistry
Yale University
New Haven, Connecticut

VOLUME 29

1975

ACADEMIC PRESS • *New York San Francisco London*

A Subsidiary of Harcourt Brace Jovanovich, Publishers

ACADEMIC PRESS, INC.
111 Fifth Avenue, New York, New York 10003

United Kingdom Edition published by
ACADEMIC PRESS, INC. (LONDON) LTD.
24/28 Oval Road, London NW1

LIBRARY OF CONGRESS CATALOG CARD NUMBER: 44-8853

ISBN 0-12-034229-4

PRINTED IN THE UNITED STATES OF AMERICA

CONTENTS

Experimental and Theoretical Aspects of Protein Folding

C. B. ANFINSEN AND H. A. SCHERAGA

CONTRIBUTORS TO VOLUME 29

Numbers in parentheses indicate the pages on which the authors' contributions begin.

C. B. ANFINSEN, *National Institute of Arthritis, Metabolism and Digestive Diseases, Bethesda, Maryland* (205)

GUNTER BLANKENHORN, *Fachbereich Biologie, Universität Konstanz, Konstanz, West Germany* (135)

HENRY B. F. DIXON, *Department of Biochemistry, University of Cambridge, Cambridge, England* (135)

ROBERT E. FEENEY, *Department of Food Science and Technology, University of California, Davis, California* (135)

N. MICHAEL GREEN, *National Institute for Medical Research, Mill Hill, London, England* (85)

H. A. SCHERAGA, *Department of Chemistry, Cornell University, Ithaca, New York* (205)

GREGORIO WEBER, *Roger Adams Laboratory, School of Chemical Sciences, University of Illinois, Urbana, Illinois* (1)

ENERGETICS OF LIGAND BINDING TO PROTEINS

By GREGORIO WEBER

Roger Adams Laboratory, School of Chemical Sciences,
University of Illinois, Urbana, Illinois

I. Introduction

The interaction of proteins with small ligands constitutes one of the more important chapters of biochemistry, and indeed of molecular biology. It embodies the fundamental physicochemical principles that are operative in the physiological regulation of enzyme activity, the specific interactions among proteins and of proteins with nucleic acids and other macromolecules. Briefly, the study of the interactions of proteins and small ligands provides the basis for the understanding of biological specificity at the molecular level.

The preoccupation with the structural origin of the interactions has lead to neglect of the energetic aspects as if these were of secondary importance or even trivial. In this review I shall attempt to show that, quite to the contrary, knowledge of the structural basis of the molecular interactions, however detailed, cannot be given an appropriate physical interpretation without the energetics. Knowledge of the average distances between the interacting groups or atoms, or their macroscopic representation in a molecular model, is insufficient, without specification of the distance–energy relations to explain the dynamics of interaction among the partners. The problem is not peculiar to macromolecules, but on account of the large number of weak, elementary interactions between a protein and a ligand made up of even a few atoms it imposes itself for consideration from the very beginning. The individual groups or residues in the protein are bound together by forces that are of the same order as those operating between proteins and ligands. As a consequence, changes in internal protein interactions cannot be separated from protein–ligand interactions, and when more than one ligand is bound the modification of the protein structure makes it impossible for the binding of the ligands to be independent in any rigorous sense. The magnitude, and therefore the significance of these effects cannot be appreciated without a detailed knowledge of the energetics. It is somewhat puzzling that in so many of the publications in which equilibria are studied by spectroscopic and other methods, it is often the case that consideration of the energetic aspects is neglected in favor of elaborate discussions on the probable structural origin of the effects, although the methods employed are suitable to cover the former aspects and quite unsuitable to provide structural information of any kind. One purpose of the present review is to present the case for the central role of the energetics in ligand–protein interactions, and the necessity of better studies to characterize it, particularly in the cases of multiple ligand binding. I shall present here a unified treatment of such multiple equilibria, and establish its consequences in the domains of protein–protein interactions, physiological regulation, and the interconversion of chemical and osmotic energies in the organism. I will try to show that in this area, as in every other area of physics and chemistry, the energetic aspects are fundamental and provide the indispensable background for the understanding of the structural and dynamic properties of the system.

Thermodynamic Characterization of the Equilibrium

We shall be concerned here with the energetics of the system formed by the protein and an indefinite number of ligands in solution. Our principal aim will be to describe the changes in free energy that occur when association of protein and ligands take place. Because the protein is itself a complex molecule, these free-energy changes have characteristics that are not encountered, to anything like the same extent, in the chemical association of smaller molecules. The simple, basic principles are, however, the same as in the simplest cases. The free energy of a system of molecules in solution can be established following principles laid down by Willard Gibbs.

At a fixed temperature and pressure, the free energy results from the addition of contributions from the various components: the protein, ligands, and solvent in our case. To each of these components, a partial free-energy or chemical potential is assigned, of the form

$$\mu_i = \mu_i^\circ + RT \ln a_i c_i \tag{1.1}$$

where a_i is the thermodynamic activity coefficient and c_i the molar concentration. For c_i sufficiently small, the activity coefficient will necessarily be unity (e.g., Lewis and Randall, 1923; Glasstone, 1947) and since fortunately this is the case for virtually all the equilibria that will interest us, the above equation can be simplified to,

$$\mu_i = \mu_i^\circ + RT \ln c_i \tag{1.2}$$

Accordingly, we shall not use activity coefficients in our discussion and will from now on use the chemical potential as the simple function of the concentration shown in Eq. (1.2). It follows from it that the chemical potential is made up of two contributions, one of which has the same value for all components with the same concentration c. The specific properties associated with the chemical reactions must therefore be contained in the term μ°, the so-called standard chemical potential. At unit concentration we must have $\mu = \mu^\circ$. It is clear then that the second term, $RT \ln c$ is simply the difference between the chemical potential of the actual component and that of a solution of unit concentration. It is in fact nothing more than the *osmotic work* required to bring the solution from its actual concentration c to the standard concentration ($c = 1$).

The change in free energy ΔF in a chemical reaction resulting from the conversion of reactants into products is given by

$$\Delta F = \sum_i \mu_i \text{ (products)} - \sum_i \mu_i \text{ (reactants)} \tag{1.3}$$

In the reaction

$$\mathrm{A + B \rightarrow C + D}$$

carried out reversibly at arbitrary, but fixed, concentrations of the reactants (A), (B), (C), (D),[1] introduction of Eq. (1.2) into the last gives:

$$\Delta F = \Delta F^\circ + RT \ln \frac{(\mathrm{C})\,(\mathrm{D})}{(\mathrm{A})\,(\mathrm{B})} \tag{1.4}$$

where ΔF° is the standard free-energy change, which in general equals

$$\Delta F^\circ = \sum_i \mu_i^\circ \text{ (products)} - \sum_i \mu_i^\circ \text{ (reactants)} \tag{1.5}$$

and in our case,

$$\Delta F^\circ = \mu_{\mathrm{D}}^\circ + \mu_{\mathrm{C}}^\circ - \mu_{\mathrm{A}}^\circ - \mu_{\mathrm{B}}^\circ \tag{1.6}$$

For a reaction carried out at the thermodynamic equilibrium concentrations of products and reactants, evidently $\Delta F = 0$, and Eq. (1.4) gives

$$\Delta F^\circ = - RT \ln \frac{[\mathrm{D}]\,[\mathrm{C}]}{[\mathrm{A}]\,[\mathrm{B}]} = - RT \ln K \tag{1.7}$$

where K is the thermodynamic equilibrium constant.

The difference in the osmotic work necessary to bring 1 mole of the reaction products and of the reactants, respectively, from the

[1] Throughout this review we shall use the convention that concentrations within parentheses ( ) are arbitrary quantities, and those within square brackets [ ] denote thermodynamic equilibrium concentrations.

standard concentration to the thermodynamic equilibrium concentrations is what may be called the "standard osmotic work" for the particular reaction.

$$\Delta F_{\mathrm{OS}} = -(RT \ln [\mathrm{C}] + RT \ln [\mathrm{D}] - RT \ln [\mathrm{A}] - RT \ln [\mathrm{B}]) \quad (1.8)$$

and according to Eq. (1.7),

$$\Delta F^\circ = \Delta F_{\mathrm{os}} \quad (1.9)$$

The change in the standard chemical potential in the reaction equals the standard osmotic work, which can provide therefore a direct experimental measure of the former. Moreover, Eq. (1.4) shows that the chemical and osmotic energies are directly coupled in the sense that a free-energy source (another chemical reaction, an electric field, radiation) that introduces an energy ΔF into the system can bring about changes in osmotic energy, and reciprocally that osmotic energy that results in $\Delta F \neq 0$ can be transformed, by suitable couplings into chemical energy capable of promoting a further chemical reaction. We shall examine these relations in detail in Section IV, where we shall show that such suitable couplings are provided by the protein molecules.

Conservation of the Standard Free Energy

In discussing multiple binding equilibria in proteins, we shall exclusively deal with ΔF°, the standard free energy change that takes place when ligands are bound to protein. Every complex formed by protein and ligands may be assigned a fixed standard chemical potential, and the standard free-energy change upon the change of one complex into another may be measured by the difference in such standard chemical potentials. In carrying out, in a reversible fashion, the reactions that convert one protein ligand complex A into another, B the standard free-energy change is

$$\Delta F^\circ (\mathrm{A} \rightarrow \mathrm{B}) = \mu_{\mathrm{B}}^\circ - \mu_{\mathrm{A}}^\circ \quad (1.10)$$

The formation of a unique molecular complex from two partners can be carried out in only one way. One can say that there is a single path from reactants to products. Ligand–protein complexes can involve more than one type of ligand, so that there are several possible *equilibrium* paths by which the reaction can be carried out, by performing the ligand additions in different order. Whatever the way in which the reactions are carried out the standard free-energy change ΔF° ($\mathrm{A} \rightarrow \mathrm{B}$) has a unique value determined by the standard chemical potentials μ_{A}° and μ_{B}°. We shall refer to this property as the *con-*

servation of the standard free energy. Evidently no new principle is involved here. It is only a case of giving a name, for the sake of convenience, to a characteristic property of thermodynamic potentials. We shall have cause to refer often to this property in the course of this review. Examples of its usefulness may be found in the derivation of Adair's equation, given in the Appendix, and in the introduction of conditional free energies and free-energy couplings in the case of multiple ligand binding by proteins, which we shall discuss next.

II. Multiple Ligand Binding by Proteins

The chemist who investigates complexes between small molecules has usually to contend with equilibria in which no more than two components, and perhaps the solvent are involved. Not so the protein chemist: Proteins are polyelectrolytes and they cannot be found in solution free from counterions. Their properties are critically dependent upon the ionic strength and specific ionic composition of the solvent, particularly the concentration of hydrogen ions. The simplest system encountered in practice is composed of the protein molecule interacting with solvent and at least two kinds of ions, besides any added specific ligand. We can well imagine the oversimplification involved in considering such system as virtually comprising one single component, "the protein" in solution, with a unique chemical potential, without reference to any internal complexity. This kind of oversimplification has been taken so much for granted that those cases which could no longer be described as resulting from a unique protein species in solution have been put in a special category of "allosteric phenomena" and *ad hoc* properties of the protein molecules invented to explain them. The distinction of such cases from those of apparent simplicity of behavior, and most of the proposed explanations, become superfluous if it is admitted from the start that we *never* deal with the binding of a single ligand to a protein, but that simultaneous, interacting equilibria are the rule when proteins are considered. We need then no apologies for introducing the simultaneous binding of two ligands as the simplest, as well as the standard, case. Those phenomena also explained on the "one protein, one ligand" hypothesis will then appear as limiting cases which will very likely become rarer as experimental methods improve, and smaller degrees of ligand–ligand interaction become detectable.

A. *Simultaneous Binding of Two Ligands. Free-Energy Coupling*

Consider the chemical reactions I to IV below,

$$\begin{aligned} &P + X \rightarrow PX && (I) \\ &P + Y \rightarrow YP && (II) \\ &PX + Y \rightarrow YPX && (III) \\ &YP + X \rightarrow YPX && (IV) \end{aligned} \qquad (2.1)$$

The standard free-energy changes of reactions I and II will be called $\Delta F°$ (X) and $\Delta F°$ (Y), respectively. Reactions III and IV correspond, respectively, to the binding of Y when X is already bound, and to the binding of X when Y is already bound. Their standard free-energy changes I shall write respectively as $\Delta F°$ (Y/X) and $\Delta F°$ (X/Y) and will call them "conditional free energies of binding" (Weber, 1972a) or simply "conditional free energies." They stand in contrast to $\Delta F°$ (X), $\Delta F°$ (Y), the unconditional free energies, which presuppose the absence of the second ligand. It will be noticed that the same final product, YPX is formed as a result of two successive equilibria corresponding to reaction I followed by III or reaction II followed by IV. We can therefore apply the principle of free-energy conservation in the form

$$\Delta F° \text{ (X)} + \Delta F° \text{ (Y/X)} = \Delta F° \text{ (Y)} + \Delta F° \text{ (X/Y)} = \Delta F° \text{ (XY)} \qquad (2.2)$$

$\Delta F°$ (XY) designates the free-energy change in the formation of YPX from the reactants Y, P, and X in their standard states. Equation (2.2) is graphically represented in Fig. 1. The free energy of the upper level is given by the sum of the standard chemical potentials of P, X, and Y. The lowest state is given by the standard chemical potential of the doubly liganded protein, YPX. The sums in Eq. (2.2) correspond to the free-energy changes depicted in Fig. 1.

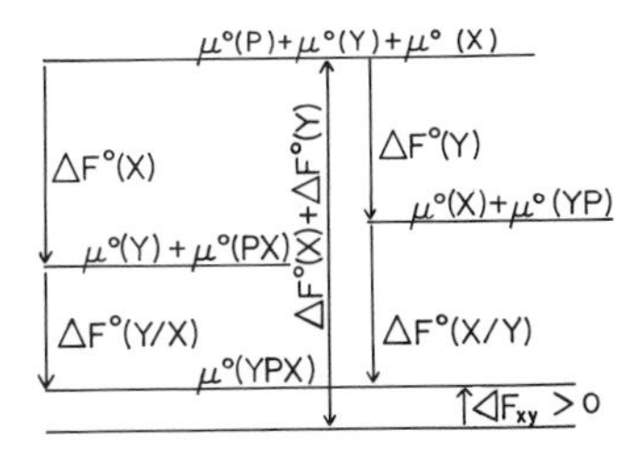

FIG. 1. Free-energy level scheme for the binding of two ligands, X and Y, to the protein P. $\Delta F°$(X), $\Delta F°$(Y): free energies of binding of X, Y to P. $\Delta F°$(X/Y), $\Delta F°$(Y/X): conditional free energies. ΔF_{xy}: coupling free energy.

There is evidently no requirement that ΔF° (X) = ΔF° (X/Y), or ΔF° (Y) = ΔF° (Y/X), but free-energy conservation imposes the condition

$$\Delta F^\circ\,(X/Y) - \Delta F^\circ\,(X) = \Delta F^\circ\,(Y/X) - \Delta F^\circ\,(Y) = \Delta F_{xy} \tag{2.3}$$

Thus the difference between the unconditional free energy of binding for X and its conditional free energy when Y is bound must be the same as the difference between the unconditional free energy of binding for Y and its conditional free energy when X is bound. In other words, the free-energy conservation condition leads directly to *reciprocity* between X and Y, in the sense that energetically the binding of X produces upon the binding of Y the same effects that the binding of Y produces upon the binding of X. The unique difference between the corresponding conditional and unconditional free energy changes given by Eq. (2.3) will be called the free-energy coupling between X and Y symbolized by ΔF_{xy}. From the last two equations,

$$\Delta F_{xy} = \Delta F^\circ\,(X/Y) - \Delta F^\circ\,(X) \tag{2.4}$$

$$\Delta F^\circ\,(X/Y) = \Delta F^\circ\,(XY) - \Delta F^\circ\,(Y) \tag{2.5}$$

and eliminating ΔF° (X/Y)

$$\Delta F_{xy} = \Delta F^\circ\,(XY) - \Delta F^\circ\,(X) - \Delta F^\circ\,(Y) \tag{2.6}$$

The free-energy coupling is therefore also the difference between the sum of the unconditional free energies of binding and the standard free energy of formation of the biliganded species from the protein and the two ligands.

In general ΔF_{xy} may be equal to zero or greater or smaller than zero. When $\Delta F_{xy} = 0$ there is no interaction between the ligands and the free energy of formation of the biliganded species equals the sum of the unconditional free energies of binding. If $\Delta F_{xy} \neq 0$ its sign is determined by the definition according to Eq. (2.3) or (2.6), which indicate that ΔF° (Y) is to be subtracted from ΔF° (Y/X), or that ΔF° (X) + ΔF° (Y) are to be subtracted from ΔF° (XY), and *not the reverse,* ΔF° (Y/X) subtracted from ΔF° (Y) or ΔF° (XY) subtracted from ΔF° (X) + ΔF° (Y). Without any loss of generality for the case of dilute solutions, it can be assumed that ΔF° (X), ΔF° (Y), ΔF° (Y/X), and ΔF° (XY) *are all negative.* It then follows that if $\Delta F_{xy} > 0$ there is *antagonism* between the ligands, which oppose each other's binding, but that if $\Delta F_{xy} < 0$ there is *cooperation* between the ligands so that they facilitate each other's binding.[2]

[2] It will be noticed that a negative free-energy coupling ($\Delta F_{xy} < 0$) corresponds to a stabilizing effect and therefore could be spoken of as a "positive interaction" among

1. Chemical Interpretation of the Free-Energy Coupling

Consider the change in standard chemical potential in the conditional and unconditional bindings (reactions III and I, respectively)

$$\Delta F^\circ\,(\mathrm{X}) = \mu^\circ\,(\mathrm{PX}) - \mu^\circ\,(\mathrm{P}) - \mu^\circ\,(\mathrm{X}) \tag{2.7}$$

$$\Delta F^\circ\,(\mathrm{X/Y}) = \mu^\circ\,(\mathrm{YPX}) - \mu^\circ\,(\mathrm{YP}) - \mu^\circ\,(\mathrm{X}) \tag{2.8}$$

and therefore, from 2.3

$$\Delta F_{\mathrm{xy}} = \mu^\circ\,(\mathrm{YPX}) + \mu^\circ\,(\mathrm{P}) - \mu^\circ\,(\mathrm{YP}) - \mu^\circ\,(\mathrm{PX}) \tag{2.9}$$

From the last equation it follows that ΔF_{xy} is the standard free energy change in the reaction

$$\mathrm{PX} + \mathrm{YP} \rightarrow \mathrm{P} + \mathrm{YPX}$$

the equilibrium constant of which is given by

$$K_{\mathrm{xy}} = \frac{[\mathrm{YPX}][\mathrm{P}]}{[\mathrm{YP}][\mathrm{PX}]} = \exp(-\Delta F_{\mathrm{xy}}/RT) \tag{2.10}$$

If $\Delta F_{\mathrm{xy}} < 0$, $K_{\mathrm{xy}} > 1$ and YPX and P predominate over PX and YP. The binding of the two ligands being cooperative, Y and X are more likely to be bound to the same molecule than occupying separate molecules of protein. Precisely the opposite situation occurs if $\Delta F_{\mathrm{xy}} > 0$. In this case PX and YP predominate over YPX and P: as a result of antagonism the ligands are more likely to be found in different protein molecules. To compare the distributions obtaining as a function of ΔF_{xy}, we consider in greater detail the situation that arises when the concentrations of the free ligands, [X] and [Y] are so adjusted as to produce occupancy of half the sites that they can potentially fill. Then the overall saturations for X and Y, which we shall call respectively S_{x} and S_{y}, will equal $\frac{1}{2}$. If the total protein concentration is P_0

$$S_{\mathrm{x}} = ([\mathrm{YPX}] + [\mathrm{PX}])/\mathrm{P}_0 = \tfrac{1}{2} \tag{2.11}$$

$$S_{\mathrm{y}} = ([\mathrm{YPX}] + [\mathrm{YP}])/\mathrm{P}_0 = \tfrac{1}{2} \tag{2.12}$$

$$[\mathrm{YPX}]/\mathrm{P}_0 = S_{\mathrm{xy}} \tag{2.13}$$

where S_{xy} is the fraction of the protein molecules bound to both ligands. Using the last three equations, Eq. (2.10) becomes

---

ligands. Conversely a positive free-energy coupling ($\Delta F_{\mathrm{xy}} > 0$) corresponds to a destabilizing effect and could be referred to as a "negative interaction." This language difficulty seems inevitable if one is to preserve the convention, firmly established, of considering negative free energies as stabilizing the system and positive free energies as destabilizing it.

$$K_{xy} = \frac{S_{xy}^2}{(\frac{1}{2} - S_{xy})^2}; \qquad S_{xy} = \frac{1}{2}\left[\frac{(K_{xy})^{1/2}}{1 + (K_{xy})^{1/2}}\right] \tag{2.14}$$

Thus for $K_{xy} = 1$, $S_{xy} = \frac{1}{4}$ as expected for the random distribution which evidently must be reached in this case since each ligand binds independently of the other. If $K_{xy} \neq 1$ the ligand molecules are no longer distributed at random among the protein molecules. From (2.14) we have

$$\frac{\Delta F_{xy}}{4.6RT} = \log \frac{2S_{xy}}{1 - 2S_{xy}} \tag{2.15}$$

Equation (2.15) plots like an ordinary titration curve (Fig. 2). The free energy coupling of zero corresponds to $S_{xy} = \frac{1}{4}$. For values of S_{xy} symmetrically disposed about $\frac{1}{4}$, ΔF_{xy} has equal absolute values, positive and negative, and becomes infinite for $S_{xy} = 0$ and $S_{xy} = \frac{1}{2}$. If $\Delta F_{xy} = \pm 2.76$ kcal and $T = 300°$K, the values of S_{xy} are 0.045 and 0.455, respectively. Thus for a free-energy coupling with an absolute value of 2.76 kcal, a good fraction of a "strong" binding energy—calling "strong," a binding energy in the region of 10 kcal—we can obtain a good, though not yet perfect, correlation between the binding of both ligands, X and Y. In the case of negative interactions ($\Delta F_{xy} > 0$) they are bound to different protein molecules in

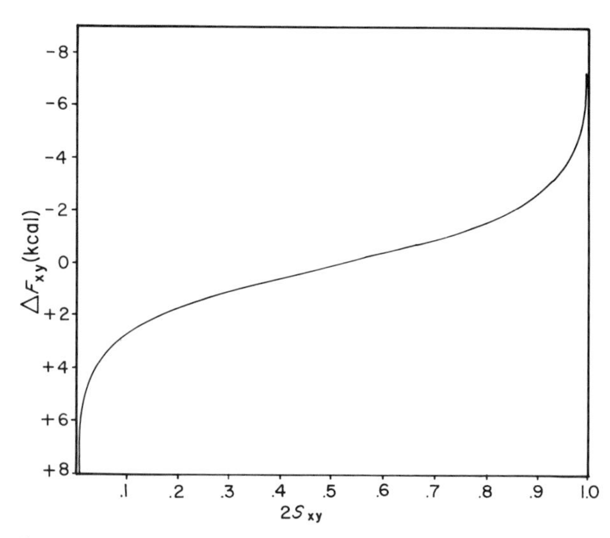

FIG. 2. Relation between degree of double saturation S_{xy} and free-energy coupling ΔF_{xy} of the bound ligands, when the degree of saturation of the ligands (S_x,S_y) equals 1/2.

90% of the total cases, while for a positive interaction ($\Delta F_{xy} < 0$) they find themselves bound to the same protein molecule in similar proportion. Figure 2 shows that, for a value of ΔF_{xy} of approximately 2.5 kcal, the plot of S_{xy} against ΔF_{xy} shows almost an inflection, so that beyond this value of S_{xy} the free energy required to enhance the binding correlation increases steeply. For 99% correlation $\Delta F_{xy} = \pm 5.52$ kcal, etc. We encounter here for the first time a situation that presents itself regularly in considering the biological properties of proteins. Certain functions of proteins are preferentially associated with the singly or the doubly liganded species. In these cases the perfection of the correlation between the binding of both ligands, given by Eq. (2.15) will be directly related to the efficiency of the biological function in question. A modest free-energy coupling of the magnitude just discussed, that is 1–2 kcal, will be required to achieve what may be considered a satisfactory correlation, but a *perfect* correlation could be achieved only by an *infinite* free-energy coupling. Very often the chemist or physical-chemist, in an attempt to visualize the effects postulates a "mechanical" molecular model which virtually ensures a perfect correlation, either negative or positive between the binding of the two ligands. Evidently such perfect mechanical contrivances are ruled out by the infinite free-energy couplings that they theoretically require, as well as by the very modest values of ΔF_{xy} encountered in practice. It is remarkable that this far-reaching consequence as regards the unsuitability of mechanical models of molecular behavior may be deduced, without any kind of doubt, from the very simple properties of the chemical potentials contained in Eq. (1.1).

2. *Determination and Magnitude of Free-Energy Couplings*

To the free energies ΔF° (X) and ΔF° (X/Y) correspond dissociation constants K(X) and K(X/Y) which can be experimentally determined by a study of the degree of dissociation of PX as a function of free-ligand concentration. K(X) is determined in the absence of Y and K(X/Y) is determined in the presence of saturating concentrations of Y. The saturation S_x observed is given in general by the equation (Weber, 1972a)

$$S_x = \frac{[X]}{[X] + K(X)K_{yx}\left[\dfrac{(\epsilon + 1)}{(\epsilon + K_{xy})}\right]} \tag{2.16}$$

with $\epsilon = [Y]/K(Y)$, and a similar equation for S_y. If [Y] is kept constant, the last equation may be written,

$$S_X = \frac{[X]}{([X] + C)} \tag{2.17}$$

where the apparent dissociation constant of the PX complex is

$$C = K_{xy}\, K(X) \left[\frac{(\epsilon + 1)}{(\epsilon + K_{xy})}\right] \tag{2.18}$$

Evidently for $[Y] \ll K(Y)$ we obtain $C = K(X)$, and for $[Y] \gg K(Y)$, $C \simeq K(X)$. $K_{xy} = K(X/Y)$, with intermediate values of C for Y comprised within the stated range. The experimental titration curves, obtained by plotting log [X] against S_X (see Appendix, Section A, and Fig. 3) will show no indication by their shape that any interaction between X and Y exist. In actual titration experiments the total concentration of Y is kept constant, while that of X is varied. In these cases the free concentration [Y], and therefore ϵ, may be considered

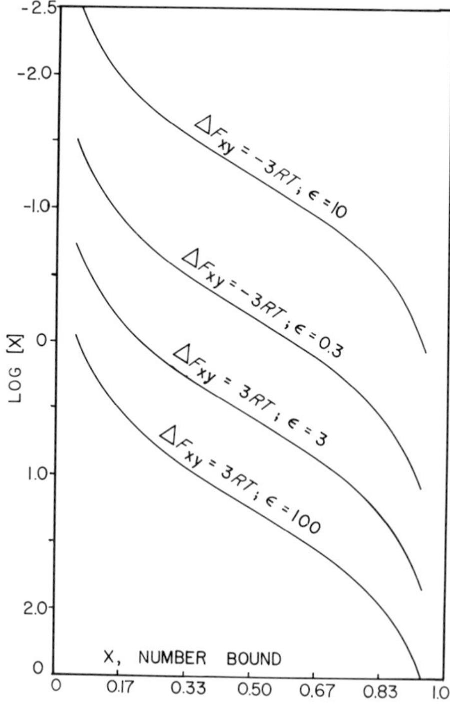

FIG. 3. Titration curves of P with X (Bjerrum plots) for positive free energy couplings ($\Delta F_{xy} = 3RT$) and negative free-energy coupling ($\Delta F_{xy} = -3RT$) and values of Y comparable to $K(Y)$ ($\epsilon = 3,0.3$) or much larger than $K(Y)$ ($\epsilon = 100,10$). For $\Delta F_{xy} = 0$ the midpoint of the titration curve is at log $[X] = 0$.

practically constant only if the protein concentration is negligible in comparison with that of Y, so that formation or dissociation of the YP complex does not change appreciably the value of [Y]. Antagonism of the ligands will be indicated by a displacement of the titration curve toward higher concentrations of [X] and will, in this respect be indistinguishable from a simple competitive effect in which X and Y vie for the same binding site (Fig. 3). The experimental distinction is, however, very simple. In case of a true competitive effect between X and Y,

$$S_x = \frac{[X]}{[X] + K(X)(1 + \epsilon)} \tag{2.19}$$

Evidently the dissociation constant for X increases *without limit* as [Y] is increased, while in the case of a finite free-energy coupling it tends to a fixed value $K(X)K_{xy}$ as [Y] increases indefinitely. In the case of cooperative action in the binding of X and Y, ($\Delta F_{xy} < 0$) there is no possibility of confusion: As [Y] increases the apparent dissociation constant for X decreases, tending to a limit when $[Y] \gg K(Y/X)$ (Fig. 3).

3. *Free-Energy Couplings and the Linkage Concept*

Wyman has shown in a series of classical papers (1948, 1964, 1965) how the reciprocity of the effects between bound ligands may be derived from the notion of the chemical potential. For any two ligands, X and Y, Wyman reciprocity relations may be written

$$\frac{d\mu_x}{dS_y} = \frac{d\mu_y}{dS_x} \tag{2.20}$$

or

$$\frac{d \ln [X]}{dS_y} = \frac{d \ln [Y]}{dS_x} \tag{2.21}$$

Relations (2.20) and (2.21) convey information similar to that of relations (2.3), (2.6), and (2.10), which define the free-energy coupling between the ligands. Wyman's equations are readily derived from the concept of a thermodynamic potential, whereas the free-energy coupling concept is more of an intuitive chemical notion. For this reason, it is easier to grasp and to follow through the actual experimental behavior. Additional advantages reside in a simple graphical representation of the effects (Figs. 1, 4, 13, and 19) and in suggesting a direct way of measuring the energetic coupling between linked binding processes.

4. *Literature Values of Free-Energy Couplings*

Relatively few acceptable data are found in the literature for this quantity. The deficiency is due undoubtedly to the lack of an appropriate formalism like the one described above, to measure the interactions between the ligands. But it is also due in good measure to the orientation of much research on the subject toward proving, or disproving, structural theories of binding rather than toward achieving a satisfactory description of the energetics. The few values of Table I are sufficient to show that free-energy couplings between the ligands both positive and negative have been observed, and that they are all small, in the neighborhood of 1 kcal or less in all cases. The explanation for the small values must be looked for, without doubt, in the general features of the internal interactions in proteins, as we shall later discuss.

Apart from the Bohr effect (antagonism of H^+ and oxygen binding in hemoglobin) few cases have been reported in which, by the measurement of the four constants $K(X)$, $K(Y)$, $K(X/Y)$, $K(Y/X)$, the reciprocity of the effects has been demonstrated. Mildvan (1972) has shown the reciprocity of effects between PEP and K^+ in pyruvate kinase. Benesch *et al.* (1968) showed for the first time the reciprocity of effects of oxygen and DPG in hemoglobin.

B. *Multiple Interactions of Two Kinds of Ligands*

To maintain the simplicity and directness of the exposition of free energy couplings we postulated [Eq. (2.1)] that only 1 mole of X and

TABLE I
Free-Energy Coupling between Ligands

Protein	Ligand couple[a]	ΔF_{xy}(kcal)	Reference
Hemoglobin	Oxygen, 2,3 DPG	+1.3	Tyuma *et al.* (1971a)
Hemoglobin	Oxygen, IHP	+2.3	Tyuma *et al.* (1971b)
Bovine serum albumin	ANS, 3,5-dihydroxybenzoate	+1.5	Kolb and Weber (1972)
Pyruvate kinase	Phosphoenol pyruvate, K^+	−1.2	Mildvan (1972)
Pyruvate kinase	K^+, Mn^{2+}	−1.4	Mildvan (1972)
Pyruvate kinase	Phenylalanine, Mn (2+)	+0.8	Kayne and Price (1972)
Aspartate transcarbamilase	CTP, succinate	+0.5	Changeux *et al.* (1968)
Chicken heart lactate dehydrogenase	NADH, oxalate	−1.5	Kolb (1974)

[a] IHP, inositol hexaphosphate; CTP, cytidine triphosphate; DPG, 2,3-diphosphoglycerate; ANS, 1-anilinonaphthalene 8-sulfonate.

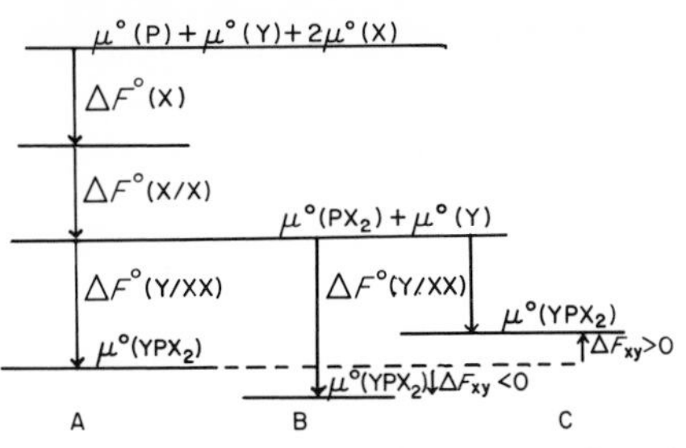

FIG. 4. Free-energy level scheme for binding of 2 moles of X and 1 mole of Y by the protein P. A: independent ligands; B: "positive" X-Y interactions ($\Delta F_{xy} < 0$); C: "negative" X-Y interactions ($\Delta F_{xy} > 0$).

1 mole of Y were bound per mole of protein. An interesting new aspect is seen on considering the binding of several moles of X and Y to 1 mole of protein, *if there are multiple interactions between them,* so that binding of one ligand affects two or more ligands of a different type. The simplest case of this type would be represented by a molecule that binds a single molecule of Y and two of X. The analysis of this case is facilitated by the diagram of the standard chemical potentials, or free-energy levels, shown in Fig. 4. As depicted on the figure, binding of the Y ligand affects equally both X ligands, so that ΔF_{xy} is the same for both interactions. This restriction is not necessary, and possibly this case is less likely in practice than the general case of multiple, unequal interactions. However, it lends itself best to an understanding of the effects. The equilibria and the dissociation constants[3] involved are shown in the accompanying scheme:

$$\begin{array}{ccccc} & Y + P & \overset{K(Y)}{\rightleftharpoons} & YP & \\ K(X) & \upharpoonleft\downharpoonright & & \downharpoonleft\upharpoonright & K(X)\cdot\beta \\ & Y + PX & & YPX & \\ K(X) & \upharpoonleft\downharpoonright & & \downharpoonleft\upharpoonright & K(X)\cdot\beta \\ & Y + PX_2 & & YPX_2 & \end{array}$$

where $\beta = \exp - (\Delta F_{xy}/RT) = K_{xy}$.

The saturation for X is given by

$$S_x = \frac{[PX] + [YPX] + 2[PX_2] + 2[YPX_2]}{2[[P] + [YP] + [PX] + [YPX] + [PX_2] + [YPX_2]]} \qquad (2.22)$$

[3] The equilibrium constants $K(Y/X)$ and $K(Y/XX)$ for the binding of Y to PX and PX_2, respectively, are determined by those shown through the principle of conservation of the standard free energy [Eq. (2.3)].

substituting the equilibrium concentrations by their value as function of the dissociation constants, we find

$$S_X = \frac{\dfrac{[X]}{K(X)}\left(\dfrac{1+\epsilon\beta^{-1}}{1+\epsilon}\right) + \dfrac{[X]^2}{K(X)^2}\left(\dfrac{1+\epsilon\beta^{-2}}{1+\epsilon}\right)}{1+\dfrac{2[X]}{K(X)}\left(\dfrac{1+\epsilon\beta^{-1}}{1+\epsilon}\right) + \dfrac{[X]^2}{K(X)^2}\left(\dfrac{1+\epsilon\beta^{-2}}{1+\epsilon}\right)} \tag{2.23}$$

The last equation has the form that corresponds to the binding of 2 moles of X with dissociation constants,

$$K_1(X) = K(X)\left[\frac{(1+\epsilon)}{(1+\epsilon\beta^{-1})}\right]$$

$$K_2(X) = K(X)\left[\frac{(1+\epsilon\beta^{-1})}{(1+\epsilon\beta^{-2})}\right] \tag{2.24}$$

Evidently for $\beta = 1$, S_X reduces to the simple, independent binding of 2 moles of X for any value of ϵ, namely

$$S_X = \frac{[X]}{[X] + K(X)} \tag{2.25}$$

If there is interaction between X and Y ($\beta \neq 1$) the ratio of the difference $K_1(X) - K_2(X)$ to the unconditional dissociation constant for X, $K(X)$ equals

$$\frac{K_1(X) - K_2(X)}{K(X)} = \frac{\epsilon(1-\beta^{-1})^2}{(1+\epsilon\beta^{-1})(1+\epsilon\beta^{-2})} \tag{2.26}$$

Both numerator and denominator of (2.26) are positive, whether β is larger or smaller than 1. It follows that *always* $K_1(X) > K_2(X)$ so that whether the effects between the ligands X and Y are positive or negative, they must give rise to an apparent cooperative binding of X. We examine now how such a "forced cooperativity" of X binding arises. If ϵ is made sufficiently large so that $\epsilon\beta^{-1} \gg 1$, then

$$K_1(X) = K_2(X) = \beta \cdot K(X) \tag{2.27}$$

and

$$S_X = \frac{[X]}{([X] + \beta K(X))} \tag{2.28}$$

so that cooperativity disappears if either $\epsilon\beta^{-1} \ll 1$ or $\epsilon\beta^{-1} \gg 1$. Considering ϵ as a parameter, the value of β at which $(K_1(X) - K_2(X))/K(X)$ reaches a maximum may be obtained by differentiating the right-hand side of Eq. (2.26) with respect to ϵ and setting the derivative to zero. It is then found that maximum difference between the con-

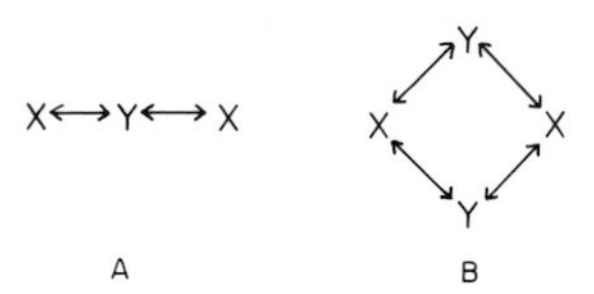

FIG. 5. Two modes of homogeneous couplings of ligands (X and Y) discussed in the text. All couplings shown are assumed of equal value.

stants $K_1(\mathrm{X})$ and $K_2(\mathrm{X})$ occurs when

$$\epsilon^2 = \beta^3$$

or

$$\epsilon\beta^{-3/2} = 1 \tag{2.29}$$

On introduction of Eq. (2.29) it is found that

$$\frac{K_1(\mathrm{X})}{K_2(\mathrm{X})} = \frac{(1 + \beta^{3/2})(1 + \beta^{-1/2})}{(1 + \beta^{1/2})^2} \tag{2.30}$$

According to Eq. (2.30)

$$\begin{aligned} &\text{if } \beta \gg 1 \qquad K_1(\mathrm{X})/K_2(\mathrm{X}) = \beta^{1/2} \\ &\text{if } \beta \ll 1 \qquad K_1(\mathrm{X})/K_2(\mathrm{X}) = \beta^{-1/2} \end{aligned} \tag{2.31}$$

Therefore the maximum cooperativity is equivalent for any value of β and its reciprocal. The free-energy couplings between X and Y that give rise to the effects described by Eqs. (2.22) to (2.30) are those shown in Fig. 5A. Figure 5B shows a more complex case, in which 2 moles of X and 2 moles of Y are coupled symmetrically and

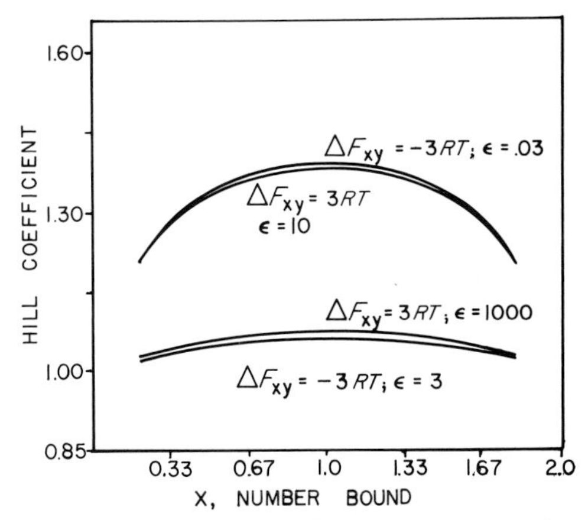

FIG. 6. The Hill coefficient (see Appendix, Section D) as a function of X-number bound, for the couplings of Fig. 5A. Note similarities for equal negative and positive couplings at optimal values of ϵ.

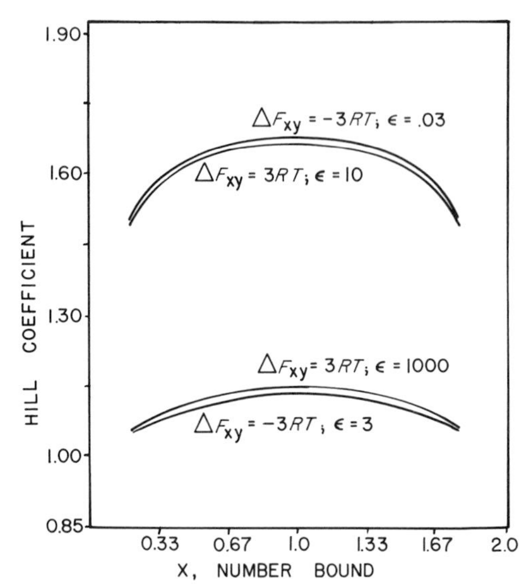

FIG. 7. Similar to Fig. 6, for the energy couplings shown in Fig. 5B. Note the increased cooperativity resulting from the larger numbers of couplings between X and Y.

homogeneously. The analysis of this latter case shows that the binding of X is characterized by two apparent dissociation constants

$$K_1(\mathrm{X}) = K(\mathrm{X})\left[\frac{(1+\epsilon)}{(1+\epsilon\beta^{-1})}\right]^2$$
$$K_2(\mathrm{X}) = K(\mathrm{X})\left[\frac{(1+\epsilon)}{(1+\epsilon\beta^{-2})}\right]^2 \qquad (2.32)$$

so that the ratio $K_1(\mathrm{X})/K_2(\mathrm{X})$ can reach a value which is the square of that obtaining for the case of Fig. 5A.

These effects of the Y concentration upon the cooperativity are shown in Figs. 6 and 7 by means of plots of the Hill coefficient (see Appendix, Section D) against the degree of X-saturation in the different cases.

1. The Origin of Cooperativity. Presence in Single-Chain Proteins

We have made a formal examination of the origin of the cooperativity, but this can also be seen intuitively: In the case of antagonism between X and Y we shall suppose that we fix the concentration of the latter, so that there will be almost, though not quite, complete saturation of its site in the absence of X. Such a concentration will be intermediate between $K(\mathrm{Y})$ and $K(\mathrm{Y/X})$. Upon addition of X, no appreciable binding of this ligand will be reached until [X] is close to $K(X/Y)/10$, and as X becomes bound it will result in a release of Y since its free concentration is less than $K(Y/X)$. The second site for X will be now more readily filled since the second constant, $K_2(X)$ will roughly correspond to $K(X)$ rather than $K(Y/X)$, as was the case for the first mole of X bound. The cooperativity resulting from positive X–Y interactions is similarly explained. In this case $K(Y/X) < K(Y)$ and a concentration of Y intermediate between these two will then be insufficient to produce appreciable saturation in the absence of X. On [X] reaching a value close to $K(X/Y)$ simultaneous binding of both ligands will readily result in the addition of the second mole of X. It follows from the description that cooperativity due to the presence of the effector Y would disappear if its concentration is kept greater than $K(Y/X)$, and that in these cases no change in Y-saturation will take place upon addition of X. Thus cooperativity requires a change in Y-saturation upon addition of X, this change resulting from breakdown of the YP complex (if $\Delta F_{xy} > 0$) or formation of it (if $\Delta F_{xy} < 0$). A number of examples in the literature are readily explained on these principles. Among them is the observation of Tyuma *et al.* (1971a) that addition of 2,3-diphosphoglycerate (DPG), which is antagonistic to oxygen in the O_2–hemoglobin equilibrium, produces an enhancement in the cooperativity which is reflected in a change in the Hill coefficient from 2.5 to 3. Such an effect must be dependent upon DPG concentration (Tomita and Riggs, 1971; Weber, 1972a), or rather upon the ratio of the concentration of DPG to the dissociation constant of the hemoglobin–DPG complex. Replacement of DPG by a similar concentration of the more strongly bound inositol hexaphosphate (IHP) results according to Tyuma *et al.* (1971b) in disappearance of the enhancement of the cooperative effect, as would naturally be expected from our previous considerations.

Excessive concern with certain theories of allosteric binding has resulted in a neglect of the study of ligand interactions in proteins made up of a single peptide chain. Yet there is every reason to believe that the physical basis for the appearance of energetic coupling among ligands must be a property of all compact globular polymers. In particular, neighboring ions must exercise these effects by virtue of their charges, without any need for postulating in the protein structure a property other than a resistance to relaxation, which would, if present, destroy the greater part of the negative interactions. With these ideas in mind, Kolb and Weber (1972) have studied by means of fluorometric titrations the binding of anilino-

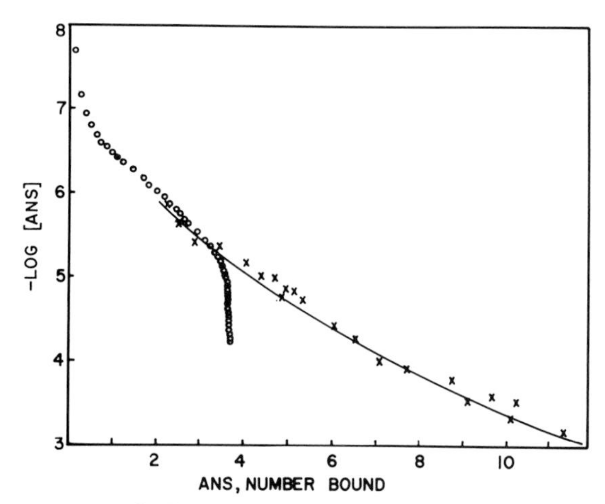

FIG. 8. Titration curves for bovine serum albumin monomer with 1 anilinonaphthalene 8-sulfonate (ANS). ○—○: Fluorescence enhancement measurements (Kolb, 1974); ×—× equilibrium dialysis (Pasby, 1969). Note the complete character of the fluorescence titration curve, which shows about 4 sites per albumin molecule as against the incomplete equilibrium dialysis curve, which shows an indefinite number of additional sites of lesser affinity.

naphthalene sulfonate (ANS) to bovine serum albumin (BSA) in the presence and in the absence of 3,5-dihydroxybenzoate. Pasby (1969) had observed that 4 moles of ANS are bound to albumin sites that are characterized by strong fluorescence and have larger free energy of binding, as compared with an indefinite number of sites showing much less, or even negligible fluorescence enhancement in comparison with the former group. The existence of these two classes of sites is evident in Fig. 8, which compares the titration curves of BSA with ANS, using fluorescence enhancement and equilibrium dialysis. By fluorescence enhancement a *complete* titration curve (see Appendix, Section A) is obtained, while equilibrium dialysis experiments yield an incomplete titration curve. It is evident that there are four strong binding sites in which ANS is protected from the fluorescence quenching action of the water, and an indefinite number of weaker sites, which must be in much more direct contact with the solvent. It was judged that a very hydrophilic anion such as dihydroxybenzoate would occupy mainly the hydrophilic sites and have an antagonistic effect, probably by electrostatic repulsion, to the occupation of the hydrophobic sites of ANS. Figure 9 conforms to this prediction and indicates the appearance of cooperative binding of ANS at intermediate concentrations of dihydroxybenzoate. The

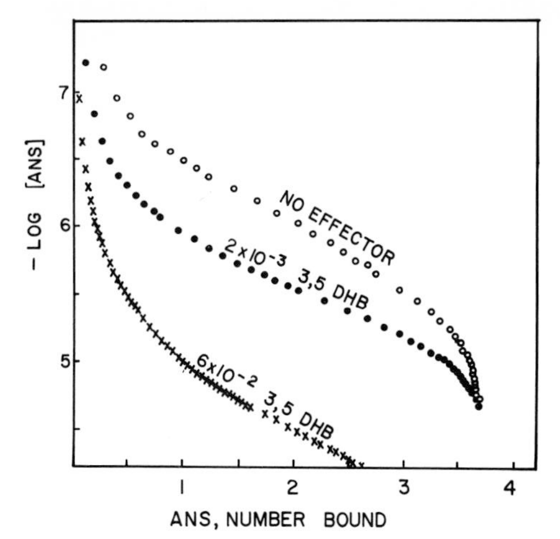

FIG. 9. Fluorescence enhancement titration curves as in Fig. 8, but with different concentrations of 3,5-dihydroxybenzoate (DHB) present. At DHB $= 2 \times 10^{-3}\ M$ a cooperative curve (●●●) is seen, but the cooperativity is almost absent when excess DHB ($6 \times 10^{-2}\ M$) (×××) is added. ○○○, No effector. Data of Kolb (1974).

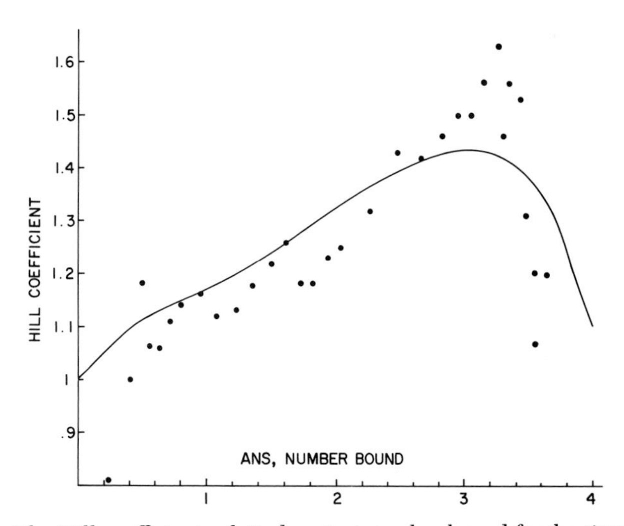

FIG. 10. The Hill coefficients plotted against number bound for the titration in Fig. 9, in the presence of $2 \times 10^{-3}\ M$ dihydroxybenzoate (DHB). The solid curve is calculated for 2 moles of 1-anilinonaphthalene 8-sulfonate coupled to DHB with a free energy $\Delta F_{xy} = 3RT$ and 2 other moles coupled with an energy of $1RT$. $[\mathrm{DHB}]/K_{\mathrm{DHB}} = 10$. Data of Kolb (1974).

maximum Hill coefficient is reached at 3 moles bound and has a value close to 1.5. From these facts, and by comparison with theoretical titration curves obtained by varying the conditions and the value of the free-energy coupling, Kolb and Weber (1972) concluded that satisfactory simulation of the experimental results requires free-energy couplings of +0.5 kcal for two of the four ANS ligands with a dihydroxybenzoate molecule and +1.5 kcal for the other two (Fig. 10). These results would lead to us to expect the demonstration of allosteric properties in enzymes made up of a single peptide chain. According to Panagou *et al.* (1972), such is the case with ribonucleotide reductase from *Lactobacillus leishmannii*.[4]

2. *Ligand Distribution and Protein Conformation*

Figures 11 and 12 show how, depending upon the concentration of Y the distribution of liganded species varies with the saturation of one of the ligands. For $\Delta F_{xy} \neq 0$ and the smaller value of ϵ the binding of X results in the release or uptake of Y, but for the larger values of ϵ the binding of X does not lead to an appreciable change in Y saturation. It is then clear that a one-to-one correlation cannot exist between protein conformation and binding of a given ligand. For any given complex YPX, YP, YPX_2, etc. there must exist differences in protein conformation, albeit small. The reciprocity of effects between the bound ligands denies also the possibility that one ligand imposes a conformation to the exclusion of the effects owing to the other ligands bound. It is true that the reciprocal effects among ligands do not compel us to accept that the conformation of YPX is an intermediate between those of YP and PX, because of the alternative possibility that complementary fractions of YPX exist in the conformations of YP and PX, respectively. In the latter case the additional hypothesis is to be made that conformations intermediate between those of YP and PX are excluded. For such supposition there is no experimental evidence. Moreover, cases of three different interacting ligands X, Y, Z bound by the same protein are known (e.g., protons, DPG, and oxygen to hemoglobin, or the different ligands bound by pyruvate kinase or ATCase). In these cases the assumption that conformations intermediate to those stabilized by each separate ligand are not physically possible would appear even less justified.

[4] The observations of Panagou *et al.* have been recently confirmed by Chen *et al.* (1974).

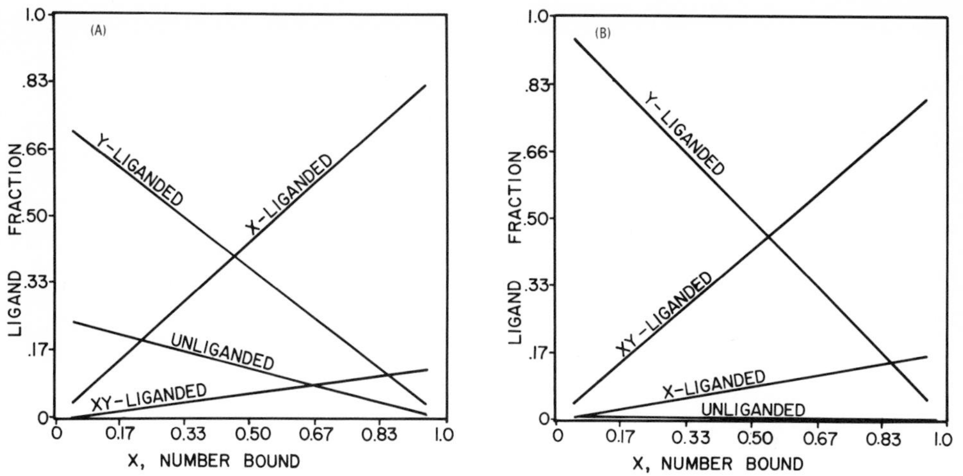

FIG. 11. Ligand distributions. Graphs (A) and (B) compare the ligand distributions obtained for the value of $\Delta F_{xy} = 3RT$ when 1 mole of X and one of Y are bound, when (A) $\epsilon = [Y]/K(Y) = 3$ and (B) $\epsilon = 100$. In (A) a linear release of Y takes place as X is added. In (B) no displacement of Y takes place when X is bound.

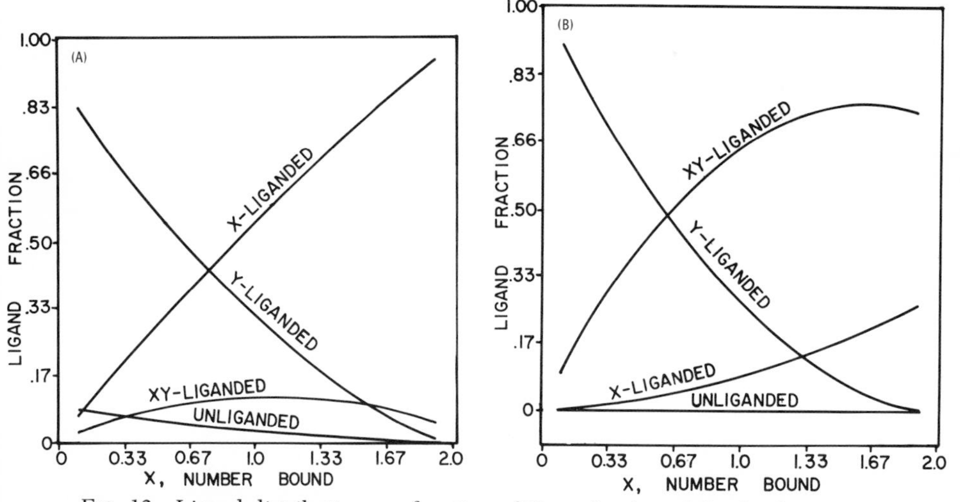

FIG. 12. Ligand distributions as function of X-number bound for the free-energy couplings shown in Fig. 5A. $\Delta F_{xy} = 3RT$ as in Fig. 11. As in the former case displacement of Y when X is bound takes place when $\epsilon = 10$ (A), but simple addition of X takes place at $\epsilon = 1000$ (B). Notice that the linear character of the displacement of Y shown in Fig. 11 is replaced by a more complex behavior.

C. *Physical Basis of Ligand Interactions in Proteins*

Two common features of protein structure, as it is known to us from a variety of experiments, seem particularly favorable for the appearance of interligand effects. In the first place proteins are compact globular structures, little penetrated by solvent and kept in their native conformation by *a large number of independent, or almost independent, small energy interactions.* (On this point more is said in Section V.) As a result the addition of a ligand must always produce some changes in the energetic balance of its immediate surroundings. There is no need to postulate any major structural rearrangements when the free energies of interaction are of the order of 1 kcal, as has repeatedly been observed. In fact, any rearrangement or relaxation of the structure following the addition of ligands will be in the direction of reducing any antagonism between them, so that the existence of measurable repulsive interactions between ligands indicates that the intervening structure must be capable of withstanding it without undergoing relaxation, and that the maximum value of such negative interactions must be limited by the magnitude of the forces that maintain the protein structure. We expect these forces to be generally weak, in consonance with the small values observed for the free-energy couplings. We can imagine the protein as a globule of plastic material capable of flowing under an applied stress, but doing so only after a "yield value" has been reached, just as is the case for a non-Newtonian, or Maxwellian fluid (Houwink, 1958).

In the second place, theoretical speculations, experiments on the unfolding of proteins at interfaces (Bull, 1964), and the direct observations of the structure of protein crystals by X-rays (Richards *et al.*, 1972; Richards and Lee, 1971), all lead to the conclusion that the amino acid residues out of contact with the solvent provide a core of relatively low polarizability, in contrast to the amino acid residues in direct contact with the solvent. Binding of two charged ions will result in much stronger electrostatic interactions across the core of the protein than when the same ions are free in solution (Epstein, 1971) in a highly polar solvent like water. Similar remarks apply to the interactions between a charged ligand and charged groups or polarizable protein groups involved in the binding of a second, perhaps uncharged, ligand. There is every reason to suppose that the negative effects between oxygen and DPG [or inorganic phosphate or IHP] are of electrostatic origin since many anions are capable of exercising similar effects (Rossi-Fanelli *et al.*, 1961), and since IHP has decidedly the strongest interaction as compared with other anions. We have here an interesting proof that in the specific biological effects the specific part is due to the binding affinity while the actual effects are due to ordinary molecular forces to which no specificity can be attributed.

In summary, the protein structure would appear particularly suited to the establishment of weak interactions between bound ligands to the point that we may well ask whether there are any cases of bound ligands indifferent to each other. The reason why so many cases of binding appear to be independent is that quite apart from the necessity of very precise observations to detect the weaker interactions, it is indispensable to carry out the measurements over an appropriately large range of ligand concentrations and saturations if the existence of interaction is to be rigorously excluded. As our operating procedures increase in accuracy and ease of execution and the observations are specially designed to reveal the effects of ligands upon each other it seems likely that the cases of independent binding and of competition for the same binding site will be found to be rarer than the phenomena that we have just discussed.

III. Binding by Multimer Proteins

Intracellular proteins are often found to consist of several peptide chains, identical or somewhat different in amino acid sequence, that associate into multimers of fixed composition. Dimers, trimers, tetramers, hexamers, and dodecamers have been isolated and characterized. This prevalence, and the belief that the more sophisticated molecular functions of proteins are, and perhaps can only be, performed by them have excited the interest of many biochemists. The progress of X-ray structural analysis of protein crystals (Dickerson, 1964, 1972) has had the greatest impact upon these matters by showing that in the particular case of hemoglobin the liganded form, oxyhemoglobin, has distinct differences in the folding of the peptide chains and their relative disposition as compared with the unliganded form, deoxyhemoglobin (Muirhead *et al.*, 1967). The fact that hemoglobin shows cooperative ligand binding as well as a variety of oxygen-linked effects, with protons and with diphosphoglycerate in particular, has given rise to much speculation as regards the relation between ligand binding and structural changes in multimer proteins. I shall not attempt here to analyze in detail the relations of structure and function in these proteins, but will confine myself to those aspects that can be clarified by the study of the binding process alone, and in particular to the relations that according to thermodynamics must exist between different functions, particularly the association of the protomers to form the multimer, and the binding of specific ligands. I shall do this by a simple extension of the method already used in dealing with multiple ligand binding to a single peptide chain.

A. Relation between Free Energy of Protomer Association and Ligand Binding

The simplest case will be that of association of two identical monomers. In dealing with it we shall need to distinguish clearly between two different kinds of molecular associations: On the one hand, there is the association of the monomers to form a dimer, on the other the association of a ligand—a small molecule—with a specific binding site of the monomers. The former type will be called the "macroassociation," the latter type the "microassociation." Macro- and microassociation processes may be considered equivalent: a dimer that binds a mole of ligand can also be considered a monomer that undergoes association with two ligands, the small molecule and the second monomer. It follows that the description of the process will not entail anything fundamentally different from the case in which two different ligands, X and Y are bound to the same monomer, which we have already discussed in detail. However, from the descriptive point of view we must recognize here one new feature not explicitly present before. While the relations between the protein molecule with ligand shown in Eq. (2.1) allow for only two states, liganded and unliganded, we must allow here that the free energy of macroassociation can take as many distinct values as the number of ligands that the dimer can bind, besides a free energy of macroassociation in the unliganded state. The free-energy level scheme for a dissociating dimer binding at most 2 moles of X, is shown in Fig. 13. The quantity ΔF°_{11} is the standard free energy of microdissociation of the monomer while ΔF°_{21} and ΔF°_{22} are the free energies of microdissociation of the dimer. The superscripted free energies ${}^2\Delta F^\circ_0$, ${}^2\Delta F^\circ_1$, ${}^2\Delta F^\circ_2$ are the free energies of macroassociation of the unliganded, uniliganded, and biliganded dimer. Applying the principle of conservation of the standard free energy we find,

$$\begin{aligned} \Delta F^\circ_{21} &= \Delta F^\circ_{11} + {}^2\Delta F^\circ_1 - {}^2\Delta F^\circ_0 \\ \Delta F^\circ_{22} &= \Delta F^\circ_{11} + {}^2\Delta F^\circ_2 - {}^2\Delta F^\circ_1 \end{aligned} \tag{3.1}$$

If the free energy of microassociation of the isolated monomers are the same, as recognized in Fig. 13 and in Eq. (3.1) by the unique value ΔF_{11}, ΔF_{21}, and ΔF_{22} can only differ when the free energies of

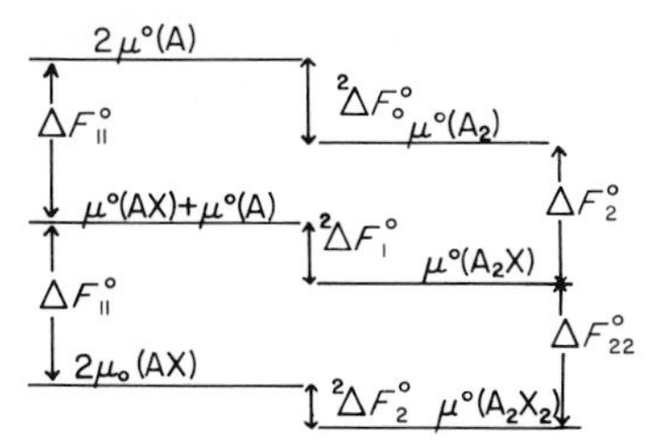

FIG. 13. Energy level scheme for a dimer A_2 binding 2 moles of ligand X. Note that, in agreement with Eq. (3.3), the case shown is one of dependence of macroassociation upon X binding, although the free energy of X binding by the dimer has a unique value ($\Delta F^\circ_{21} = \Delta F^\circ_{22}$).

macroassociation fail to obey the equality

$$^2\Delta F^\circ_2 - {}^2\Delta F^\circ_1 = {}^2\Delta F^\circ_1 - {}^2\Delta F^\circ_0 \tag{3.2}$$

or

$$^2\Delta F^\circ_1 = \tfrac{1}{2}({}^2\Delta F^\circ_2 + {}^2\Delta F^\circ_0) \tag{3.3}$$

Equation (3.2) implies that the free energy of association of the uniliganded dimer is the mean between those of the unliganded and biliganded states. One trivial case in which Eq. (3.2) must hold is when

$$^2\Delta F^\circ_0 = {}^2\Delta F^\circ_1 = {}^2\Delta F^\circ_2 \tag{3.4}$$

In this case the free energies of micro- and macroassociations are independent, a case that presents no interest to us, for then the dimer will behave, as regards microassociations, as would the independent monomers. If Eq. (3.3) holds, but not Eq. (3.4), it follows that binding of the small ligand has an effect upon the energy of macroassociation, but that the binding of the first and second molecules of X produce identical changes, either positive or negative, in the free energy of macroassociation (Fig. 13). The important point follows

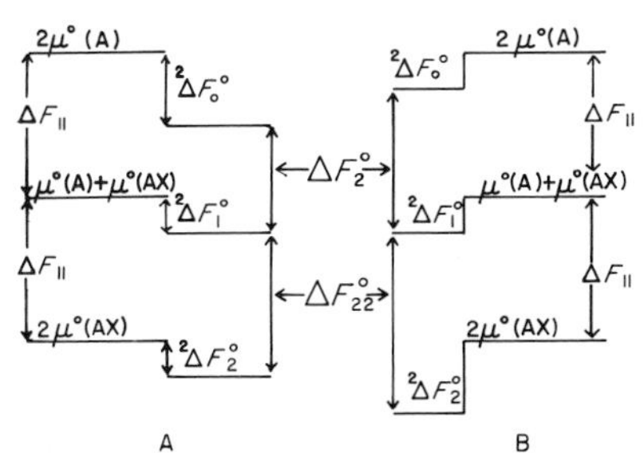

FIG. 14. The two cases discussed in the text in which cooperativity of ligand binding arises from changes in macroassociation. A: X addition promotes dimer dissociation; B: X addition promotes dimer association.

that if an analysis of the microassociation shows that $\Delta F^\circ_{21} = \Delta F^\circ_{22}$, it is not possible to conclude that macro- and microassociations are independent. A considerable number of cases have been reported in which addition of a ligand produces evident dissociating or associating effects upon an oligomer protein, without a similar report of differences in the free energies of microassociation. Although most of these cases require a more thorough investigation before being put in this category, theory certainly has a place for them, and they may well turn out to be the most common in practice.

Cooperative and Antagonistic Character of Microassociations

Evidently in those cases in which $\Delta F^\circ_{21} \neq \Delta F^\circ_{22}$, Eq. (3.1) is not obeyed. In the case of cooperative binding of X, we must have that, in absolute value, $\Delta F^\circ_{22} > \Delta F^\circ_{21}$, and therefore,

$$^2\Delta F^\circ_2 - {}^2\Delta F^\circ_1 > {}^2\Delta F^\circ_1 - \Delta F^\circ_0 \tag{3.5}$$

As shown by Fig. 14 cooperative behavior can arise both when the greater part of the change in macroassociation free energy takes place upon the binding of the first or of the second molecule of X. These two possibilities are easily distinguished in practice: If binding of X promotes dissociation $^2\Delta F^\circ_0$ is greater than $^2\Delta F^\circ_2$ and we must of necessity have $\Delta F^\circ_{21} < \Delta F^\circ_{11}$, a decreased affinity of the dimer for the first mole of X bound, as compared to the affinity of X for the isolated monomer. On the other hand, if the binding of X promotes association then $^2\Delta F^\circ_2 > {}^2\Delta F^\circ_0$ and therefore $\Delta F^\circ_{22} > \Delta F^\circ_{11}$. The second mole of X must be more strongly bound by the dimer than by the isolated monomer. In the case of hemoglobin such comparison can readily be made since the affinity of the isolated chains for oxygen has been measured (Brunori *et al.*, 1968) and found to be much greater than that of the intact tetramer for the first molecule of oxygen bound. It follows then that the binding of oxygen must facilitate dissociation, a fact that has also been experimentally ascertained (Guidotti, 1967) by a direct study of the molecular weight of hemoglobin at different degrees of saturation. The presence or absence of cooperativity in ligand binding rests upon small differences in the effects that the occupation of the binding sites produce at the subunit boundaries. If the intersubunit bonds broken upon microassociation are distinct for each binding site, no cooperative effect is possible. However, if the same bonds are broken as a result of occupation of one or the other binding site, cooperative binding must follow. These effects are shown in a very schematic fashion in Fig. 15. It is then easy to see how deletion or change of a single amino acid residue, by modifying the energetic balance at the boundary, can en-

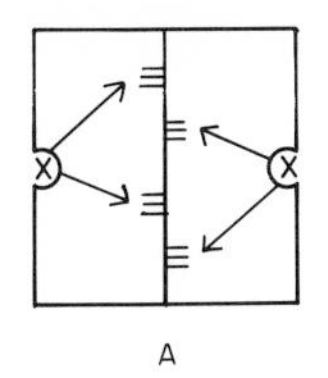

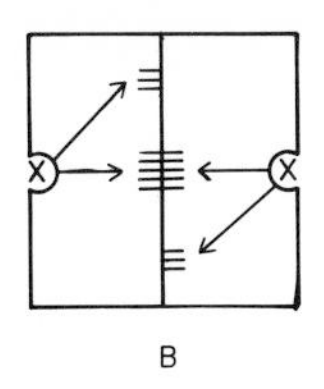

FIG. 15. The origin of cooperativity resulting from the effect of X binding. In (A) addition of X produces independent effects ($\Delta F_{21} = \Delta F_{22}$), upon subunit boundary bonds. In (B) addition of X breaks or makes some common bonds at the boundary ($\Delta F_{21} \neq \Delta F_{22}$).

hance, reduce, or abolish the cooperativity. The observations that have been carried out in hemoglobin mutants broadly conform to this concept.

The rules for the effects of the antagonistic binding of X upon macroassociation, which demand

$$^{2}\Delta F_{2}^{\circ} - {}^{2}\Delta F_{1}^{\circ} < {}^{2}\Delta F_{1}^{\circ} - {}^{2}\Delta F_{0}^{\circ} \tag{3.6}$$

can be derived in a similar way. True antagonistic binding in which the first molecule bound decreases the affinity for the second is not easily distinguished experimentally from independent binding of 2 moles of X with different intrinsic affinities. The structural changes required to introduce appreciable changes in affinity are indeed minimal, and it is not often that the structure of an oligomer protein is known in such detail that the existence of sites with different intrinsic affinities can be excluded. For this reason, although antagonistic binding has often been postulated (Conway and Koshland, 1968; Cook and Koshland, 1970), no conclusive proof against independent multiple binding has been given in these cases. Lewitzky *et al.* (1971) have deduced from the extended titration curve of cytosine triphosphate (CTP) synthetase with CTP that negative interactions between the ligands take place, but the structural knowledge of CTP synthetase is too meager yet to exclude the alternative explanation. Bernhard and McQuarrie (1972), who examined the behavior of glyceraldehyde-3-phosphate dehydrogenase, concluded in this case from indirect experiments that there must be, among the binding sites, preexistent differences that are responsible for the apparent antagonistic ligand binding.

B. The Monomer as Second Ligand. The Physical Basis for the Relations of Micro- and Macroassociations

Considering the monomer as a second ligand, which takes the place of Y in Eqs. (2.1), we can derive the equivalences shown in Table II.

TABLE II
The Monomer as Second Ligand: Equivalences with Free-Energy Level Scheme for Binding for Dimer A_2

Reaction	Standard free-energy change	Equivalent in Fig. 13
1. $A + X \rightarrow AX$	$\Delta F°(X)$	ΔF°_{11}
2. $A + A \rightarrow A_2$	$\Delta F°(A)$	${}^2\Delta F^{\circ}_{0}$
3. $AX + A \rightarrow A_2X$	$\Delta F°(A/X)$	${}^2\Delta F^{\circ}_{1}$
4. $A_2 + X \rightarrow A_2X$	$\Delta F°(X/A)$	ΔF°_{21}
5. $AX + AX \rightarrow A_2X_2$	$\Delta F°(AX/X)$	ΔF°_{2}
6. $A_2X + X \rightarrow A_2X_2$	$\Delta F°(X/AX)$	ΔF°_{22}

The conservation relations

$$\begin{aligned} \Delta F°(A) + \Delta F°(X/A) &= \Delta F°(X) + \Delta F°(A/X) \\ \Delta F°(A/X) + \Delta F°(X/AX) &= \Delta F°(X) + \Delta F°(AX/X) \end{aligned} \tag{3.6}$$

are, respectively, equivalent to those in Eq. (3.1). These equations define the free-energy couplings $\Delta F_{A.X}$ and $\Delta F_{AX.X}$, respectively. The first is the free-energy coupling between the binding of an *unliganded monomer* and a small ligand, to a monomer, the second is the free-energy coupling in the same monomer between a *liganded monomer* and a small ligand. From the last equations and Table II, we see that

$$\begin{aligned} \Delta F_{A.X} &= {}^2\Delta F_1 - {}^2\Delta F_0 = \Delta F^{\circ}_{21} - \Delta F^{\circ}_{11} \\ \Delta F_{AX.X} &= {}^2\Delta F_2 - {}^2\Delta F_1 = \Delta F^{\circ}_{22} - \Delta F_{11} \end{aligned} \tag{3.7}$$

Eq. (3.7) describes the reciprocity of the effects between the free-energy changes at the subunit boundary and at the binding site for the small ligand. The reciprocity is formally equivalent to that between two ligands bound to the same peptide chain and we expect the physical basis of the reciprocity to be the same. When a ligand is bound to a peptide chain the interactions of the contacting groups in the peptide chain and in the ligand will set up forces that are transmitted and felt at many points in the protein structure (Klapper, 1973). These forces will produce the largest effects where the bonds linking the chain are the weakest and such a region is provided by the boundary between subunits, as witnessed by the fact that dissociation and hybridization of peptide chains takes place far more readily than unfolding and denaturation. Although there is in principle no difference between the effects that are set up in the case of a monomer or multimer protein, it is easy to see why evolution has led to proteins where the interacting ligands occupy binding sites in different peptide chains. In this way, the influences can be exercised by the effects upon a common boundary which is the region most likely to be affected by the small changes in internal free energy due to ligand binding. In other words, the reason for the prevalence of multimer proteins may be found in the convenient energetic requirements that they fulfill. (On this point, more in Section IV,D.) The intervention of the subunit boundaries in the interaction between ligands not only presents possibilities that have been developed in evolution, but introduces considerable difficulties in the way to simple structural interpretations of the effects. Consider in this respect the number and nature of the boundary contacts of the subunits in hemoglobin. According to Perutz *et al.* (1968) and Bolton and Perutz (1970), the interchain atomic contacts number about a hundred per pair of chains. From the experimental data on oxygen binding (Saroff and Minton, 1972), the difference in subunit interaction energy between deoxygenated and oxygenated stripped hemoglobin amounts to approximately 1200 calories. This small change in free energy is to be explained by alterations in the balance of forces between the hundred or so atomic contacts at the boundary. The drawing up of such a balance requires not only the determination of the atomic coordinates with a precision well beyond that obtained in present-day protein X-ray analysis (Watenpaugh *et al.*, 1972), but also on a very precise knowledge of the relations between mutual displacements of the structure and energy, so that a convincing quantitative analysis of the energetic changes at the subunit boundaries in hemoglobin is unlikely to be accomplished in the near future. On a more general note, the boundary changes that follow ligand binding present us with a problem where the uncertainties may be of more fundamental origin than the technical shortcomings of the moment. The classical indetermination of physics arises in cases of large energies and very small particles. Here we seem to be at the other end of the scale, where very small energies are distributed over many groups belonging to a very large particle. It would not be therefore surprising that the impossibility of measuring all the distances and energies with the required precision would limit our knowledge of the phenomena in question in a permanent fashion. A discussion of the uncertainties involved in the measurement of very small distances is given by Brillouin (1962). Although there is no strict upper limit to the precision that may be attained, the "cost" of the information—and the chance of loss of information by irreversible damage to the structure—rise prohibitively as the distances fall well below atomic dimensions. It may well be that the simultaneous determination of a large number of such distances, with the precision necessary to draw up a precise energetic balance, will be outside our possibilities, or may be our wish. In any case, there is very little hope that examination of a relevant mutant protein, or a particularly clear experimental result on a bulk property, will show us the precise "mechanism" of the interaction between ligands in multimer proteins. Evidently the possibility of a reasonable guess as to the ligand–protein interactions responsible for a given effect by simple considerations of best fit of molecular models increases with the total energy of the interactions. Two recent examples of such attempts worth contrasting in this respect refer to the changes in subunit contacts in hemoglobin (Perutz and TenEyck, 1972) which involve a free energy of the order of 1 kcal per subunit pair and the complex between chymotrypsin and the specific inhibitor (Blow *et al.*, 1972) which involves an energy of interaction of 15 kcal (Green, 1957).

C. Macroassociation and Quaternary Structure

Since the protein dimer has a finite free energy of association, distinctly different binding phenomena will be observed depending upon the protein concentration of the solutions. At dilutions such that both liganded and unliganded forms are completely dissociated, the only relevant binding free energy of the ligand will be ΔF°_{11}. At concentrations of protein intermediate between the macrodissociation constants of the liganded and unliganded dimers, the free energies of binding of the ligand X will be a weighted average of ΔF°_{11}, ΔF°_{21}, and ΔF°_{22}. In this region of protein concentration the dimer can dissociate into monomers by addition or withdrawal of X, according to whether ${}^2\Delta F^\circ_0 > {}^2\Delta F^\circ_2$ or ${}^2\Delta F^\circ_0 < {}^2\Delta F^\circ_2$, respectively. Computations of the Hill coefficients to be observed during titration of a dimer with X, for different degrees of dissociation in the absence of ligand, and therefore for different total concentrations of protein are shown in Fig. 16. The cases A and B are those schematized in Figs. 14A and B, respectively. It is noticeable that only in case B, in which X ligation promotes macroassociation, it is possible for the Hill coefficient to rise conspicuously above unity in a solution in which dimers are only a minor fraction (10–20%) of the total protein. The origin of this difference is seen quite clearly in Fig. 17A and B, which give the fractional distributions of monomers and dimers as ligation proceeds. It is only when addition of X promotes the formation of the species with the higher cooperativity that the effect can be observed. The dependence of the Hill coefficient upon the degree of protein dissociation in the absence of ligand, for the case of a tetramer with the

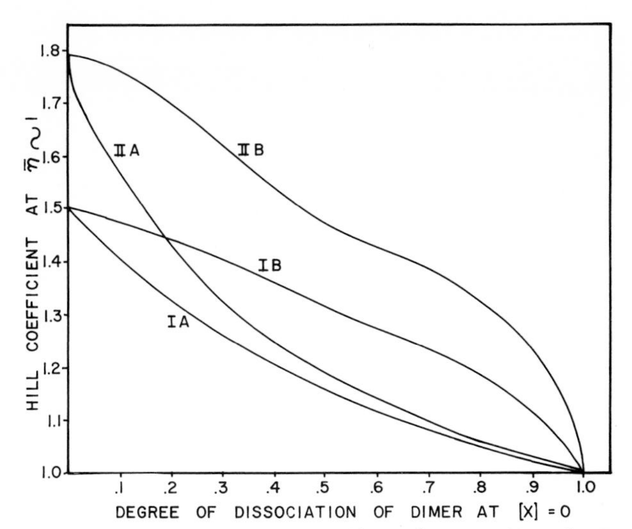

FIG. 16. The Hill coefficients to be observed in a dimer at the midpoint of the titration curve as a function of the degree of dissociation in the absence of ligand. The maximum Hill coefficients of 1.5 and 1.8 correspond to two ratios of K_{21}/K_{22}, respectively, equal to 10 and 100. Curves A and B are for the cases in which X addition promotes, respectively, dissociation (Fig. 14A) and association (14B). Notice that curves A are concave and curves B are convex, but that in both cases the Hill coefficient decreases monotonically with degree of dimer dissociation. I: $K_{21}/K_{22} = 10$; II, $K_{21}/K_{22} = 100$; A, $K_{22} = 1$; B, $K_{21} = 1$; $K_{11} = 1$.

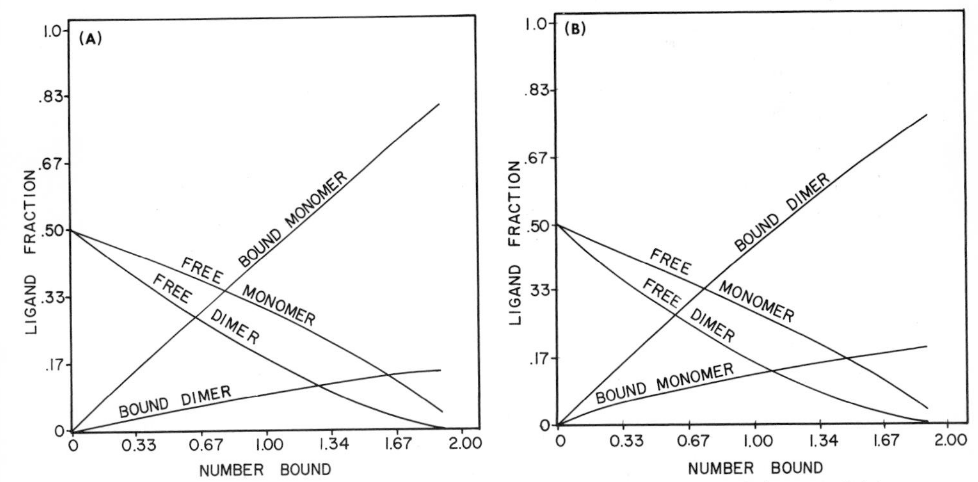

FIG. 17. (A) The ligand distribution for the case IA of Fig. 16 and degree of dissociation of dimer = 1/2 in the absence of X. Addition of X favors appearance of the form without cooperativity, hence the much lower values of the Hill coefficient in comparison with case IB (Fig. 16). $K_{21} = 10$; $K_{22} = 1.0$; $K_{11} = 1.0$.

(B) Ligand distribution for the case IB in Fig. 16. Appearance of the cooperative form as X binding proceeds explains the difference with case IA. $K_{21} = 1.0$; $K_{22} = 0.1$; $K_{11} = 1.0$.

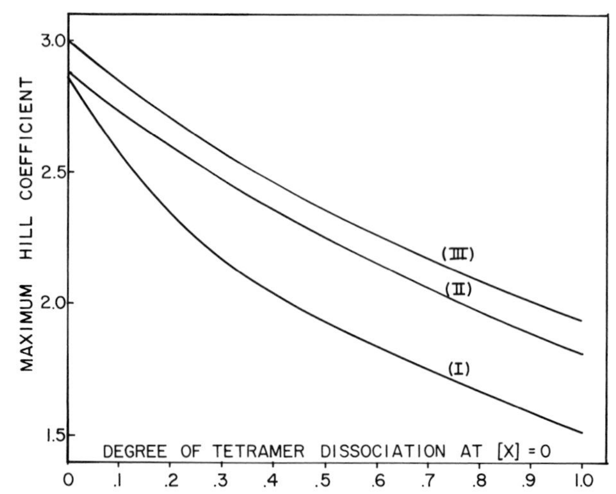

FIG. 18. The effect of dissociation of a tetramer into dimers upon the Hill coefficient observed at the midpoint of the titration curve. The oxygen dissociation constants for the tetramer ($K_{41} \ldots K_{44}$) are those given by Tyuma *et al.* (1971a) for dissociation of the oxygen from the hemoglobin-2,3-diphosphoglycerate (DPG) complex at optimal DPG concentration (see Appendix, Section B). Three different ratios of the oxygen–dimer dissociation constants (K_{21}, K_{22}) were assumed. Although no measurements have been reported, we expect the actual case to be somewhere between cases I and II. $K_{41} = 545$; $K_{42} = 118$; $K_{43} = 218$; $K_{44} = 1$; $K_{21} = 1000$ (III); $K_{21} = 100$ (II) $= K_{21} = 10$ (I): $K_{22} = 1$.

oxygen dissociation constants of the oxygen–hemoglobin–DPG complex is shown in Fig. 18. Realistic values of the oxygen–dimer dissociation constants would confine the results to be observed to those between curves I and II. Thus a value of the Hill coefficient greater than 2 could be observed only if the dissociation of deoxyhemoglobin is conspicuously less than 50%. The computations do not support the contention of Guidotti (1967) that the Hill coefficient conspicuously greater than 2 observed by Rossi-Fanelli *et al.* (1961) in concentrated salt solutions could be due to tetramer–dimer equilibria in solutions in which the deoxy dimers form only a minor fraction of the total protein. The recent experimental results of Tomita *et al.* (1972), who found a limiting value of 2 for the Hill coefficient under conditions of nearly complete dissociation of the oxygenated form is in agreement with the results expected from Fig. 18. At concentrations of protein sufficiently large compared with the macrodissociation constants of both liganded and unliganded dimer, addition or withdrawal of X will not cause any apparent change in the state of aggregation of the protein. The free energies of ligand binding experimentally determined will correspond to ΔF°_{21} and ΔF°_{22}. In actual practice, the physical techniques to determine the protein mass can be applied only over a restricted range of protein concentrations, which thus determines the aggregation phenomena to be observed upon ligand binding. In most cases described in the literature *either* complete change in the state of macroassociation, *or* no change in state of aggregation has been observed upon binding. In hemoglobin a dissociation into dimers upon oxygenation is detected only at the highest dilutions that can be experimentally handled (10^{-6} *M*), and at all higher concentrations no apparent change in aggregation is seen. This latter type must be the case in many oligomeric enzymes, since the catalytic kinetics appears independent of protein concentration (Deal *et al.*, 1963). The apparent prevalence of this case, and in particular the observation of marked crystallographic differences between the liganded and unliganded forms of hemoglobin, has lead Monod *et al.* (1965), to interpret the effect of ligand binding as producing a change in the quaternary protein structure, which is supposed to exist in two main structural states (often called T and R) characteristic of the fully liganded and unliganded forms of the protein. A similar interpretation, including the possibility of intermediate quaternary structures, is that of Koshland *et al.* (1966). Evidently any change in quaternary structure in those cases in which ligand binding does not change the state of aggregation, is included in the scheme of Fig. 13, since some structural change, albeit a minor one, will be necessary to account for the different free energy of association of the liganded and unliganded oligomers. The advantages of the treatment based upon the free energies of macroassociation rather than upon free energies of tautomerization is not limited to the fact that it shows how the phenomena observed (tautomerization or change in state of aggregation) are determined primarily by the protein concentration. One should consider also the following points:

1. The characterization of tautomerizations in molecules involving as large a number of degrees of freedom as proteins is confined to some structural detail that happens to be observable. Tautomerizations involving regions of one protomer away from the boundary cannot be separated in any simple way from those involving two neighboring subunits, a distinction that is crucial for the recognition of changes in subunit interaction as distinct from others.

2. It appears unlikely that any method can furnish us with the tautomerization constants necessary to characterize even a simple system. In contradistinction, the free energies of association of the

protein subunits are well-defined and measurable thermodynamic quantities. With progress in the physical methods of measurement of molecular size at high dilution, it is not unreasonable to expect that free energies of association of the order of 15 kcal will become measurable.

3. The tautomerization constants are first-order quantities in protein concentration and apply only when the protein concentrations exceed appreciably the macrodissociation constants of the system. They cannot describe the behavior of the system in those cases in which ligand binding produces appreciable changes in protein aggregation.

Extension of Dimer Equilibria to Oligomers of Any Order

An oligomer of N, equal or different, subunits can be considered as consisting of $N(N-1)/2$ interacting pairs to each of which the scheme of Fig. 13 applies. To perform the necessary computations for such aggregate of N protomers, we must specify (Weber, 1972a): (1) the free energies of microassociation of the N unliganded protomers $[\Delta F^\circ_{11}(i), 1 \leq i \leq N]$; (2) for each interacting pair of subunits, $i\,j$, four free energies of macroassociation: $\Delta F^\circ_{ij}(00)$, $\Delta F^\circ_{ij}(01)$, $\Delta F^\circ_{ij}(10)$, $\Delta F^\circ_{ij}(11)$. The first and fourth quantities correspond, respectively, to the unliganded and biliganded pair. The free energy of association of a dimer with liganded j and unliganded i subunit is $\Delta F^\circ_{ij}(01)$ and $\Delta F^\circ_{ij}(10)$ is the free energy of macroassociation of i liganded and j free.

With these specifications it is a simple matter to calculate the free energy of aggregation of the system at any state of ligation. It must be remembered that if J out of N monomers are bound by the ligand there are $\binom{N}{J}$ site-isomers possible, and the average free energy of association of the system $\bar{F}_J$ is determined by those of the site-isomers, $F_J(i)$ $(1 \leq i \leq \binom{N}{J})$ according to the equation

$$\bar{F}_J = \sum_i^{\binom{N}{J}} F_J(i)\exp(F_J(i)/RT) / \sum_i^{\binom{N}{J}} \exp(F_J(i)/RT) \qquad (3.8)$$

The free energy of binding of the jth ligand is given by

$$\Delta F^\circ_{NJ} = \bar{F}_J - \bar{F}_{J-1} \qquad (3.9)$$

If an arbitrary quantity is added to both $\bar{F}_J$ and $\bar{F}_{J-1}$ the value of $\Delta\bar{F}_{NJ}$ in Eq. (3.9) remains unchanged. Such added quantity is simply any part of the subunit affinities that does not depend upon ligand binding. Consider, as a limiting case, that of a protein in which there are distinct globular regions formed by different portions of the *same, unique peptide chain*, a description that appears to fit serum albumin (Foster, 1960; Weber and Young, 1964). Ligands bound to each of these regions could influence each other through their boundaries in much the same way as in proteins made of separate subunits. The only difference results from the very large free energy of the covalent bond holding the peptide chain together, which is represented in both F_J and F_{J-1} in Eq. (3.9). Here, as in most other respects we find complete equivalence of the energetic situation in single- and multiple-chain proteins. It follows from these considerations that from the experimentally determined free energies of microassociation it is possible to determine only the *average differences* of subunit association in the two liganded states, but not their absolute values. The latter can be measured only by some experimental method that determines the size of the protein aggregates. The computations by Eqs. (3.8) and (3.9), following the specifications of the energies of interaction, correspond to a lattice calculation of a particularly simple type (e.g., Hill, 1960) on account of the usually small value of N involved. An example of such calculations is given by Weber (1972a). They do not differ in essence from original calculations of Pauling (1935), except for the use of the weighting factors in Eq. (3.8). These weighting factors are important when the free-energy couplings between macro- and microassociations are of the order of RT, which is the actual case in hemoglobin and probably in most other proteins. To obtain relative values of the four oxygen dissociation constants that match the experimentally determined ones (Roughton *et al.*, 1965; Roughton and Lyster, 1965; Gibson, 1970; Tyuma *et al.*, 1971a), it is only necessary to assume a simple square lattice with a unique value of 2.7 RT units for the four coupling free energies between macro- and microassociation (Weber, 1972a). It is then evident that the oxygen dissociation curve of hemoglobin can be readily expressed as a two-parameter equation without undue difficulty. The only merit of calculations like this one is to provide a demonstration that the relevant physiological properties of hemoglobin are dependent upon *very small* free-energy couplings between macro- and microassociation. Various mutants of hemoglobin are known, particularly hemoglobin H, in which the cooperative character of oxygen binding is lost with concomitant increase in affinity, and even two cases [hemoglobins Capetown (Charache and Jenkins, 1971) and Little Rock (Bromberg *et al.*, 1973)] in which, while the overall cooperativity is preserved, the affinity is much increased. If ${}^2\Delta F_0^\circ$ and ${}^2\Delta F_2^\circ$ in our hypothetical dimer exchange relative values by one decreasing and the other increasing (Fig. 14A,B), the cooperativity is preserved, and the affinity rises, as seen in the cited mutants. Such an exchange in the values of the energies of macroassociation in a protein is not likely unless the difference between the operative free energies is small. This is the case in hemoglobin, and on these considerations it is predictable that these two mutant hemoglobins, contrary to hemoglobin A, will dissociate more readily in the unliganded form.

D. Interactions between Micro- and Macroassociations when Two Kinds of Ligands Are Bound

A further degree of complexity is reached in these cases. We may notice that proteins made up of subunits are particularly likely to present such complexity, since a ligand Y bound at the boundary between subunits will necessarily affect the binding of more than one ligand X, if Y interacts with those groups at the boundary, the free energy of interaction of which is modified when X is bound. (For a relevant crystallographic observation, see Adams *et al.*, 1973.) A case of this type, now well understood, is that of 2,3-diphosphoglycerate (DPG) and hemoglobin. After the finding by Benesch and Benesch (1967) that DPG is a normal regulator of the affinity of hemoglobin for oxygen, a large number of experiments have been carried out to determine the effects of DPG upon the oxygen–hemoglobin equilibrium. From the energetic point of view, the situation appears quite clear: In human hemoglobin a single molecule of DPG is bound with high affinity ($K \sim 10^{-5}\ M$) to the tetramer. Its effect is to decrease the oxygen binding affinity at the four heme binding sites to an equivalent or almost equivalent extent. Evidently the binding site for DPG must be at the boundary between the subunits, and it must necessarily change the free energy of macroassociation. The depression of the oxygen affinity by DPG must tend to a limit as its concentration increases, a limit that is determined by the finite free-energy coupling between the two ligands concerned (Eq. 2.3). With excess DPG the oxygen concentration for half-saturation of hemoglobin is increased by nearly an order of magnitude, so that the free-energy coupling between oxygen and DPG is close to 1.3 kcal. We have already commented upon the cooperative effects that appear in the binding of a ligand X when it is negatively or positively coupled to the binding of another ligand Y [Eq. (2.23) and Figs. 6 and 7]. In accordance with this simpler case, we expect DPG to be able to increase the cooperativity of oxygen binding by hemoglobin when present in optimal concentrations, but not when present in large excess. In fact, according to the careful measurements of Tyuma *et al.* (1971a), at millimolar concentrations of DPG the Hill coefficient of 2.5 of stripped hemoglobin rises to 3.0. On the other hand, recently Benesch *et al.* (1972) prepared a derivatized hemoglobin with permanently reduced affinity for oxygen by treatment with pyridoxal phosphate followed by borohydride reduction. The Hill coefficient is that of stripped hemoglobin, corresponding to the case $\epsilon\beta^{-1} \gg 1$, in Eq. (2.23), when no additional cooperativity can be induced by the second ligand (Weber, 1972a, and Fig. 18). The specificity of attachment of the pyridoxal phosphate seems certain from the report that no further depression of affinity was obtained on addition of DPG. As indicated by Eq. (2.23) any enhancement of the cooperativity requires the dissociation of Y from the AX_2Y complex. Evidently any structural effects caused by the binding of oxygen which are judged indispensable for the energy coupling between the hemes must be able to take place just as well in the hemoglobin–DPG complex as in stripped hemoglobin. Since it is not conceivable that DPG binds to hemoglobin without introducing some structural modifications, at least three structural forms of oxygenated hemoglobin can arise after the cooperative binding of O_2: stripped hemoglobin, stripped hemoglobin–DPG complex, and hemoglobin derivatized with pyridoxal phosphate.

IV. Extension of the Concept of Ligand Interaction to Covalent Bond Exchange in Proteins. Interconversion of Chemical and Osmotic Energies

From the thermodynamic viewpoint no distinction is to be made between the effects among ligands that are covalently coupled and those that are noncovalently attached to the protein. Such differences as exist arise from the larger free energies of binding in the former case, which restrict the observed effects to those due to protein fully saturated with the covalently bound ligand. This restriction is removed if the covalently bound ligand enters into chemical equilibrium with a third partner, usually another small molecule, which makes it now possible for a state of partial saturation of the protein with the covalently bound ligand to exist. Many cases are known in which groups attached to a protein partake of a chemical equilibrium which is appreciably shifted by changes in the concentration of the reagents. We need only mention the acyl intermediates in hydrolytic enzymes or the equilibrium between phosphoproteins and phosphate esters catalyzed by phosphoprotein kinases (Rabinowitz and Lipmann, 1960).

We shall examine first in some detail the simplest case, the influence of a noncovalently bound ligand Y upon the chemical equilibrium of a group X covalently attached to the protein P, with an external acceptor A. This chemical equilibrium may be formulated as

$$\text{I:}\quad \text{P-X} + \text{A-R} \rightarrow \text{P-R} + \text{A-X} \tag{4.1}$$

To this reaction we attach a standard free-energy change $\Delta F^{\circ}(X)$ and a dissociation constant $K(X)$ just as in the case in which X is noncovalently bound. As usual the free energy $\Delta F(Y)$ corresponds to the reaction

$$\text{Y} + \text{P} \rightarrow \text{YP} \tag{4.2}$$

Evidently the conditional standard free-energy change $\Delta F^{\circ}(X/Y)$ is that of the reaction

$$\text{II:}\quad \text{YP-X} + \text{A-R} \rightarrow \text{YP-R} + \text{A-X} \tag{4.3}$$

and depends upon the free-energy coupling ΔF_{xy} described in Eq. (2.3). A positive interaction between X and Y will stabilize the form YP-X while a negative interaction between X and Y will shift the chemical equilibrium in the opposite sense making X a better "leaving group." The free-energy scheme linking reactions I and II with the value of ΔF_{xy} is shown in Fig. 19. It is seen to be entirely equivalent to the free-energy scheme shown in Fig. 1. It represents, like the latter, a case of negative interaction between X and Y, and, in the case shown in Fig. 19, it leads to an inversion of the usual sense in which the reaction of P-R with A-X proceeds. If X is a characteristic chemical group, like the phosphate group in the various phosphate esters and phosphoproteins, it is possible to set up an *X-transfer potential* scale (Dixon, 1951) which depends upon the standard free energy of a common reaction in which the group X is

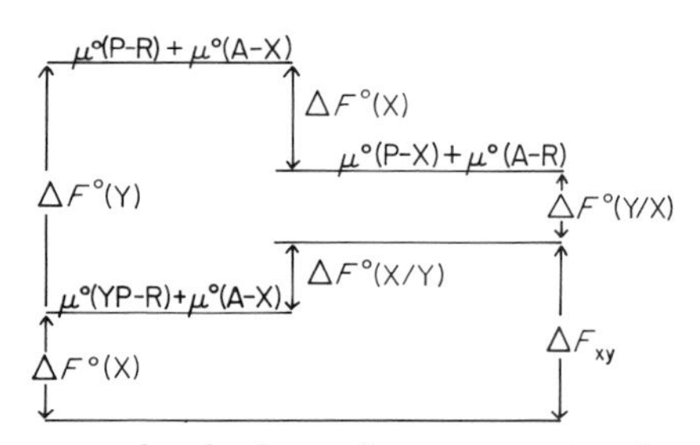

FIG. 19. The free-energy level scheme for a reaction involving covalent exchange between a protein P-X and a small molecule A-R, and the influence upon it of a second equilibrium of P with the effector Y. As depicted in the figure, ΔF_{xy} is positive and sufficient to invert the direction in which the reaction P-X + A-R → P-R + A-X would proceed in the absence of Y.

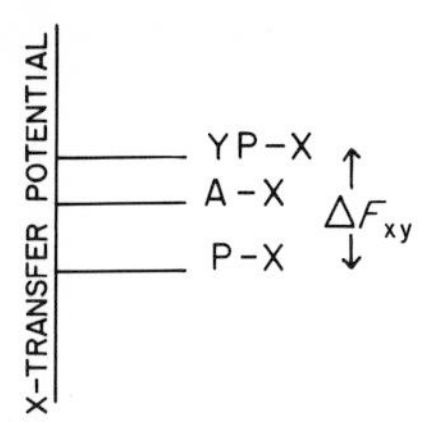

FIG. 20. X-transfer potential (Dixon, 1951) of YP-X, A-X, and P-X associated with the scheme of Fig. 19.

exchanged. (In the case of the phosphate group this common reaction is the hydrolysis of the group.) The X transfer potential for the X-derivatives considered here is depicted in Fig. 20, with the potential of P-X somewhat below that of A-X and that of YP-X a little above. Essentially the effect of binding Y will be to change the transfer potential by the amount ΔF_{xy}. It is then reasonable to expect that the binding of small ligands by proteins will be found, as indicated by Fig. 19 and 20 to modify the chemical equilibria in which proteins take part and therefore to have considerable importance in metabolism. In fact, the simple scheme presented can be extended to provide a formulation for the interconversion of osmotic and chemical energies (Weber, 1972b, 1974).

The form of the chemical potential (Eq. 1.1) and the general equation (1.4) tell us at once that it must be possible to convert directly chemical energy into osmotic energy and vice versa. Tissue concentration gradients are common to living organisms. Some of them result from accumulation of metabolites that are required by the cell at higher concentrations than those in the surrounding medium (active uptake) and represent therefore the result of conversion of *chemical into osmotic energy.* The reciprocal process, the conversion of *osmotic into chemical energy* has not been conclusively demonstrated in any particular case, but its existence and importance have been postulated by Mitchell (1966, 1972a,b) in his famous chemiosmotic theory of oxidative phosphorylation. The reciprocal character of the free-energy coupling indicates that the same processes by which chemical energy is converted into osmotic energy are potentially capable of performing the opposite conversion, of osmotic into chemical energy. The theories that attempt to account for active uptake of metabolites postulate, in one form or another, a specific carrier protein that exists in two states, one of "high," the other of "low," affinity for the ligand (e.g., Kaback, 1970, 1972). It is omitted that these relative affinities are *not* arbitrary quantities, independent of the reaction that the protein undergoes in the conversion of one state into the other. The transformation reaction and the change in affinity for the ligand are always coupled by a *finite free energy*, the reciprocal character of which determines the extent of the reverse process, the conversion of osmotic into chemical energy. The simplest system for the interconversion of chemical and osmotic energies consists of two compartments α and β, separated by a selectively permeable boundary. Some asymmetry must be postulated in the properties of the system to break up the equivalence of the diffusion from the two compartments, and this can be done equally well by supposing that certain components are present in only one of the two compartments or by endowing the boundary itself with polarity so that its properties are different at the two sides in contact with α and β, respectively. Either type of assumption can be used in setting up a description of the energetics of interconversion of chemical and osmotic energies. I shall exclusively use the supposition that certain components are restricted to one of the two compartments, while others diffuse freely from one to another. In this way the separation of the two compartments may be considered as a simple boundary, without postulating specific structural properties as must exist in real membranes. Figure 21 shows schematically the postulated arrangements. The components D-X and D-R are confined to the α compartment, and A-X and A-R are confined to the β compartment. The small ligand Y, which we shall call the effector, is present in both compartments. It is not supposed to diffuse freely from one to the other, but only through the intermediate protein complexes YP-R and YP-X. Figure 21B gives an equivalent arrangement with three compartments (Weber, 1972b) in which the D forms are in α, the A forms in β, and the P forms in an intermediate compartment M. This scheme, which assumes the protein components to be restricted to the M compartment, may be more like the actual biological situation, in which certain transporting components are expected to be an integral part of the membrane (Singer and Nicolson, 1972). From the point of view of the thermodynamic description the two schemes of Fig. 21, A and B, are equivalent. In what follows, I use the scheme in Fig. 21A, but the reader should have no difficulty in interpreting the same events in the alternative scheme. Besides the reactions I and II [Eqs. (4.2) and (4.3)], we have reactions III and IV below, which take place between the P and D components, namely

$$\text{III:} \quad \text{D-R} + \text{P-X} \rightarrow \text{D-X} + \text{P-R} \tag{4.4}$$

$$\text{IV:} \quad \text{D-R} + \text{YP-X} \rightarrow \text{D-X} + \text{YP-R} \tag{4.5}$$

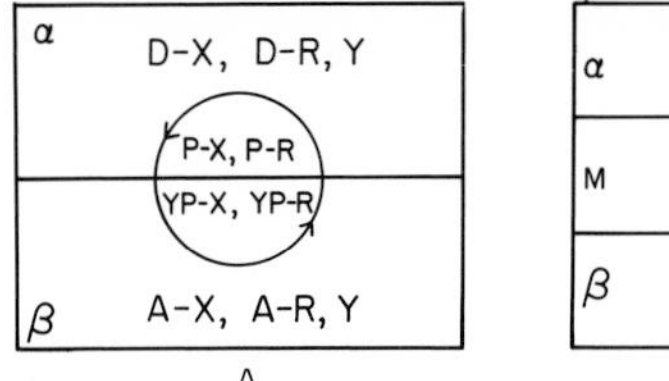

FIG. 21. Two equivalent schemes for the interconversion of chemical and osmotic energies. (A) Two-compartment system. D forms exist only in compartment α; A forms exclusively in compartment β; free effector Y and the P forms in both. Only P forms are allowed to cross the boundary. (B) Three-compartment scheme. The P forms are restricted to M, the middle compartment. Reactions of P with A and D occur at the boundaries of M with α and β.

The standard free-energy changes in the reactions I to IV are all specified by the relative positions of the four components D-X, A-X, P-X, YP-X in the X-transfer potential scale. It will be noticed that reactions I and III fix also the value of $\Delta F°(\mathrm{V})$, the standard free-energy change in the reaction

$$\mathrm{V:}\quad \text{D-X} + \text{A-R} \rightarrow \text{D-R} + \text{A-X} \tag{4.6}$$

We consider now separately the energetics of the conversion of chemical into osmotic energy and its reciprocal.

A. *Conversion of Chemical into Osmotic Energy*

We assume the set of X-transfer potentials shown in Fig. 22A, namely that the potentials of A-X and Y-PX are equal and lie above those of D-X and P-X which are also equal. It will be observed that the effect of this choice is to make

$$\Delta F_{xy} = \Delta F°(\mathrm{V}) \tag{4.7}$$

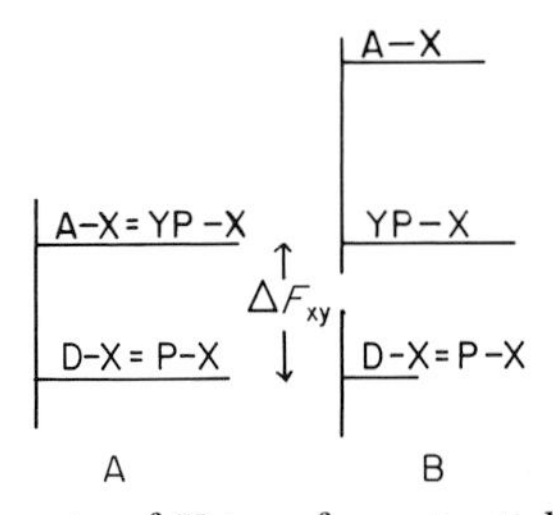

FIG. 22. Two alternative sets of X-transfer potentials with identical free-energy couplings between X and Y, which show differences in reversibility of chemiosmotic interconversion. In (A) the interconversion is wholly reversible. The set of potentials (B) gives rise to virtually irreversible conversion of chemical into osmotic energy.

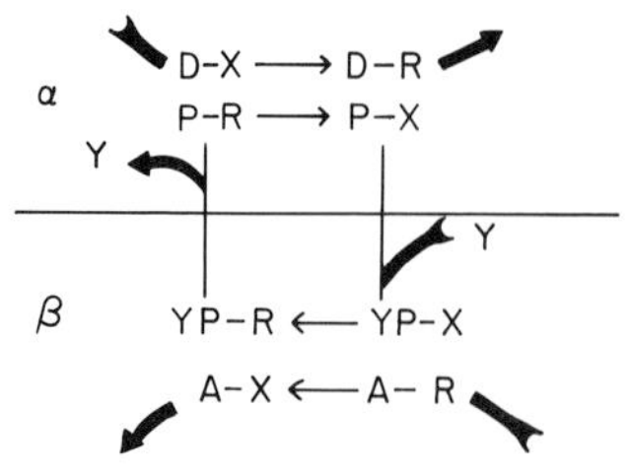

FIG. 23. The movement of Y across the boundary from β (high Y concentration) to α (low Y concentration) is associated with promotion of the reaction A-R + D-X → A-X + D-R. When the entrance and exit of the D and A forms is that shown by the arrows conversion of osmotic into chemical energy takes place. Reversing the sense of the arrows gives the condition for conversion of chemical into osmotic energy (active transport).

The chemical reactions must result in the maintenance of a difference in concentration of the ligand Y between the α and β compartments. To establish the balance between the chemical and osmotic energies, we suppose that a stationary state of chemical reaction is maintained by the entrance in α of a certain amount of D-R and withdrawal of an equal quantity of D-X, and the entrance to and exit from β of similar amounts of A-X and A-R, respectively (Fig. 23). This supposition corresponds to the physiological case in which an actual steady state is maintained. We start, however, with a concentration [Y] of effector equal in both compartments and intermediate between K(Y) and K(Y/X). From the scheme of X-transfer potentials of Fig. 22A, we see that as a result of the chemical reactions in β, P-R will decrease and P-X will increase over the respective concentrations in α. Since the predominant species are P-X and YP-R, the equalizing diffusion of the components will transport the ligand Y from α to β and at the same time drive the reaction D-R → D-X in α by the increase in P-X in this compartment. The directional transport of Y will continue so long as the liganded forms of P are at different concentrations in the two compartments, and this will be the case until $(\mathrm{Y})_\alpha = K(\mathrm{Y})$ and $(\mathrm{Y})_\beta = K(\mathrm{Y/X})$. At this point, if we assume that a compensating "leakage flux" of Y from β to α is set up, and that this leakage amounts to exactly 1 mole of Y for each mole of A-X, or D-R, entering the system, the free-energy change in the reaction V will balance exactly the osmotic work performed in the transport of Y, so that

$$\Delta F(\mathrm{V}) = -RT \ln \frac{(\mathrm{Y})_\alpha}{(\mathrm{Y})_\beta} = \Delta F_{\mathrm{xy}} \qquad (4.8)$$

The exact balance between the free-energy changes in the system must include the stationary concentrations (A-R), (A-X), (D-R), and (D-X) as discussed elsewhere (Weber, 1972b). Nevertheless Eq. (4.8) is valid under the assumption that (A-R) = (A-X) and (D-R) = (D-X), which give the system maximum stability. According to Eq. (4.8), under such optimal conditions a ratio $(Y)_\alpha/(Y)_\beta = 1/100$ can be maintained by a chemical reaction with a free-energy coupling of the two binding processes equal to 2.75 kcal.

B. Conversion of Osmotic into Chemical Energy

We assume now the existence of a difference in the concentration of Y in the compartments α and β, which is maintained at a fixed level by an external source, so that $(Y)_\alpha$ is kept smaller than K(Y), while $(Y)_\beta$ is kept equal to or greater than K(Y/X). The forms of P present in α are the unliganded forms, and the forms present in β are the liganded forms. According to the set of transfer potentials of Fig. 22A, entrance of D-X and removal of D-R from the α compartment will increase the ratio PX/PR in it, while entrance of A-R and removal of A-X will drive the reaction in β to an increase in YP-R and a decrease in YP-X. Diffusion of the P forms will produce a net transport of the effector Y down the gradient, so that the difference $(Y)_\beta - (Y)_\alpha$ will become smaller. When, in this ideal system 1 mole of products (A-X and D-R) are formed at the expense of the reactants A-R and D-X, respectively, in the steady state, 1 mole of Y has been transferred from the higher concentration $(Y)_\beta$ to the lower concentration $(Y)_\alpha$. The free energy necessary to promote reaction V, against the "normal" direction, comes from the equalization of the concentrations of Y, originally different in both compartments. The equivalence of the chemical and osmotic energies is that given by Eq. (4.8), but with the sign of both members changed.

C. Reversibility and Reciprocity of Osmotic-Chemical Interconversion

It is evident from the considerations of the last paragraph that the difference in Y concentration to be obtained from a given reaction, or conversely, the contribution to the chemical energy from a fixed concentration gradient is restricted to the amount ΔF_{xy}, the free-energy coupling for the two ligands bound to the protein P. When the X-transfer potentials are those depicted in Fig. 22A, Eq. (4.7) obtains and the reciprocity of the interconversion of chemical and osmotic energies is complete. Under the set of X-transfer potentials shown in Fig. 22B this will no longer be the case. Equation (4.7) takes then the general form

$$\delta\mu = \Delta F^\circ(\mathrm{V}) + \Delta F_{\mathrm{xy}} \tag{4.9}$$

The condition for complete reversibility is $\delta\mu = 0$. If it is obeyed by an actual system, experiments should demonstrate that an exchange in the concentrations $(\mathrm{Y})_\alpha$ and $(\mathrm{Y})_\beta$ influences considerably the driving chemical equilibrium to which the transport of Y is coupled. On the other hand, if $\Delta F^\circ(\mathrm{V})$ is much larger than ΔF_{xy} (Fig. 22B), this will no longer be the case. The chemical reaction V will not then be significantly modified by an exchange in the concentrations $(\mathrm{Y})_\alpha$ and $(\mathrm{Y})_\beta$, and the process will have all the appearances of an *irreversible conversion of chemical into osmotic energy.* A moment's consideration of the scheme implied by Fig. 22B will indicate that the larger the free-energy change in reaction V, the better the "kinetic" efficiency of the process that results from the reduced concentration of the unwanted forms: YP-R in the α compartment and P-X in the β compartment. This improvement is paid off by the excess energy $\delta\mu$, which not being converted into osmotic energy, is lost to the process. In biological systems we may expect some compromise to be achieved between ideal thermodynamic efficiency which requires $\delta\mu = 0$, and "kinetic" efficiency, that increases with $\delta\mu$. In the case of conversion of osmotic into chemical energy the improvement in "kinetic" efficiency cannot well be expected to be purchased by a much larger gradient, since those that we know hardly ever exceed 1:100, and to be of any value they would therefore require conversion into chemical energy as complete as possible. The "kinetic" requirements of the process and also the energetic ones may be assured in a very different way, and this may well turn out to be one of the major advantages in metabolism of proteins made up of similar subunits.

D. Effect of Protein Subunit Structure and Cooperative Ligand Binding upon Chemiosmotic Interconversion

The importance of these factors may be shown by comparing the addition of chemical and osmotic energies in the case of two proteins: One of them, P_1 is a monomer, binding a single mole of X and a mole of Y; the other P_4 is a tetramer binding 4 moles of the effector Y, each of them coupled to a single, covalently attached X-group in the tetramer (Weber, 1974). Such a model would have appeared unrealistic only a few years ago, but in fact it simply reproduces the oxygen–DPG coupling in hemoglobin for the case of a covalently attached group. Under stationary conditions the saturation of the proteins with Y in the compartments α and β would reach values S_α and S_β, respectively. As already discussed the binding of Y in the α compartment is determined by the value of $K(\mathrm{Y})$, while in the β compartment it is determined by the value of $K(\mathrm{Y/X})$. Therefore

$$S_\alpha = [\mathrm{Y}]_\alpha^j/([\mathrm{Y}]_\alpha^j + K(\mathrm{Y})) \tag{4.10}$$
$$S_\beta = [\mathrm{Y}]_\beta^j/([\mathrm{Y}]_\beta^j + K(\mathrm{Y/X}))$$

where j is the Hill coefficient, or effective order of the reaction of Y with P. For P_1 necessarily $j = 1$. If we suppose that P_4 binds Y cooperatively, we can use $j = 2.7$, the value that applies to the prototypic stripped hemoglobin. We may realistically demand S_α to equal 0.1 and S_β to equal 0.9 to ensure the prevalence of the required species in each compartment. The last equations give then for the ratio

$$(\mathrm{Y})_\beta/(\mathrm{Y})_\alpha = 81^{1/j} \exp\,(\Delta F_{\mathrm{xy}}/nRT) \tag{4.11}$$

where n is the number of effectors coupled to X with a mean free-energy coupling ΔF_{xy}. The equation shows that the ratio $(\mathrm{Y})_\beta/(\mathrm{Y})_\alpha$ required to provide the energy ΔF_{xy} decreases with a power equal to the number of effectors coupled to X and with the Hill coefficient j.

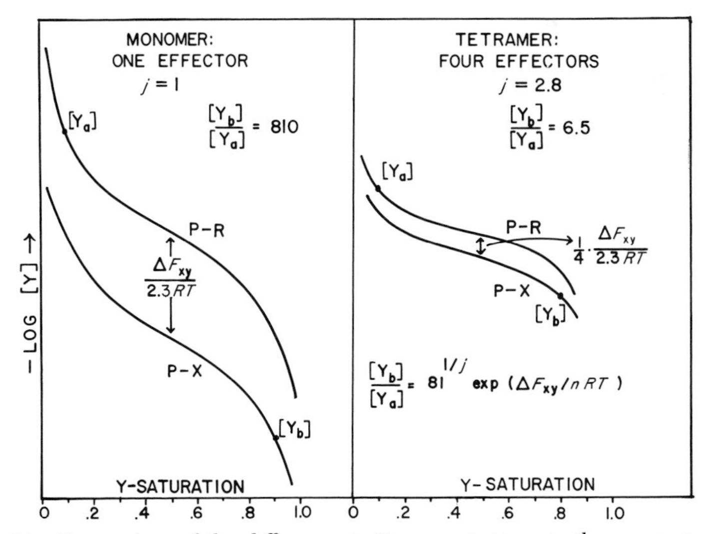

FIG. 24. Comparison of the difference in Y concentrations in the compartments α and β, for the same value of ΔF_{xy} (1.4 kcal), when the protein involved in chemiosmotic conversion is a monomer P_1 (left) and a tetramer P_4 that binds Y cooperatively with a Hill coefficient (j) of 2.8 (right).

To take a realistic example, for a value of $\Delta F_{xy} = 1.4$ kcal, if $n = 1$, $j = 1$, shown on the left of Fig. 24. $(Y)_\beta/(Y)_\alpha = 810$, for the tetramer P_4, as shown in the right of Fig. 24, with $j = 2.8$ and $n = 4$, $(Y)_\beta/(Y)_\alpha = 6.5$. The difference is dramatic enough, and on this basis we fully expect that if the conversion of chemical into osmotic energy proceeds as we have described in this chapter, the proteins involved as mediators will exhibit cooperative binding of the effector Y.

E. The Results of Positive Effects between Ligands

Active transport of Y can occur equally well if we assume the existence of positive interactions between X and Y in the protein P, leading to stabilization of the form YP-X and to a decrease in the transfer potential as compared to P-X. In that case transport of Y would occur from the compartment with excess P-X (that is, the one into which A-X is being introduced) to the compartment with excess P-R (that is, the one into which D-R enters). The main difference between the two cases is that the Y-transporting species is YP-R in the case of negative interactions and YP-X in the case of positive interactions. In theory reaction V can be driven by the osmotic energy in the gradient just as well by postulating positive as negative interactions in the YP-X complex. In the former case the low effector concentration is required in the compartment with the A components and the high effector concentration in the compartment with the D components. It will be noticed that with positive interactions the effector binding decreases the transfer potential of P-X, as opposed to the case of negative interactions. This implies that an X donor exists with a high potential capable of generating P-X, at a level with A-X. On the other hand, if the transfer potential of P-X is low, and only subsequently raised by the negative X-Y interactions the requirement in the generation of P-X is only for an energy donor of the level of D-X. We believe that the importance of the gradient is in generating a compound of *higher* X-transfer potential from a lower one, like ATP from ADP and phosphate. This evidently requires negative interactions between effector and leaving group and would naturally exclude the case of positive ligand–effector interactions as physiologically unwanted in this particular case.

F. Chemiosmotic Conversion by Proteins and Conformational Changes

From Section IV,E, it follows that negative interactions between effector and leaving group are what makes it possible to shift the equilibrium in a chemical reaction by energy from a gradient, and that the negative interactions are those that permit raising the X-transfer potential of P-X by an amount ΔF_{xy}. One can describe the effects by saying that, in a stationary state like that depicted in Fig. 23, as a result of the negative X-Y interactions the protein acts as a free-energy adder performing the addition of chemical and osmotic energies. We have already commented on the possible origin, conformational or electrostatic, of the interactions themselves in section II,C. At this point we shall attempt to give an answer to the question: What limits the negative interactions, and therefore the maximum amount of energy, that can be added by the protein? The limit must evidently be set by the capacity of the intervening protein to relax to a more stable conformation that would minimize the negative interactions between X and Y. The fact that the few values of ΔF_{xy} known to us are all of the order of 1 kcal indicates indirectly that the protein structure is unable to withstand local strains of larger magnitude without compensatory relaxation of the structure. The only means to increase substantially this amount would be to influence a single group X by several ligands Y attached to different peptide chains, and perhaps this single factor, more than any other, is responsible for the prevalence of oligomeric proteins. In recent years it has become fashionable to attribute all the effects that one ligand exerts over another to "conformational changes in the protein." The previous discussion shows that in the present case conformational changes in the protein could only provide for the minimization of the negative interactions and therefore diminish or altogether destroy the very effect that we consider physiologically valuable. This remark would apply to every case in which strain induced in the protein structure would make an energetic contribution to whatever process. In energy transfer of any kind the mediator molecule must not absorb the energy that is being transferred, and that is what a conformational change would precisely do.

G. Electrical Effects and Chemiosmotic Energy

Mitchell, who was the first to postulate the conversion of osmotic into chemical energy in living organisms (1966, 1972a) assumed that suitable couplings existed, capable of operating this conversion, and wrote the equivalence relation:

$$\Delta F = -RT \ln \frac{(\mathrm{Y})_\beta}{(\mathrm{Y})_\alpha} + \Delta\psi_{\alpha\beta} \tag{4.12}$$

The first term on the right side corresponds to our Eq. (4.8) giving the ideal case of complete reciprocal conversion of the two forms of energy. The second term represents the contribution from the electrical potential difference between the two compartments, as Mitchell assumed that, in principle, conversion of the electrostatic energy in the field into chemical energy was just as likely as that of osmotic into chemical energy. To evaluate this possibility it is necessary to keep clearly in mind the distinction between the chemical effects associated with an actual electric current and the direct effects from an electric field. The chemical effects of an electric (ion or electron) current, authenticated already by Faraday, are not in question. The same cannot be said of direct electrostatic effects due to the field. An electric field cannot per se modify the equilibrium constant of a chemical reaction. Energy conservation requires that the chemical energy be gained at the expense of a reduction of the field strength, which will then under stationary conversion conditions attain a value $\Delta\psi'_{\alpha\beta}$ smaller than $\Delta\psi_{\alpha\beta}$, the value in the absence of electrochemical conversion. The difference $\Delta\psi_{\alpha\beta} - \Delta\psi'_{\alpha\beta}$ would correspond to the net gain of chemical energy and $\Delta\psi'_{\alpha\beta}$ will be the part of the electrostatic energy not available for chemical work. The reduction of the electrostatic potential can only be achieved by movement of electrical charges, that is by an actual electric current. Such electrically driven ion fluxes are actually observed (Witt, 1971). In terms of the two-compartment scheme of Fig. 21, if our ligand Y is charged the difference in concentration in the two compartments, α and β, will generate a potential difference $\Delta\psi_{\alpha\beta}$ unless passive diffusion of an accompanying counterion prevents this from arising. With the same value of ΔF_{xy} a smaller ratio $(Y)_\beta/(Y)_\alpha$ will be reached when $\Delta\psi_{\alpha\beta} \neq 0$ that when $\Delta\psi_{\alpha\beta} = 0$ since in the former case the chemical energy has to pay for the generated potential difference as well as for the concentration gradient. On the other hand, the concentration gradient of the counterion, the diffusion of which prevented the generation of an electric potential or decreased its strength, could presumably be utilized in the same manner as the actively transported ions to promote a chemical reaction. We thus conclude that the *stationary electric field energy* $\Delta\psi_{\alpha\beta}$ associated with a concentration gradient of ions does not represent energy that can be usefully converted into chemical work.

V. Cooperativity and Ligand Correlation

In dealing with the interactions of two different ligands bound to the same protein, we touched upon the relation between ligand correlation and free-energy coupling (Eq. 2.15). The conclusion there

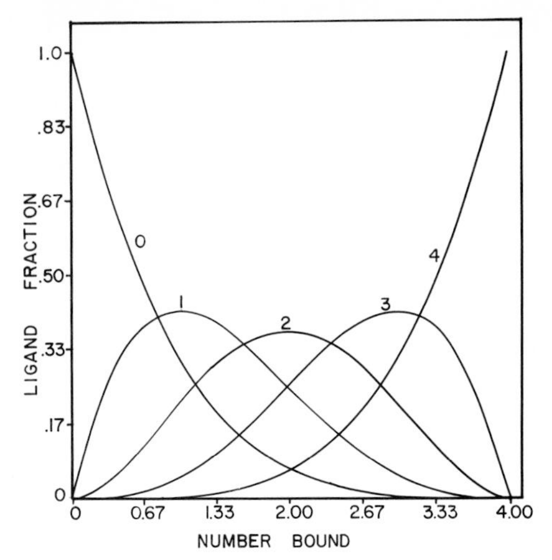

FIG. 25. Normal distribution of protein liganded forms in binding to the four sites of a protein with equal affinity. Ordinate: Fraction of protein molecules carrying 0, 1 · · · 4 ligands. K_1, K_2, K_3, K_4 each = 1.0.

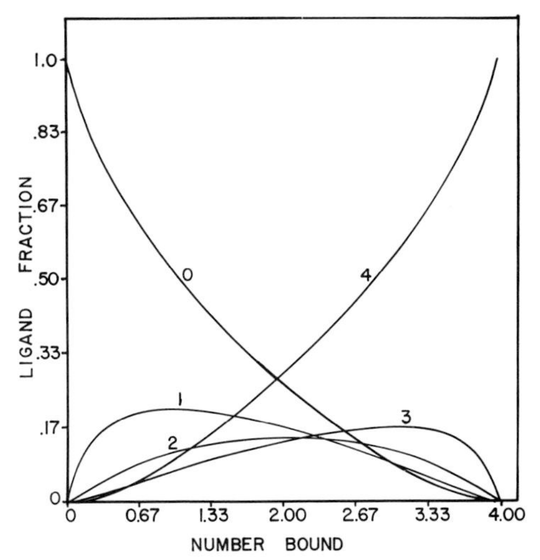

FIG. 26. Distribution of liganded protein species as function of number bound for a protein with ligand dissociation constants like those of stripped hemoglobin (Tyuma *et al.*, 1971a; Gibson, 1970). Notice that at $n = 2$, the molecules with four ligands represent fewer than half of the liganded molecules. $K_1 = 55.0$; $K_2 = 15.0$; $K_3 = 6.0$; $K_4 = 1.0$.

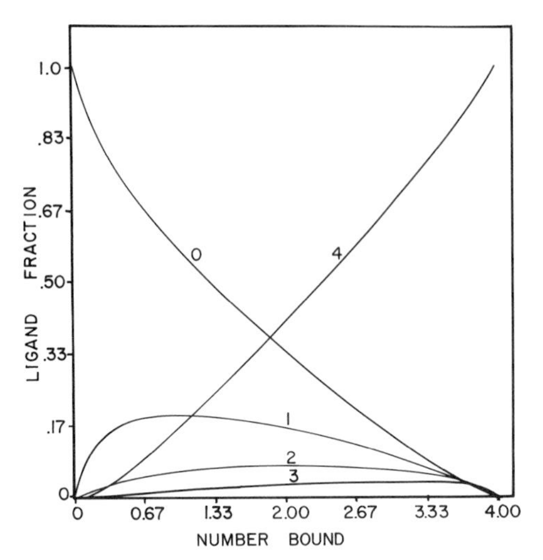

FIG. 27. Distribution of liganded protein species for a protein with dissociation constants like those oxygen in the oxygen-hemoglobin-DPG complex (Tyuma *et al.*, 1971a). Notice the progress toward a bimodal distribution by comparison with Fig. 26. At $\bar{n} = 2$, more than half of the liganded species are molecules of protein with four ligands. $K_1 = 544.0$; $K_2 = 218.0$; $K_3 = 118.0$; $K_4 = 1.0$.

reached is indeed a general one: The normal distribution of ligands among the protein molecules characterize their independent binding and conversely a departure from the normal distribution presupposes the existence of free-energy couplings, positive or negative among the bound ligands. Figures 25–27 show the changes in ligand distribution with number bound introduced by the free-energy couplings operating in hemoglobin in the absence of DPG or in the presence of optimal concentrations of DPG. It is clear that we are still considerably far from a bimodal distribution in which the only species present are the unliganded and fully liganded forms. It is also to be noticed that the ligand distribution is completely specified by the dissociation constants of the system (see Appendix, Section C) and therefore that they provide only alternative, not additional, information to that contained in the free energies of ligand binding. However, a small energy of interaction can give rise to a conspicuous departure from the normal distribution [Eq. (2.15) and Fig. 11], and it follows that a direct study of ligand distribution would be a convenient and sensitive method of revealing free-energy couplings among ligands. Unfortunately not many physical techniques are capable of giving the necessary information. Electronic energy transfer among ligands is a conspicuous exception and has been used to study the distribution of identical fluorescent ligands, because the resulting fluorescence depolarization increases with the number of ligands bound to a protein molecule (Weber and Daniel, 1966). In the case of different ligands bound the appearance of sensitized fluorescence fulfills the same function as it depends upon the presence of the two different ligands, donor and acceptor on the same protein molecule (Brewer and Weber, 1968; Steinberg, 1971).

A. *External or Contingent Correlation*

Although the existence of ligand correlation requires the presence of free-energy couplings among the ligands, the effects can be greatly enhanced by circumstances that are *bound to be present* in actual living organisms. I am referring to the existence of an external correlation, or contingent correlation among the concentrations of the ligands bound to the protein. The effects can be best illustrated by reference to the simple system considered in Section II,A where 1 mole of X and one of Y are bound to the protein. Equation 2.16 describes the effects when [X] is allowed to change and [Y] is kept constant at a value fixed by the parameter $\epsilon = [Y]/K(Y)$. The effects that can be expected under concomitant changes of [X] and [Y] can be simply computed under the assumption that an external correlation is imposed upon their values. We can then write

$$[X] = f(t); \qquad [Y] = g(t) \tag{5.1}$$

making [X] and [Y] functions of a parameter t. In a living organism t will be the time and the functions $f(t)$ and $g(t)$ will be those that describe possible biological processes through which [X] and [Y] are changed. For example, if [X] is being created through metabolic reactions and Y is being destroyed, a valid description of their variation would be given by the equations

$$\begin{aligned} [Y] &= [Y_0]\exp(-\lambda t) \\ [X] &= [X_0] + at \end{aligned} \tag{5.2}$$

where a, Y_0, and λ are parameters and t is restricted to the range $0 < t < t_{max}$ so as to keep both [X] and [Y] within a realistic range of values at all times. It is interesting then to compare the results with those for $\lambda = 0$, in which $[Y] = [Y_0]$, which are described by Eqs. (2.16), (2.23), or (2.32). This comparison gives the magnitude of the changes that can be expected by introduction of an external correlation, and indirectly of the differences to be expected between the results of a simple *in vitro* titration and the *in vivo* behavior. We have carried out computations of this type for a variety of systems,

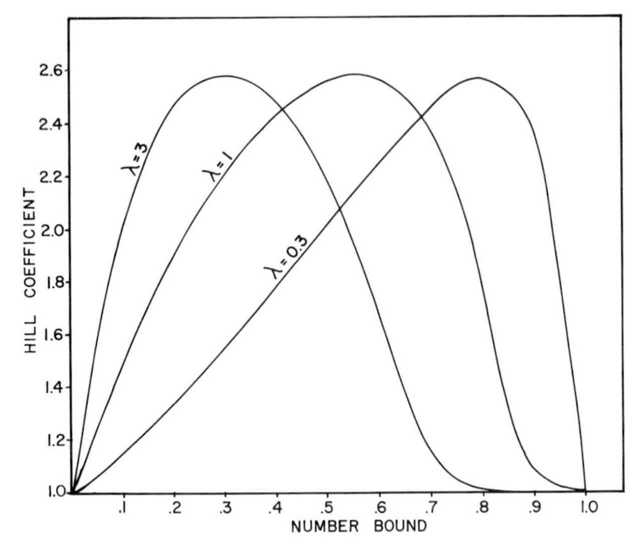

FIG. 28. The Hill coefficient as a function of the number bound for the binding of X, when 1 mole of X and 1 mole of Y can be bound when an external correlation is imposed upon the concentrations of X and Y. The λ's are the rate of disappearance of the negative effector Y. Notice that decrease in λ does not alter the maximum value of the Hill coefficient reached but displaces it toward higher saturation. $\Delta F_{xy} = 3RT$ $[Y_0]/K(Y) = 30$.

the simplest being the one in which 1 mole of X and 1 mole of Y are bound [Eq. (2.16)] and the most complex being a tetramer binding 4 moles of X and 4 moles of Y, and in which each Y ligand is energetically coupled with the two neighboring X ligands. The results are presented in Figs. 28–32. Several general points can be made:

1. External correlation can introduce the appearance of an apparent cooperativity in the binding of X when a single mole of X is bound. This is to be taken to mean that the apparent dissociation constant for X shows a change with degree of X-saturation. If the concentration of free Y changes in the appropriate range as X is added, it will be close to $K(X)$ for some degree of saturation of X, and close to $K(X/Y)$ for another value of X-saturation. Thus introduction of an external correlation can lead to a Hill coefficient very different from 1 (Fig. 28).

2. The presence of an external correlation can lead to a large increase in the value of H, the Hill coefficient at midsaturation, over that obtained when Y is maintained constant. For example, in the case of a tetramer with four Y effectors (Fig. 30) with an intrinsic cooperativity for X leading to a Hill coefficient of 3, for [Y] = constant at an optimal value, the introduction of an external correlation

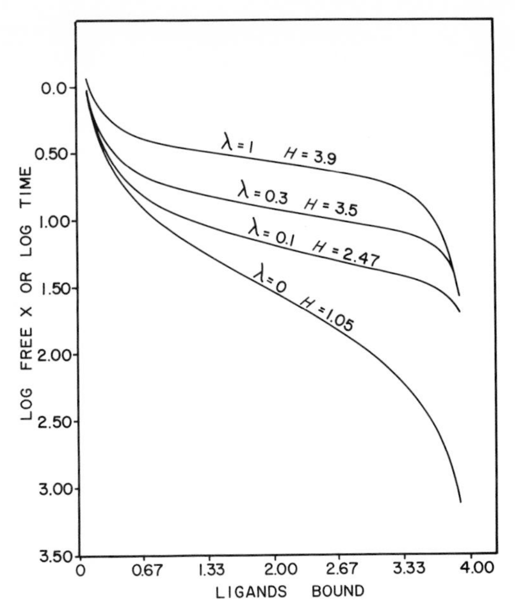

FIG. 29. Effect of external correlation between X and Y for a tetramer binding 4 moles of X and 4 moles of Y, and in which X is coupled to two Y's and Y to two X's. Titration curves in the absence of intrinsic cooperative binding of X ($\Delta F_{XX} = 0$), when $\lambda = 0$ (no external correlation) and when λ is increased. H = Hill coefficient. $\Delta F_{XX} = 0$; $\Delta F_{xy} = 2RT$; $[Y_0]/K(Y) = 100$.

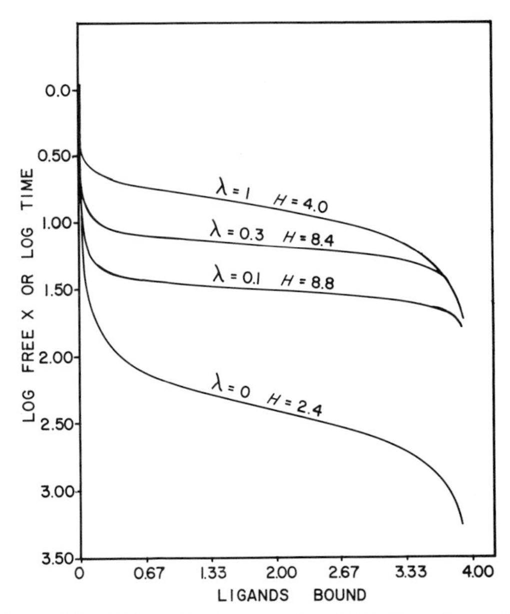

FIG. 30. Similar to Fig. 29 but for the case in which there is intrinsic cooperativity for the binding of X in absence of Y($\Delta F_{XX} = 2RT$). Notice the much higher values of the Hill coefficient (H) reached for this case, indicating the importance of intrinsic cooperativity. $\Delta F_{XX} = 2RT$; $\Delta F_{xy} = 2RT$; $[Y_0]/K(Y) = 100$.

can lead to $H = 11$ to 13. The very sharp transition from the form with no X bound to the fully saturated form takes then all the appearances of an "all or none" transition as envisioned by classical physiology. It is interesting that in order to reach these very high degrees of cooperativity it is necessary for the ligand X to be bound cooperatively in the absence of Y. Figure 32, which compares the degree of cooperativity in X binding as a function of the rate of effector decay for various values of ϵ, shows this feature quite clearly.

3. The very high degrees of cooperativity reached through external correlation are not critically dependent upon initial concentration of effector, Y_0, nor upon λ the rate of effector decay. Figures 31 and 32 show that both the initial effector concentration and the rate of decay can vary within an order of magnitude ($2 < \log \epsilon < 3$; $0.3 < \lambda < 4$) without the characteristic Hill coefficient falling below

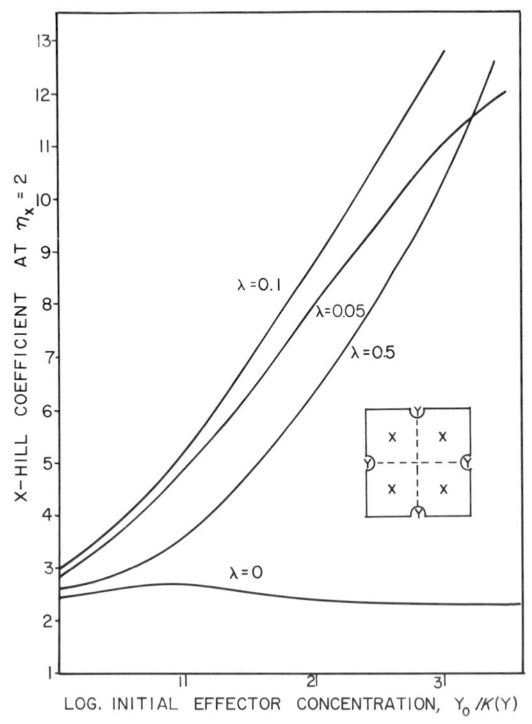

FIG. 31. Cooperativity due to external correlation of ligand and effector concentration. The effect of changing initial effector concentration Y_0 upon the case shown in Fig. 30. Notice that for a 10-fold range in λ and a 10-fold range in Y_0, the Hill coefficient remains above 6. Tetramer with 4 effectors. $\Delta F_{XX} = 2RT$; $\Delta F_{xy} = 2RT$; $[Y] = Y_0 \exp(-\lambda t)$; $[X] = t$.

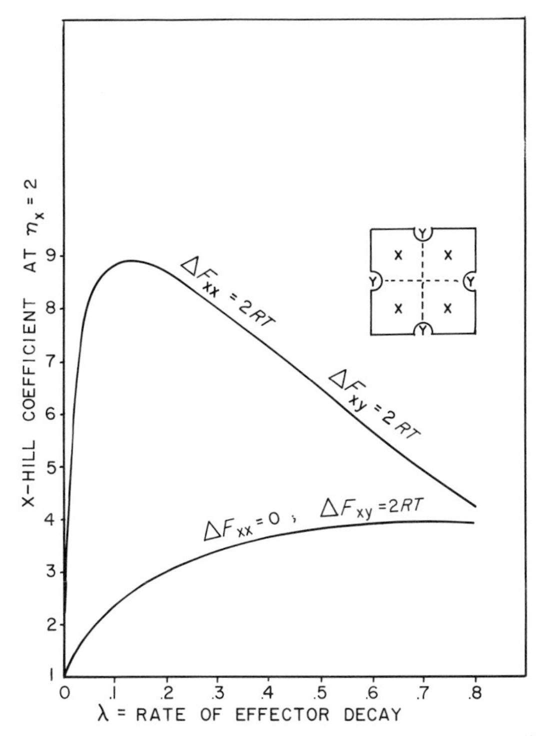

FIG. 32. Cooperativity due to external correlation of ligand and effector concentration. The effect of changing the rate of effector decrease upon the system of Fig. 30. Tetramer with 4 effectors. $[Y] = Y_0 \exp(-\lambda t)$; $[X] = t$; $Y_0/K(Y) = 100$.

7. From this kind of behavior it seems that our phenomenon has every chance of being present and important in living systems.

B. Ligand Correlation and the Two-State Hypothesis of Binding by Proteins

The interactions among bound ligands and the changes in free energy of subunit association with ligand binding can both lead to the appearance of cooperative ligand binding by virtue of the "miniature" lattice interactions that are set up in these cases, as described in Sections II and III. Increased effects are obtained when both these effects are present, and as the number of ligands and the value of the free-energy couplings among them arise, a continuous smooth increase in cooperativity of binding takes place. When both of the mentioned causes contribute to the phenomenon, and the ensuing cooperativity is very pronounced, intermediate states of ligation become rarer, and the unliganded and fully liganded species predominate, although in the known cases they are still not close to the limit, in which they are the only discernible ones (Figs. 26 and 27). It is in the light of these considerations that the hypotheses as to the existence of abrupt structural transitions in protein structure with ligand binding need to be seen. It is important in this context to distinguish between a description of the phenomenon by the assumption of a limited number of states and the actual existence of physically demonstrable unique molecular species. The postulation of two unique states (T and R) with different binding constants (Monod *et al.*, 1965) to describe the cooperativity of binding in hemoglobin is a legitimate oversimplification well founded in many successful analogous precedents. Thus, it is known that even changes in continuous distributions can be represented by discontinuous changes in a few, or even only two species, witness the assumption of dipoles pointing either along or against the field in computations of electric polarization (e.g., Böttcher, 1952) or the use of rotational transitions of 90° in nuclear magnetic resonance (Bloch, 1946) or fluorescence polarization (Spencer and Weber, 1970). These two-state or few-states descriptions are successful in predicting actual experimental results, so long as the methods of observation determine only the linear average of some property of the population, and cannot make by themselves a distinction between cases that yield the same average from different distributions of elements. This is precisely the case with populations of molecules studied by the usual spectroscopic methods (Weber, 1970). It is another matter as to the existence of demonstrable unique molecular states in proteins. The ready identification of the crystal structures of unliganded and fully liganded hemoglobin with the T and R forms of Monod *et al.* (Perutz, 1970) has lent seriousness to this possibility. In Section VI, I shall discuss the connection of protein structure and ligand binding, and it will then be apparent that such close structural identifications are not realistic. In the case of hemoglobin in particular, a third partner, diphosphoglycerate, is critically involved in the binding of oxygen. Efforts to accommodate all the experimentally observed equilibria among these three within the scheme of two unique forms, soon lead to contradiction (Antonini and Brunori, 1970). The suggestion of Perutz (1970) that DPG occupies "a central cavity between the subunits" and that, upon oxygenation, "the subunits close up so that DPG is expelled from the central cavity" is a good example of the erroneous picture that can be conjured up if both the finite character of the free-energy couplings among ligands and the reciprocity of the effects are disregarded and the physical reality of two unique determinant structures is put in its place.

VI. Biological Specificity and Ligand Binding

The complete description of molecular complexes involves three separate yet closely interrelated aspects: structure, energetics, dynamics. Although in this review I am primarily concerned with the energetics it is indispensable to pay some attention to the other two. A reasonable understanding of biological specificity requires the formulation of the necessary relations among the three aspects.

A. *Range of Binding Energies*

A very large spectrum of free energies of interaction of protein–ligand complexes is observed in practice. Table III shows a short list of such values. They vary from 22 kcal/mole to 3 kcal/mole, which correspond to dissociation constants from 10^{-15} to 10^{-2} M. Kinetically the dissociation constant equals the ratio k_-/k_+ of the dissociation and association rates. Direct measurements, particularly by stopped-flow techniques have shown that k_+ never differs greatly from the rate of a diffusion-controlled reaction; in other words, there is no large entropy or enthalpy of activation associated with ligand association. One would expect this to be the case, for the opposite

TABLE III

Free Energies of Formation of Some Protein–Ligand Complexes[a]

Protein	Ligand	ΔF° (kcal)	Observer
Avidin	Biotin	−20	Green (1963)
Lac repressor	DNA	−22	Muller-Hill *et al.* (1968)
			Jobe and Bourgeois (1972)
Anti-DNP-globulin	DNP-lysine	−12, −8[b]	Eisen and Siskind (1964)
Horse liver ADH	NADH	−9	Theorell and McKinley-McKee (1961)
Beef heart LDH	NADH	−8.8	Anderson and Weber (1965)
Serum albumin	ANS	−7.9	Daniel and Weber (1966)
Yeast ADH[c]	Ethanol	−2.5	Mahler and Douglas (1957)
	Acetaldehyde	−4.7	
Fumarase	l-Malate	−2.5	Massey (1953)
	Fumarate	−3	

[a] Abbreviations: DNP, dinitrophenyl; ANS, 1-anilinonaphthalene 8-sulfonate. LDH, lactate dehydrogenase.

[b] Anti-DNP fractions with free energies of associations too large to be determined by the methods employed, but believed to be in the range −12 to −16 kcal, were also present.

[c] Binding of substrates to alcohol-dehydrogenase and fumarase have been selected as representative because they correspond to cases in which the chemical equilibrium permits observations of the forward and backward reactions catalyzed by the enzymes.

would imply that protein molecules exist in a conformation inappropriate for binding and the ligand would have to wait for the gate to open, so to speak. Such a device would be plainly less efficient and would be eliminated by selection in favor of forms that are conformationally available most of the time, and consequently display rates of association close to the diffusion-controlled limit. Thus the 12 or more orders of magnitude difference observed in the dissociation constants must reflect primarily the rates of dissociation. Taking 10^7 $mole^{-1}$ sec^{-1} as a typical value for k_+ we see that k_- would vary in the examples of Table III from 10^{-5} sec^{-1} (or one dissociation per month) to 10^6 (or one dissociation per microsecond). On considering the wide range of values of the dissociation constant the conclusion is inescapable that they are determined by functional requirements. The role of avidin and of the antibodies is to reduce the concentration of biotin and of the antigens, respectively, to the lowest limits. Hemoglobin and myoglobin are stable functional proteins, in comparison with the unstable, nonfunctional apoproteins. The heme–protein complexes owe their stability to the multiple points of attachment of the heme which also serve to transmit to the protein the effects on the electronic distribution in the porphyrin ring caused by oxygen binding. The free energies of formation of enzyme–coenzyme complexes are probably larger than those of common enzyme–substrate complexes for similar reasons. The enzyme–substrate complex has to dissociate in a short time to allow for efficient functioning of the catalyst, and this would be prevented by the slow rate of dissociation that a smaller dissociation constant would imply.

B. Forces Responsible for the Specific Complexes

The early speculations of Ehrlich on the complementary fit of antigens and antibodies, and the postulation of specific complexes as intermediates in enzyme catalysis (Michaelis and Menten, 1913) were finally given a precise formulation in 1948 by Pauling, according to whom biological specificity is always due to short-range interactions, particularly van der Waals dispersion forces between the partners. No specific differences in these forces can be expected in different complexes, apart from the dependence on the molecular polarizability (Hirschfelder, 1965), so that the total energy of interaction depends upon close approach and thus critically upon molecular shape. This simple interpretation of molecular complexes cannot go unqualified. In the first place there are the effects of permanent electric charges, which must be of paramount importance in many cases: The differences in the free energy of binding of NAD^+ and

NADH by the dehydrogenases, or the strong binding of anions, weaker binding of neutral molecules, and very weak binding of cations by serum albumin are two among many examples. Also the solvent is always present as a third component and binding can hardly be accomplished without solvent displacement. In comparing the effects observed upon binding of various ligands to protein the possible intervention of the solvent is often disregarded although from the energetic point of view ligand binding could be more realistically viewed as the partition of the ligand between two media, solvent and protein surface, or as the displacement of one ligand, the solvent, by another. In the simple molecular complexes between two small molecules water often acts as an intermediate binding agent, as shown by Ballshmitter and Katz (1969) for the formation of chlorophyll dimers and by Shefter (1968) for the crystallized complex of thymidine and acridine. Although this mode of behavior has not been reported for protein ligand interactions it would be surprising if it is not found to occur. The importance of the solvent in ligand binding phenomena as well as its contribution to the internal bonding in proteins has been much discussed (Kauzmann, 1959; Némèthy and Scheraga, 1962; Sinanoglu, 1968) but no methodology to achieve a clear distinction of solvent involvement in particular cases has evolved. As a result we must content ourselves with a rather hazy picture, in the sense that although the main sources of molecular interaction have been enumerated and can be recognized in particular cases, their relative importance is in doubt in almost any one. It must be acknowledged that this vagueness about the contribution of different factors is rather in the nature of the problem in that differences in binding must result from sources none of which is truly dominant, if there is to be binding discrimination among close molecular species. Only in this way can a fine degree of molecular recognition by the protein be achieved. One single contribution, however strong, cannot evidently confer the desired specificity. The crystallographic analysis by X-rays of protein–ligand complexes (Phillips, 1967) confirms the existence of multiple points of attachment and a rigorous spatial configuration, both of which characteristically attest to the presence of many energetically equivalent contributions in the molecular complexes. The fact that the binding energy depends upon many small independent contributions has some important consequences: Since no atom or group in the ligand or protein is all-important, it follows that molecules differing from a specific ligand by a single feature, like the replacement of a group, will in many cases be bound as well, and sometimes better, than the specific ligand. Thus the very origin of molecular recognition implies that it cannot be perfect in any sense. These considerations form part of much rational enzymology and pharmacology, and the effects in question must have played a considerable part in the evolution of proteins: If protein specificity of binding is the result of many independent effects none of which is dominant, the resultant property, the binding energy, will have a normal distribution with respect to those changes in the enzyme that give different weights to the individual interactions, an example of Liapunoff's theorem (e.g., Gnedenko and Kintchine, 1964) that applies to such normal distributions. Evolution, playing on the protein modifications has finally adjusted the binding energy to the optimum value for each particular case. It is evident that our knowledge of such weak, multiple interactions is at present very inadequate, and it seems that progress in this area can come only from two sources: a much improved knowledge, of "weak" molecular interactions, both as regards theory and experimental measurements on the very simplest molecular complexes, and structural studies by X-ray crystallography of complexes of diverse proteins with similar ligands, to reveal through invariance those structural features that are important in the particular cases (Rao and Rossmann, 1973).

C. Dynamics of Protein Molecules

Many observations show the mutual influence of ligands bound to a protein, and it is hardly possible to envision such mutual influence taking place without some concomitant change in the protein structure. The X-ray crystallographic observations inform us as to the average position of the atoms in the crystal, but do not contain direct information as to the possible fluctuations in structure either as regards their frequency or energy characteristics. The formation of a crystal itself may already modify the average solution structure or restrain some of the possible fluctuations. Although at present we cannot clearly assess the importance of such crystal forces, they cannot be lightly dismissed for, after all, without them there would be no crystal. The observation that ligand binding produces sometimes shattering of the crystals of the unliganded form (Haurowitz, 1938) and in some others it does not (Chance and Rumen, 1967) indicates the existence of variable reciprocal effects between inter- and intramolecular interactions. Various methods have been used to compare the properties of the molecules in the crystal and in solution. One of the most general is that of low-angle X-ray scattering of the solutions (Kratky and Pilz, 1972). This method appears to be particularly useful in the cases in which the crystal structure has been determined with a high degree of resolution, which permits then calculation of the solution scattering and comparison with the experimental results. A recent study of myoglobin by Stührmann (1973) has led him to conclude that there are differences in the scattering predicted from the crystal and experimentally found, and these represent contributions from scattering centers at distances of the order of 5 Å. The distances between these do not appear as fixed as suggested by the crystal studies, and the solution scattering appears to be made up from contributions belonging to several different conformations. Among investigations employing spectroscopic techniques, a recent study of the crystals and solutions of carboxypeptidase deserves special mention. Johansen and Vallée (1971, 1973) have shown that the α, β, and γ forms of carboxypeptidase A treated with the diazonium salt of arsanilic acid yield almost uniquely the tyrosine 248-azo derivatives. While the solutions are red on account of the formation of the zinc complex of the azo dye, the crystals are uniformly yellow, indicating that the conformation selected by crystallization prevents the interaction of the zinc atom with the azo chromophore. The observation of Quiocho *et al.* (1972) that crystals of similarly labeled carboxypeptidase A_α can be prepared which display the red color of the zinc complex both in solution and in the crystal stresses the fact that the conformations that allow or prevent appearance of the red complex cannot have very different energies and that the crystal forces (Lipscomb, 1973) may tip the balance one way or another depending on various external conditions. Kinetic methods have been used to investigate directly the time-dependent fluctuations of the protein molecules. The evidence from isotope exchange, summarized by Hvidt and Nielsen (1966) is conclusive in showing that a wide spectrum of structural fluctuations is operative. The isotope exchange method fixes attention on the more slowly exchanging hydrogens in the protein. The fast exchanging ones, often comprising more than half of the total, fall within the dead time of the experiments and are not measurable by this technique in real time.[5]

Direct evidence for the existence of very rapid structural fluctua-

[5] Measurable isotope exchange takes place with rate constants of 10^{-5}–10^{-2} sec^{-1}. This slow rate of exchange of many groups is not directly determined by their accessibility to solvent, but by the hydroxyl-dependent exchange rate of exposed groups, which for neutral solutions is close to 10 sec^{-1}. From these considerations it follows that the faster exchange rates *measurable* at pH 7 correspond to hydrogens that become accessible owing to fluctuations in protein structure occurring in the microsecond time range (Å. Hvidt, personal communication).

tions in proteins in the nanosecond time scale has been recently presented by Lakowicz and Weber (1973). Oxygen is an "ideal" quencher of the fluorescence of singlet excited states, in the sense that every collision with an excited fluorophore results in quenching. The cross section for oxygen quenching appears to be that of the fluorophore itself, so that a direct collision, without an interposed molecule, is required for quenching. In eleven proteins studied by Lakowicz and Weber quenching of the tryptophan fluorescence by oxygen was found 10–20 times more effective than quenching by iodide (Lehrer, 1971) although the latter substance quenches the fluorescence of free tryptophan and its derivatives almost as effectively as oxygen. The differences between oxygen and iodide all but disappear in 6*M* guanidine hydrochloride solution so that one is forced to conclude that in the native protein oxygen reaches the vicinity of the tryptophan residues with much higher efficiency than iodide, and that oxygen must penetrate the protein structure helped by its liposolubility and absence of electric charge. These conclusions are strengthened by two other observations: (1) The fluorescence lifetime of the protein decreases proportionally with oxygen quenching, so that this must be a dynamic phenomenon taking place within the lifetime of the excited state (2–6 nsec). (2) Proteins with the bluer emissions, in which the tryptophans are out of contact with the solvent (Teale, 1960), like aldolase or azurin, are just as effectively quenched as lysozyme or bovine serum albumin in which the redder-emitting tryptophans are exposed to the solvent. The rapid diffusion of oxygen to the vicinity of the tryptophans evidently requires rapid fluctuations of the protein structure in the nanosecond time scale. Separation of residues or atoms must take place to permit oxygen motion and the separating groups cannot be held by energies greater than a few kilocalories if they are to become separated by an energy fluctuation occurring in a fraction of a nanosecond. These observations, taken together with the isotope exchange results show that a variety of independent interactions covering a very large energy spectrum are found in protein molecules. From these results it can be definitely concluded that the time-average structure observed by X-ray diffraction is something of an abstraction, since it is not itself widely—or even sparsely—represented at any given time in the population of molecules. I mean by this that if we were to take an instantaneous picture of the molecular population showing us all the coordinates of the atoms for each individual molecule we would have difficulty in finding one that will match the average in *all* respects, although *most* of the molecules will have *most* features in common with it. Indeed the protein molecule model resulting from the X-ray crystallographic observations is a "platonic" protein, well removed in its perfection from the kicking and screaming "stochastic" molecule that we infer must exist in solution. The great importance of the former lies in that it has permitted us to see the origin of the "bulk properties" of the protein, which result from averaging over the whole population. However, when it comes to the conversion of one structure into another, and to render account of any dynamic function of the protein, consideration of the "stochastic" species is indispensable.

Energy–Time Relations in the Fluctuations of Structure

The "fluctuation spectrum" of the protein, that is the graph giving the dependence of the frequency of fluctuations upon the energy of activation E, can be constructed from very general considerations, at least as a first approximation. We shall assume that each nearest-neighbor interaction can be broken independently of all others and that it involves, on average a small energy $\bar{E}$. This assumption of the independence of bonds is one kind of limiting hypothesis. The directly opposite one would be that no localized change in protein conformation is possible. Such a hypothesis would demand a delocalized energy of stabilization, very much like the resonance energy of aromatic hydrocarbons, connecting all parts so as to permit either an overall rearrangement or none at all. We know of no property of proteins, or for that matter of any biopolymer that would justify postulation of such delocalized energy. In its absence, the maintenance of the protein structure must depend upon local bonds and be largely determined by nearest-neighbor interactions, although the assumption of *complete* independence of all bonds would be a great oversimplification. Accepting it as a first approximation, the rate of appearance of a conformation involving simultaneous breaking of N bonds of average energy $\bar{E}$ would be

$$Z = A \exp(-N\bar{E}/RT) \qquad (6.1)$$

The preexponential factor A may be estimated in various ways: It represents in order of magnitude, the rate of exchange of an amount of energy RT. In a tightly coupled system, already Polanyi and Wigner (1928) had estimated this factor as 10^{15} sec^{-1} and justified their choice by comparisons with the rate of unimolecular reactions of known energy of activation. Alternatively, we can think that energy transfer in a periodic, covalently bonded structure like a peptide backbone must be close to the optical vibration rate (Brillouin,

1946), which would yield again $A \simeq 10^{15}$ sec^{-1}. The last equation may then be conveniently written as

$$\log Z = 15 - (N\bar{E}/2.3\,RT) \simeq 15 - N\bar{E}/1.4 \quad (6.2)$$

if $\bar{E}$ is expressed in kilocalories per mole and $T \sim 300°K$. The average interaction energy between neighboring amino acid residues may be set at 3 kcal so that $E/2.3\,RT \sim 2$. Fluctuations that distort the protein structure sufficiently to make possible the motion of oxygen molecules within its matrix will involve a value of N of perhaps 2–3 giving $\log Z \sim 10$, sufficient to explain the observed quenching phenomena. Structural fluctuations involving simultaneous motion of eight to ten residues would take place only once a second, and those involving two or three dozen at once would occur only once in a human lifetime.

Note that the exponential nature of the dependence of Z upon N implies a relatively sharp limit between effects that can be observed in "experimental times," which for proteins are of the order of hours, or at most days, and those that cannot be observed on account of rarity. It is also to be noticed that if the high-energy side of the fluctuation spectrum is limited by the finite time of observation, the low-energy side is equally limited by our capacity to observe effects that depend upon very small energies of interaction. In the experiments on the quenching by oxygen, which on account of the very short times involved, reflect very low energy fluctuations, this difficulty is circumvented by creating first a high-energy state, the singlet state of tryptophan, and observing its extinction upon collisions with oxygen. The creation of the high-energy state is not without influence upon the surrounding protein structure: Because of the changed dipole moment of the tryptophan in the singlet excited state (Lippert, 1957; Weber, 1961; Weber and Lakowicz, 1973) and the concomitant fast relaxation phenomena in the surrounding protein structure (Brand and Gohlke, 1971; Lakowicz and Weber, 1973), we are observing the behavior of a somewhat distorted protein structure in comparison with the structure that surrounds the tryptophan in the ground state. This limitation is characteristic of our experiment, but a limitation must necessarily appear in one form or another whenever we try to investigate the fluctuations of the protein that take place within very short times and depend therefore upon very small energies.

D. Observational Limitations. "Size" of Proteins as Systems

Observational limitations are not often discussed in relation to macromolecules, the implicit assumption being that our present-day methods have not brought us yet to a point at which they would become significant. The example that I have just discussed, as well as previous remarks on the computation of the energy of interaction between protein subunits (see Section III,B), show that we should consider carefully this question particularly in relation to the usefulness of proposing models of protein behavior that are not verifiable by experiment. It is in the light of these remarks that the physical reality of two-state models (or few-states models) of protein behavior must be judged. Whether we examine the experimental evidence for the dynamic fluctuations of structure, or the requirements for the drawing up of an exact energetic balance of the interaction among protein parts, or the limitations imposed by the form of the chemical potential, we should conclude that their physical reality is beyond all possibility of proof.

In cybernetics (e.g., Ross Ashby, 1961) an important distinction is made between "small" and "large" systems. A "large" system is one in which the number of distinctions that the observer can make by the use of his methods is too large to be subjected to analysis. Evidently as our capacity of enumerating the finer details of protein structure through the application of high-resolution techniques like X-ray crystallography and nuclear magnetic resonance (NMR) has grown, the "size" of the protein system has increased enormously. Whether it is already so large that complete description for practical purposes is impossible already, or whether some general rules will be found in the future that reduce this complexity to more manageable proportions it is not yet clear, but it is at least quite certain that we cannot hope to reduce it to the very short number of states that characterize a truly small system.

E. Some General Conclusions: Structure and Function; Molecular Properties and System Properties

The mutual effects exercised by small ligands bound to a protein correspond to a free energy coupling no greater than 1.5 kcal. This value, small compared to the total energy of interaction of the amino acid residues within a globular protein is nevertheless sufficient to introduce correlations among the bound ligands that explain many observed effects. The small energies required for this satisfactory ligand correlation render very doubtful the proposition that special "mechanisms" have evolved to perform them. They appear rather as a simple and inevitable consequence of the fact that the forces binding the individual residues in the protein to each other are of the same magnitude as the forces between proteins and ligands. We can expect mutations and other alterations of the protein to change the balance of forces in what from the standpoint of present knowledge appears as virtually unpredictable fashion. This circumstance does not depend as much from our present ignorance of the nature of internal protein interactions as from their small strength and multiplicity. A further consequence of the large number and small energy, and the reciprocal character of the interactions involved, is to render meaningless the postulation of a hierarchy of elements or forces in the protein, so that the influence of some is paramount in "imposing" a conformation on the others. Many parts of the compact protein structure can and will contribute toward the shift in energy and structure taking place upon ligand binding. To give every part of the protein its precise weight in such process seems certainly beyond present day possibilities. The energies of ligand–ligand interaction, although small are nevertheless sufficient to shift the covalent equilibria in which proteins take part, to a significant extent, and this will, in our opinion, be found responsible for the interconversion of chemical and osmotic energies in metabolism.

The small observed energies of ligand–ligand interaction arise from the limited strain that can be set up in the structure of a globular protein without producing a compensatory relaxation. A means of increasing the differences in energy of the liganded and unliganded states will be to associate globular units so that the individual effects become applied to the common boundaries. It is this energetic requirement which, in our opinion, has been responsible for the evolution of multichain proteins. The observation that the effect of an external correlation among bound ligands can greatly enhance those arising from ligand–ligand interactions within the protein leads us to a very important question: The relation of molecular properties and system properties. While the ligand correlation due to the existence of free-energy coupling among the bound ligands is a *molecular property* to be explained by a study of the isolated protein, the enhancement of the effects by the simultaneous changes in ligand concentration is a *system property* which requires consideration of other entities. Many of the most interesting properties of proteins will be found to belong to this latter category, just in the same way as the lively and varied activities of a computer depend upon the mutual relations of switches and relays and are not easily predicted from their monotonous isolated behavior. It is not difficult to foresee that observations on proteins as parts of larger systems will be found ultimately to be more important in the explanation of organic functions that many present-day attempts to "read" in the simple properties of the isolated molecule the part played by the protein in the living system. The consideration of ligand–protein interactions provides us also with a model for the relative importance to be assigned to molecules in the organic functions: We can think of the protein as determining the fate of the small molecules by its catalytic activity, but we can also think of the small molecules as forcing the bigger fellows to act in their own interests by the changed behavior resulting from ligand binding, or even influencing each other through the "passive" protein medium.

APPENDIX

A. *Titration Curves*

The dependence of the equilibrium upon the ligand concentration can best be characterized by means of a plot introduced by Niels Bjerrum (1923), also used by J. Bjerrum (1944), in which the logarithm of the free-ligand concentration is plotted against the degree of saturation, or the average number bound. The resulting graph may be termed the *titration curve* of the protein with the ligand X. Several aspects make this plot greatly preferable to others; these include the following ones.

1. A *complete titration curve* is characterized by the presence of a saturation region, at which increment of free ligand does not result in any appreciable change in saturation, thus giving rise to a "vertical" segment in the titration curve.

2. An *incomplete titration curve* is shown by the absence of a saturation region. The plot does not permit a linear extrapolation, and thus compels the observer to admit the incomplete character of the experimental data. This is to be contrasted with the linear plots in which by linear extrapolation a stoichiometry and a dissociation constant can always be calculated, regardless of the adequacy of the data (Weber, 1965; Deranleau, 1969).

3. The existence of equal, independent binding sites, irrespective of their number per protein molecule, is shown by a titration curve with a logarithmic interval in X of 1.908 units, between the saturations of 0.1 and 0.9. A shortened span indicates a cooperative binding process in which certain sites increase in affinity when others are occupied, and a lengthened span indicates either the opposite situation (decrease in affinity when certain sites are occupied) or the presence of heterogeneous, independent binding sites.

B. General Formulation of Multiple Equilibria

Consider for simplicity a macromolecule made up of a single polymer chain, therefore unable to dissociate into subunits, and having N potential binding sites[6] per molecule for a ligand X. These N sites may be numbered as distinct, for example $a_1, a_2 \ldots a_N$. If J out of the N sites are occupied $\binom{N}{J}$ distinct combinations of occupied sites or site-isomers are possible. These isomers can be numbered from 1 to $\binom{N}{J}$ so that the index i appearing below in parentheses identifies a given site-isomer. The change in standard free energy upon formation of the ith site-isomer from J ligand molecules and protein is independent of the order of addition, or "path" of the reaction, and will be called

$$\Delta F_J^\circ(i) \qquad \left(1 \leqslant i \leqslant \binom{N}{J}\right) \tag{A1}$$

The standard chemical potential of the same site-isomer is

$$\mu_J^\circ(i) \tag{A2}$$

A site-isomer can be converted into any of the others by dissociation of a suitable number of ligands and addition of an equal number to other sites. The change in standard free energy upon conversion is independent of the way in which this process is carried out and for conversion of the ith into the kth site-isomer equals,

$$\Delta F_J^\circ(i \rightarrow k) = \mu_J^\circ(k) - \mu_J^\circ(i) \tag{A3}$$

The ratio of the isomers present at equilibrium is determined by the equilibrium constant $K_J(i,k)$,

$$K_J(i,k) = \exp\,(\Delta F_J^\circ(i \rightarrow k)/RT) \tag{A4}$$

Accordingly the site isomers of the species PX_J are represented in the equilibrium population with weights

$$W_J(i) = \frac{\exp\,(-\Delta F_J^\circ(i)/RT)}{\sum_{i=1}^{\binom{N}{J}} \exp\,(-\Delta F_J^\circ(i)/RT)} \tag{A5}$$

It follows that there is for each species PX_J an *average standard*

[6] It is usual to find statements, particularly in the biochemical literature, that binding sites "become available" as a result of the binding of the same or another ligand, as though N could not be in principle a fixed number. Such a statement implies only that the affinity of some of the N sites for the ligand is small enough for binding to be undetectable unless other sites are occupied. Moreover the change in affinity of these "unavailable" sites is entirely determined by the free-energy couplings among the bound ligands.

chemical potential which is determined by the existence of the fixed proportions of site-isomers, given by Eq. (A5), and that this equals

$$\overline{\mu^\circ(\mathrm{PX}_J)} = \sum_{i=1}^{\binom{N}{J}} \mu_J^\circ(i)\, W_J(i) \tag{A6}$$

A similar equation defines $\overline{\mu^\circ(\mathrm{PX}_{J+1})}$ and the standard free-energy change in the reaction

$$\mathrm{PX}_J + \mathrm{X} = \mathrm{PX}_{J+1} \tag{A7}$$

is therefore,[7]

$$\Delta F^\circ(J \to J+1) = \overline{\mu^\circ(\mathrm{PX}_{J+1})} - \overline{\mu^\circ(\mathrm{PX}_J)} - \mu^\circ(\mathrm{X}) \tag{A8}$$

In consequence the intuitive treatment of Adair (1925) is vindicated: Any equilibrium involving the addition of N moles of ligand to a macromolecule involves only N standard free-energy changes, or N equilibrium constants corresponding to the N successive additions of the ligand. The dependence of $\bar{n}$, the average number bound, upon the concentration of free ligand [X], is given by Adair's equation:

$$\bar{n} = \frac{\sum_{J=0}^{N} J\binom{N}{J} \dfrac{[\mathrm{X}]^J}{K_0 \cdots K_J}}{\sum_{J=0}^{N} \binom{N}{J} \dfrac{[\mathrm{X}]^J}{K_0 \cdots K_J}}; \qquad K_0 = 1 \tag{A9}$$

or in a compact form, useful for computation purposes,

$$\sum_{J=0}^{N} \binom{N}{J} \frac{[\mathrm{X}]^J}{K_0 \cdots K_J} (\bar{n} - J) = 0 \tag{A10}$$

The N dissociation constants appearing in Adair's equation, in the form given above, are defined by the equations,

$$K_{J+1} = \frac{(N-J)\,[\mathrm{PX}_J][\mathrm{X}]}{(J+1)\,[\mathrm{PX}_{J+1}]} \tag{A11}$$

in which the concentrations of free sites and bound sites ($(N-J)$ $[\mathrm{PX}_J]$ and $(J+1)$ $[\mathrm{PX}_{J+1}]$, respectively) are used rather than the macromolecular concentrations. If the latter concentrations are used, the values of K_J come to depend not only upon the free energy of binding to the sites but also upon N and J. In a tetramer with independent binding sites of equal affinity, the four constants would

[7] The equilibrium constants for the steps are numbered K_1 to K_N. They may stand for dissociation or association constants, which are then simply the reciprocal of each other. The corresponding free energies are equal, though of opposite sign.

not be equal but would stand in the ratios 4:3/2:2/3:1/4, respectively. It is unfortunate that the definition of the dissociation constants is often not mentioned in papers in which actual experimental values are given, so that the reader is often unable to decide on the exact quantitative significance of the findings. Binding site concentrations—not macromolecular concentrations—should be used in formulating the equilibrium whenever N is known. If $K_1 \cdots K_N$ are set equal to K in Eq. (A9), this reduces to

$$\bar{n} = \frac{N \dfrac{[X]}{K} (1 + [X]/K)^{N-1}}{(1 + [X]/K)^N} = N \frac{[X]}{[X] + K} \tag{A12}$$

the equation for simple independent binding to N sites of equal affinity. This would not be the case if the factors $\binom{N}{J}$, arising from the expression of the equilibria in terms of site concentrations, were left out.

C. Ligand Distributions

Equation (A9) may be simply written

$$\bar{n} = \sum_{J=0}^{N} f(J) \cdot J \tag{A13}$$

where

$$f(J) = \frac{\binom{N}{J} \dfrac{[X]^J}{K_0 \cdots K_J}}{\sum\limits_{J=0} \binom{N}{J} \dfrac{[X]^J}{K_0 \cdots K_J}} \tag{A14}$$

is the fraction of the macromolecules bound to j ligands. The distribution of the liganded species, irrespective of site-isomers, is thus determined by the N constants of Adair's equation. Specification of the site-isomer distributions requires also $\binom{N}{J}$ quantities for each stage of ligation J, or 2^N in all. Thus, for $N = 4$, specification of 20 independent quantities would be required. The demands for a complete description of such systems stand in contrast to the lack of appropriate present methodology to study directly the distribution of ligands.

Average Free-Energy Coupling Among Bound Ligands

Corresponding to the average chemical potential and the average free energy of formation there is an average free-energy coupling among the J bound ligands which may be defined by a simple extension of Eq. (2.6).

$$\Delta F_{XX} = \Delta F^\circ(P \rightarrow PX_J) - J\Delta F^\circ(P \rightarrow PX) \qquad (A15)$$

while $\Delta F^\circ(P \rightarrow PX)$ is calculable from the experimental equilibrium constant observed at very low degrees of saturation, when most protein molecules have one ligand bound, $\Delta F^\circ(P \rightarrow PX_J)$ cannot be directly measured, except when $J = N$, in which case it is determined by the dissociation constant for X observed as saturation approaches unity. In stripped hemoglobin the dissociation constants at low saturation and those at which saturation is almost complete (Gibson, 1970; Roughton *et al.*, 1955) differ by a factor of about 60. From this figure it follows that the average free-energy coupling between binding sites approximates 1.25 kcal. The multiple binding of two ligands, X and Y, may be described by a simple extension of the method used to characterize the binding of a single ligand. In this way average free energies of binding for the formation of mixed species may be defined, ΔF° $(P \rightarrow PX_JY_K)$ with properties identical to the corresponding quantities for a single ligand. Apart from the average free-energy couplings between X ligands on one hand, and between Y ligands on the other, it will be possible to determine experimentally the mixed or X-Y free-energy couplings in any given case.

D. The Hill Coefficient

Consider for simplicity the case in which only 2 moles of ligand are bound, so that Eq. (A9) gives,

$$S = \bar{n}/2 = \left(\frac{[X]}{K_1} + \frac{[X]^2}{K_1K_2}\right) \Big/ \left(1 + 2\frac{[X]}{K_1} + \frac{[X]^2}{K_1K_2}\right) \qquad (A16)$$

A mean dissociation constant $K = (K_1K_2)^{1/2}$ may be defined, which implies

$$K_1 = mK; \qquad K_2 = Km^{-1}; \qquad K_1/K_2 = m^2 \qquad (A17)$$

where m is an arbitrary number. Forming the quantity $R = S/(1 - S)$, after introduction of Eq. (A17) into the preceding equation,

$$R = \frac{[X]}{K}\left(1 + m\frac{[X]}{K}\right) \Big/ \left(m + \frac{[X]}{K}\right) \qquad (A18)$$

A plot of log R against log [X] is called a Hill plot, and its slope at a given saturation is H_s, the corresponding Hill coefficient. According to this definition,

$$H_s = \frac{d \log R}{d \log [X]} = \frac{[X]}{R} \frac{dR}{d[X]} \tag{A19}$$

and using Eq. (A18)

$$H_s = \left(1 + 2m \frac{[X]}{K} + \frac{[X]^2}{K^2}\right) \Big/ \left(1 + (m + m^{-1}) \frac{[X]}{K} + \frac{[X]^2}{K^2}\right) \tag{A20}$$

Evidently the Hill coefficient is uniformly unity if $K_1 = K_2$ or $m = 1$. It tends to unity, independently of m, when $[X]/K \ll 1$, and when $[X]/K \gg 1$, that is at very low and very high degrees of saturation. When $m > 1$, $H_s > 1$ and the binding is cooperative, denoting a negative free energy of ligand interaction ($\Delta F_{XX} < 0$). On the other hand, when $m < 1$, $H_s < 1$, and the binding is by intrinsically different and independent sites or by identical sites with a positive free energy of interaction ($\Delta F_{XX} > 0$). The maximum value of 2 of the Hill coefficient is approached when m is allowed to increase indefinitely, but it should be noticed that a free energy coupling of -2.8 kcal between the binding sites would already result, for the midsaturation region, in a Hill coefficient of 1.81, therefore within 10% of the maximum value. Setting in Eq. (A9),

$$a_J = \frac{\binom{N}{J}}{K_0 \cdot \cdot \cdot K_J}$$

It is easily deduced that, in the general case,

$$H_s = \frac{\sum_{J=0}^{N} a_J J^2 [X]^J}{\sum_{J=0}^{N} a_J J [X]^J} - \frac{\sum_{J=0}^{N} a_J J(N-J)[X]^J}{\sum_{J=0}^{N} a_J (N-J)[X]^J} \tag{A21}$$

The sums in the second quotient of the right-hand side of Eq. (A21) include only powers of [X] up to the $(N-1)$th, since $N - J = 0$ for $J = N$. Therefore the maximum value of H_s is reached when the coefficient a_N greatly exceeds all the others, and the limiting value becomes then that of the first quotient, which in that case evidently tends to N. When $[X]^J \ll a_J$, that is, when $\bar{n} \ll 1$, the first quotient tends to 1 and the second to zero. When [X] increases indefinitely the first quotient tends to N and the second to $N - 1$. Thus H_s tends always to unity at low and high degrees of saturation, and the same rules apply to the general case as those displayed in Eq. (A20) for the case of $N = 2$.

The Hill coefficient measured close to midsaturation, $H_{1/2}$, may be taken as an approximate measure of the order of the binding reaction, that is, the effective number of ligand molecules that combine "simultaneously" with the protein, in which case the saturation may be expressed by

$$S = [X]^{H_{1/2}}/(K + [X]^{H_{1/2}}) \tag{A22}$$

The last expression can evidently hold true only for the region of saturations close to $S = 1/2$, but as Figs. 6 and 7 shows, no gross error would be incurred by supposing this range to extend from $S = 1/4$ to $S = 3/4$.

Discussions of various properties of the Hill coefficient are given by Wyman (1964), Weber and Anderson (1965), and Weber (1965).

E. Correlation between Ligands: Computation of Binding Curves

The computations on which Fig. 28–32 are based use Eqs. (5.2) together with Eq. (A10). The apparent dissociation constants for X are functions of $\epsilon = [X]/K(Y)$ like those of Eqs. (2.24) or (2.32). In the computations t is used as the incremented variable, $[X]$ (t) and $[Y]$ (t) are computed, and with the help of the set of functions of ϵ giving the dissociation constants, and Eq. (A10), a value of $\bar{n}_x$ is obtained. An iterative procedure involving halving of t intervals is used to obtain values of $\bar{n}_x$ as close as desired to multiples of 0.1. The Hill coefficient is calculated as ratio of the finite differences $\Delta \log (N/\bar{n}_x - 1)/\Delta \log [X]$, between neighboring values of $\bar{n}_x$ and corresponding values of $[X]$, respectively.

F. Remarks on the Experimental Methods for the Study of Chemical Equilibria of Ligands and Proteins. Uniqueness of the Experimental Quantities

The equilibria between proteins and ligands are measured by direct, semidirect, and indirect methods. Direct methods are those in which the concentrations of free and bound ligands are measured by their contribution to some intrinsic property of the solution that permits to differentiate between the free and bound species. In most cases, the direct methods are spectroscopic methods in which use is made of differences in the interactions of the molecules with radiation. In the semidirect methods a further equilibrium is created by the use of a semipermeable membrane, a second solvent or a different phase like a highly cross-linked polymer, in which the activity of the bound ligand is effectively zero. In the indirect methods, use is made of *kinetics*, in one form or another, to decide upon the equilibrium values. In one type of indirect method the rates of association and dissociation are measured separately; in another the kinetics of a second process, known to depend upon the concentration of either free or bound ligand, is followed. While direct or semidirect methods depend only upon the thermodynamic properties of the system, the indirect methods involve always additional assumptions as regards the relations of the kinetic processes studied to the thermodynamic behavior of the system. The recently developed methods of fluctuation analysis (Magde *et al.*, 1972) are of the indirect type. The direct methods are the more accurate, although not always the simplest or most reliable, but under the best conditions the semidirect methods are fully comparable. The indirect methods can be relied upon only for comparative purposes or for magnitude determination, rather than for the obtention of precise values. It must be acknowledged, however, that in the cases of very high standard free energies of binding, of the order of 15 kcal or more, all the known values have been established by indirect methods. On comparison of the binding free energies obtained by different methods, one may be tempted to assume that the results—if reliable—should be identical to any order of accuracy. This, however, is not the case, and consideration of the simplest examples should be sufficient to prove the point. Consider the equilibrium leading to formation of a complex between two molecules, e.g., nucleotides, studied by two different methods, say NMR and absorption spectroscopy. The exact value of the dissociation constant, and therefore the binding free energy, will be determined by the concentration at which half the maximum effect is observed. If the interactions of the molecules were due to one single structural feature, we would expect complete agreement between the methods, but this is never the case. When two aromatic rings interact at short distances the perturbation of the absorption or emission is due to interactions of the π electrons of the two molecules. On the other hand, the NMR effects result from protons in one molecule being subjected to the ring currents in the other. Particularly if the forces holding the molecules are weak, and this is the experimental case with nucleotides (Ts'o *et al.*, 1963; Spencer and Weber, 1972) there will be contributions from a number of relative dispositions of one molecule with respect to the other, and the two methods will almost certainly give different weights to the different spatial dispositions. As a consequence, the average effects observed by both methods, and the free energies of association calculated from the experimental data, will differ, without this implying any experimental error in either usage. In other words, even in simple molecular complexes we are faced with the problem that the thermodynamic quantities represent average values that may be differently weighted by the various methods.[8] This need not be always a disadvantage, for an estimation of the heterogeneity of the population can be made only by the formation of several averages in which the members of the population are differently weighted. The situation with ligand–protein complexes is the same as for the complexes between small molecules. The parallel extends even to the uniqueness of the crystallographic structure in molecular complexes, which, if taken to imply literally the existence of a unique type of association in solution, would indeed rule out the possibility of obtaining different values for the free energy of binding by the use of different methods.

No single method of investigation can be applied in the determination of the whole range of dissociation constants shown in Table III. Each method is limited in its applicability by the concentration at which comparable proportions of free and bound ligand are found. The thermodynamic activity of proteins is simply proportional to the concentration only in the region below $10^{-4}\,M$, and in many cases even less. For all practical purposes the rule is valid that the observations must be made at concentrations at which midsaturation is attainable, which requires both protein and ligand to have concentrations close to the dissociation constant of the complex. A conspicuous exception is the binding of protons. The unique characteristics of the electrometric methods permits to measure $10^{-14}\,M$ protons in the presence of 110 M bound protons. This ratio of 10^{16} of bound to free ligand is unique. In the case of molecular complexes, it is rare that such ratio reaches the value of 10^2, hence the requirement for the measurements to be in the region in which free and bound ligand are comparable. Since the methods of measurement have usually a range of concentration that is optimum for the observation as well as clear upper and lower bounds of concentration, beyond which they cannot be applied, it follows that the magnitude of the standard free energy of binding of the complex under study will determine the method to be used in its measurement.

Equilibria with standard free energies of association of 14 kcal or more ($K < 10^{-9}\,M$) are not measurable by direct methods because the very small amount of free ligand present at equilibrium in solutions containing 10^{-4} to $10^{-7}\,M$ protein cannot be detected by

[8] The differences that I have in mind can hardly be expected to be as large as a factor of two, not the much larger differences often reported, which imply some systematic error, conceptual or experimental, in one of the methods used.

present-day spectroscopic methods. In these cases a successful examination has been carried out by measuring independently k_+ and k_-, the rates of association and dissociation. This method is originally due to Green (1963), who used it to study the avidin-biotin equilibrium. The concentration range accessible to the spectroscopic methods (light absorption, fluorescence, nuclear magnetic resonance, and electron spin resonance) is in the range of 10^{-1} to 10^{-8} M. The lower limit is determined by the sensitivity of detection in absorption spectroscopy ($\sim 10^{-6}$ M) and by the presence of emitting impurities in fluorescence. The upper limit is determined by the difficulty of detecting the small amounts of ligand bound to the protein when the latter is in the presence of 100 or more times its molarity in ligand. Low free-energy complexes are formed by proteins with inorganic monovalent anions and cations and since these low-energy complexes do not appreciably perturb the protein, they are not easily revealed by spectroscopic methods. The equilibria of proteins with inorganic ions are undoubtedly very important in the molecular description of such functions as muscle contraction and nerve conduction. The development of methods for their accurate study would fill what is now the most conspicuous gap in our knowledge of protein–ligand interactions.

Acknowledgments

Part of this review was written during a stay at the Carlsberg Laboratory in Copenhagen. The author wishes to record his gratitude to the director, Dr. Martin Ottesen, and to other members of the Laboratory for the generous hospitality accorded to him; to Professor Aase Hvidt, of the Oersted Institute, who read parts of the first version of this review and offered valued and much needed criticism; to colleagues and students at the School of Chemical Sciences of the University of Illinois for many discussions on the subject throughout the years; and finally, to Drs. J. T. Edsall and F. M. Richards, who made many suggestions that helped to remove a number of inconsistencies and obscurities from the text.

References

Adair, G. S. (1925). *J. Biol. Chem.* **63**, 529.
Adams, M. J., Liljas, A., and Rossmann, M. G. (1973). *J. Mol. Biol.* **76**, 519.
Anderson, S. R., and Weber, G. (1965). *Biochemistry* **4**, 1948.
Antonini, E., and Brunori, M. (1970). *Annu. Rev. Biochem.* **39**, 977.
Antonini, E., Wyman, J., Rossi-Fanelli, A., and Caputo, A. (1962). *J. Biol. Chem.* **237**, 2773.
Ashby, R. (1961). "Introduction to Cybernetics," p. 61. Chapman & Hall, London.
Ballshmitter, K., and Katz, J. J. (1969). *J. Amer. Chem. Soc.* **91**, 2661.
Benesch, R., and Benesch, R. E. (1967). *Biochem. Biophys. Res. Commun.* **26**, 162.
Benesch, R., Benesch, R. E., and Yu, C. I. (1968). *Proc. Nat. Acad. Sci. U.S.* **59**, 526.
Benesch, R. E., Benesch, R., Renthal, R. D., and Maeda, N. (1972). *Biochemistry* **11**, 3567.

Bernhard, S. A., and McQuarrie, R. A. (1972). *J. Mol. Biol.* **74**, 73.
Bjerrum, J. (1944). *Kgl. Dan. Vidensk. Selsk. Mat. Fys. Medd.* **21**, 4.
Bjerrum, N. (1923). *Z. Phys. Chem.* **106**, 219.
Bloch, F. (1946). *Phys. Rev.* **70**, 460.
Blow, D. M., Wright, C. S., Kukla, D., Ruhlmann, A., Steigeman, W., and Huber, R. (1972). *J. Mol. Biol.* **69**, 137.
Bolton, W., and Perutz, M. F. (1970). *Nature (London)* **228**, 552.
Böttcher, C. J. F. (1952). "Theory of Electric Polarization." Elsevier, Amsterdam.
Brand, L., and Gohlke, J. R. (1971). *J. Biol. Chem.* **246**, 2317.
Brewer, J., and Weber, G. (1968). *Proc. Nat. Acad. Sci. U.S.* **59**, 216.
Brillouin, L. (1946). "Wave Propagation in Periodic Structures." Dover, New York.
Brillouin, L. (1962). "Science and Information Theory," 2nd ed., Chapter XV. Academic Press, New York.
Bromberg, P. A., Alben, J. O., Bare, G. H., Balcerzak, S. P., Jones, R. J., Brimhall, B., and Padilla, F. (1973). *Nature (London), New Biol.* **243**, 179.
Brunori, M., Noble, R. W., Wyman, J., and Antonini, E. (1968). *J. Biol. Chem.* **241**, 5238.
Bull, H. B. (1964). "Introduction to Physical Biochemistry," Chapter 9. Davis, Philadelphia, Pennsylvania.
Chan, S. I., Schweizer, M. P., Ts'o, P.O.P., and Helenkamp, G. K. (1964). *J. Amer. Chem. Soc.* **86**, 4182.
Chance, B., and Rumen, N. (1967). *Science* **156**, 563.
Changeux, J. P., Gerhardt, J., and Schachman, H. K. (1968). *Biochemistry* **7**, 531.
Charache, S., and Jenkins, T. (1971). *J. Clin. Invest.* **50**, 1554.
Chen, A. K., Bhan, A., Hopper, S., Abrams, R., and Franzen, J. S. (1974). *Biochemistry* **13**, 654.
Conway, A., and Koshland, D. F., Jr. (1968). *Biochemistry* **7**, 4011.
Cook, R. A., and Koshland, D. R., Jr. (1970). *Biochemistry* **9**, 337.
Daniel, E., and Weber, G. (1966). *Biochemistry* **5**, 1893.
Deal, W. C., Rutter, W. J., Massey, V., and van Holde, K. E. (1963). *Biochem. Biophys. Res. Commun.* **10**, 49.
Deranleau, D. A. (1969). *J. Amer. Chem. Soc.* **91**, 4044 and 4050.
Dickerson, R. E. (1964). *In* "The Proteins" (H. Neurath, ed.), 2nd ed., Vol. 2, p. 603. Academic Press, New York.
Dickerson, R. E. (1972). *Annu. Rev. Biochem.* **41**, 815.
Dixon, M. (1951). "Multienzyme Systems," p. 93. Cambridge Univ. Press, London and New York.
Eisen, H. N., and Siskind, G. W. (1964). *Biochemistry* **3**, 996.
Epstein, H. F. (1971). *J. Theor. Biol.* **31**, 69.
Foster, J. F. (1960). *In* "The Plasma Proteins" (F. W. Putnam, ed.), Vol. 1, p. 221. Academic Press, New York.
Gibbs, W. (1948). "Collected Works," Vol. I, p. 92. Yale Univ. Press, New Haven, Connecticut.
Gibson, Q. H. (1970). *J. Biol. Chem.* **245**, 3285.
Glasstone, S. (1947). "Thermodynamics for Chemists," pp. 213 and 273. Van Nostrand-Reinhold, Princeton, New Jersey.
Gnedenko, B. V., and Kintchine, A. I. (1964). "Introduction à la théorie des probabilitiés," p. 151. Dunod, Paris.
Green, N. M. (1957). *Biochem. J.* **66**, 407.
Green, N. M. (1963). *Biochem. J.* **89**, 585.
Guidotti, G. (1967). *J. Biol. Chem.* **242**, 3704.

Haurowitz, F. (1938). *Hoppe-Seyler's Z. Physiol. Chem.* **254,** 268.
Hill, T. L. (1960). "An Introduction to Statistical Thermodynamics," Part III, p. 235. Addison-Wesley, Reading, Massachusetts.
Hirschfelder, J. O. (1965). *In* "Molecular Biophysics" (B. Pullman and M. Weissbluth, eds.), p. 325. Academic Press, New York.
Houwink, R. (1958). "Elasticity, Plasticity and Structure of Matter," 2nd ed., p. 7. Dover, New York.
Hvidt, A., and Nielsen, S. O. (1966). *Advan. Protein Chem.* **21,** 288.
Jobe, A., and Bourgeois S. (1972). *J. Mol. Biol.* **72,** 139.
Johansen, J. T., and Vallée, B. L. (1971). *Proc. Nat Acad. Sci. U.S.* **68,** 2532.
Johansen, J. T., and Vallée, B. L. (1973). *Proc. Nat. Acad. Sci. U.S.* **70,** 2006.
Kaback, R. H. (1970). *Annu. Rev. Biochem.* **39,** 561.
Kaback, R. H. (1972). *Biochim. Biophys. Acta* **265,** 367.
Kauzmann, W. (1959). *Advan. Protein Chem.* **14,** 34.
Kayne, F. J., and Price, N. C. (1972). *Biochemistry* **11,** 4415.
Klapper, M. H. (1973). *In* "Progress in Bio-organic Chemistry" (E. T. Kaiser and F. J. Kezdy, eds.), p. 55. Wiley, New York.
Kolb, D. (1974) Ph.D. diss., Univ. of Illinois.
Kolb, D., and Weber, G. (1972). *Fed. Proc., Fed. Amer. Soc. Exp. Biol.* **31,** 423 (abstr.).
Koshland, D. E., Jr., Némèthy, G., and Filmer, D. (1966). *Biochemistry* **5,** 365. The point of view presented in this paper as regards the binding of ligands by multimer proteins is somewhat similar to that adopted by us (see Weber, 1972a) in this review. It differs in that the treatment is limited to the range of protein concentrations where ligand binding does not produce changes in macroassociation, and that the effects are described by means of arbitrary dissociation constants rather than conditional and unconditional free energies of binding, as we propose.
Kratky, O., and Pilz, I. (1972). *Quart. Rev. Biophys.* **5,** 481.
Lakowicz, J. R., and Weber, G. (1973). *Biochemistry* **12,** 4161 and 4171.
Lehrer, S. (1971). *Biochemistry* **10,** 3254.
Lewis, G. N., and Randall, M. (1923). "Thermodynamics," 1st ed., Chapter XVIII. McGraw-Hill, New York.
Lewitzky, A., Stallcup, W. B., and Koshland, D. E., Jr. (1971). *Biochemistry* **10,** 3371.
Lippert, E. (1957). *Z. Elecktrochem.* **61,** 962.
Lipscomb, W. N. (1973). *Proc. Nat. Acad. Sci. U. S.* **70,** 3797.
Mahler, H. R., and Douglas, J. (1957). *J. Amer. Chem. Soc.* **79,** 1159.
Magde, D., Ellson, E., and Webb, W. W. (1972). *Phys. Rev. Lett.* **29,** 705.
Massey, V. (1953). *Biochem. J.* **55,** 172.
Michaelis, L., and Menten, M. L. (1913). *Biochem. Z.* **49,** 333.
Mildvan, A. (1972). *Biochemistry* **11,** 2819.
Mitchell, P. (1966). *Biol. Rev. Cambridge Phil. Soc.* **41,** 445.
Mitchell, P. (1972a). *J. Bioenerg.* **3,** 5.
Mitchell, P. (1972b). *J. Bioenerg.* **4,** 63.
Monod, J., Wyman, J., and Changeux, J.-P. (1965). *J. Mol. Biol.* **12,** 88. This paper has had great importance by drawing attention to the close relationship of protein structure with metabolic regulation, but there does not seem to be any reasonable physical basis for the principle of "conversation of symmetry" promulgated by Monod, Wyman, and Changeux as the basis of ligand-binding cooperativity and other properties. The overall symmetry observed in the disposition of subunits arises from their similarity and does not carry any consequences as to the chemical reactions that do not directly or indirectly modify those groups involved in boundary interactions.

Muirhead, H., Cox, J. M., Mazzarella, L., and Perutz, M. F. (1967). *J. Mol. Biol.* **28**, 117.
Muller-Hill, B., Crapo, L., and Gilbert, W. (1968). *Proc. Nat. Acad. Sci. U. S.* **59**, 1259.
Némèthy, G., and Scheraga, H. A. (1962). *J. Phys. Chem.* **66**, 1773.
Panagou, D., Dunstone, J. R., Blakley, R., and Orr, M. D. (1972). *Biochemistry* **11**, 2378.
Pasby, T. (1969). Doctoral Thesis, University of Illinois, Urbana.
Pauling, L. (1935). *Proc. Nat. Acad. Sci. U. S.* **21**, 168.
Pauling, L. (1948). *Nature (London)* **161**, 707.
Perutz, M. F. (1970). *Nature (London)* **228**, 738. This paper of Perutz is the most comprehensive attempt at a detailed description of the relations between structure and function in a protein so far published. It relies almost exclusively on two sources of information: visual impression from molecular models built according to X-ray crystallographic data; and functional changes connected either with the observed changes in molecular structure upon oxygen ligation or with replacement of certain amino acid residues in mutant hemoglobins. It is implicit in the method used that structure–function relations can be derived with some certainty without need of detailed computation of the multiple interactions among the parts. From the point of view that we have maintained in this review it would not seem possible that such a method could lead to reliable conclusions when the energy differences involved are as modest as is the case in hemoglobin.
Perutz, M. F., and TenEyck, L. F. (1972). *Cold Spring Harbor Symp. Quant. Biol.* **36**, 295.
Perutz, M. F., Muirhead, H., Cox, J. M., and Goaman, L. C. G. (1968). *Nature (London)* **219**, 131.
Phillips, D. C. (1967). *Proc. Nat. Acad. Sci. U. S.* **57**, 484.
Polanyi, M., and Wigner, E. (1928). *Z. Phys. Chem., Abt. A* **139**, 439.
Quiocho, F. A., McMurray, C. H., and Lipscomb, W. N. (1972). *Proc. Nat. Acad. Sci. U. S.* **69**, 2850.
Rabinowitz, M., and Lipmann, F. (1960). *J. Biol. Chem.* **235**, 1043.
Rao, S. T., and Rossmann, M. G. (1973). *J. Mol. Biol.* **76**, 241. The observation of Rao and Rossmann that very similar, though quite complex structures are at the basis of NAD binding by dehydrogenases and FAD binding by rubredoxin (Watenpaugh *et al.*, 1972) is indeed a crucial finding, which opens new possibilities in the understanding of biological specificity.
Richards, F. M., and Lee, B. (1971). *J. Mol. Biol.* **55**, 379.
Richards, F. M., Wyckoff, H. W., Carlson, W. D., Allewell, N. M., Lee, B., and Mitsui, Y. (1972). *Cold Spring Harbor Symp. Quant. Biol.* **36**, 35.
Rossi-Fanelli, A., Antonini, E., and Caputo, A. (1961). *J. Biol. Chem.* **236**, 397.
Roughton, F. J. W., and Lyster, R. L. J. (1965). *Hvalradets Skr.* **48**, 185.
Roughton, F. J. W., Otis, A. B., and Lyster, R. L. J. (1955). *Proc. Roy. Soc., Ser. B* **144**, 29.
Saroff, H. A., and Minton, A. P. (1972). *Science* **175**, 1253.
Shefter, E. (1968). *Science* **160**, 1351.
Sinanoglu, O. (1968). *In* "Molecular Associations in Biology" (B. Pullman, ed.), p. 427. Academic Press, New York.
Singer, J. S., and Nicolson, J. L. (1972). *Science* **175**, 720.
Spencer, R. D., and Weber, G. (1970). *J. Chem. Phys.* **52**, 1654.
Spencer, R. D., and Weber, G. (1972). *In* "Structure and Function of Oxidation-Reduction Enzymes" (Å. Åkeson and A. Ehrenberg, eds.), p. 393. Pergamon, Oxford.

Steinberg, I. Z. (1971). *Annu. Rev. Biochem.* **40**, 83.
Stühmann, H. B. (1973). *J. Mol. Biol.* **77**, 363.
Teale, F. J. W. (1960). *Biochem. J.* **76**, 381.
Theorell, H., and McKinley-McKee, J. S. (1961). *Acta Chem. Scand.* **15**, 1811.
Tomita, S., and Riggs, A. (1971). *J. Biol. Chem.* **246**, 574.
Tomita, S., Enoki, Y., Ochiai, T., Kawase, M., and Okuda, T. (1972). *J. Mol. Biol.* **73**, 261.
Ts'o, P. O. P., and Chan, S. I. (1964). *J. Amer. Chem. Soc.* **86**, 4176.
Ts'o, P. O. P., Melvin, I. S., and Olson, A. C. (1963). *J. Amer. Chem. Soc.* **85**, 1289. See Ts'o and Chan (1964) and Chan *et al.* (1964).
Tyuma, I., Shimizu, K., and Imai, K. (1971a). *Biochem. Biophys. Res. Commun.* **43**, 423.
Tyuma, I., Imai, K., and Shimizu, K. (1971b). *Biochem. Biophys. Res. Commun.* **44**, 682.
Watenpaugh, K. D., Sieker, L. C., Herriot, J. R., and Jensen, L. H. (1972). *Cold Spring Harbor Symp. Quant. Biol.* **36**, 359.
Weber, G. (1961). *In* "Light and Life" (W. D. McElroy and B. Glass, eds.), p. 82. Johns Hopkins Press, Baltimore, Maryland.
Weber, G. (1965). *In* "Molecular Biophysics" (B. Pullman and M. Weissbluth, eds.), p. 369. Academic Press, New York.
Weber, G. (1970). *In* "Spectroscopic Approaches to Biomolecular Conformation" (D. W. Urry, ed.), p. 23. Amer. Med. Ass., Chicago, Illinois.
Weber, G. (1972a). *Biochemistry* **11**, 864.
Weber, G. (1972b). *Proc. Nat. Acad. Sci. U. S.* **69**, 3000.
Weber, G. (1974). *In* "The Mechanism of Energy Transduction in Biological Systems." *Ann. N. Y. Acad. Sci.* **227**, 486.
Weber, G., and Anderson, S. R. (1965). *Biochemistry* **4**, 1942.
Weber, G., and Daniel, E. (1966). *Biochemistry* **5**, 1900.
Weber, G., and Lakowicz, J. R. (1973). *Chem. Phys. Lett.* **22**, 419.
Witt, H. T. (1971). *Quart. Rev. Biophys.* **4**, 365.
Weber, G., and Young, L. (1964). *J. Biol. Chem.* **239**, 1415.
Witt, H. T., (1971). *Quart. Rev. Biophys.* **4**, 365.
Wyman, J. (1948). *Advan. Protein Chem.* **4**, 407.
Wyman, J. (1964). *Advan. Protein Chem.* **19**, 223.
Wyman, J. (1965). *J. Mol. Biol.* **11**, 631.

Additional References

Rice, S. A., and Nagasawa, M. N. (1961). "Polyelectrolyte Solutions." Academic Press, New York. Contains a theoretical treatment of binding to polymers. From the point of view of the biochemist it amounts to a very complete treatment of "unspecific" binding.
Steinhardt, J., and Reynolds, J. A. (1970). "Multiple Equilibria in Proteins." Academic Press, New York. Extensive description and analysis of binding by proteins; it is at present the standard work on the subject.

General Reviews

Klotz, I. M., Langermann, N. R., and Darnall, D. W. (1970). Quaternary structure of proteins. *Annu. Rev. Biochem.* **39**, 25. Reviews structure, as well as energetics, the latter from a rather different viewpoint than that adopted here. The larger biological associations of interacting proteins are reviewed by:
Ginsburg, A. E., and Stadtman, E. R. (1970). *Annu. Rev. Biochem.* **39**, 429.

Laurence, D. J. R. (1972). Interactions of polymers with small molecules. *In* "Physical Methods in Macromolecular Chemistry" (B. Carroll, ed.), Vol. 2, p. 91. Dekker, New York. Reviews theory, methodology, and experimental results of binding to a large number of systems. Very complete list of references on the subject.

Matthews, B. W., and Bernhard, S. A. (1973). Structure and symmetry of oligomeric enzymes. *Annu. Rev. Biophys. Bioeng.* **2**, 257. Emphasizes the structural principles involved in the formation of protein aggregates, almost to the exclusion of energetic considerations, and presents therefore a point of view quite divergent from our own.

AVIDIN

By N. MICHAEL GREEN

National Institute for Medical Research, Mill Hill, London, England

I. Introduction

A. History

The early history of the protein avidin is part of the story of the discovery, isolation, and synthesis of biotin. Although avidin accounts for only 0.05% of the protein of egg white, its presence was betrayed by an unusual dermatitis in rats fed with dried egg white as the sole source of protein (Boas, 1927). This could be cured by a protective factor, called vitamin H by György, which was present in many foodstuffs. The history of these investigations in the 1930s and 1940s has been well summarized by György (1954). They led to the identification of vitamin H with biotin (György *et al.*, 1940), the synthesis of biotin, and the partial purification of avidin (Eakin *et al.*, 1941). Although these preparations contained only 30% of active avidin, it proved possible to obtain a crystalline protein from them (Pennington *et al.*, 1942). The activity was decreased by crystallization, and the observation was not followed up. However, in spite of their low activity the crystals appear to be isomorphous with those of pure avidin obtained many years later (Green and Toms, 1970) under rather similar conditions of crystallization.

Since that time interest in the nutritional effects of avidin has waned, but its biotin binding properties have continued to attract the attention of protein chemists and enzymologists. The dissociation constant of the avidin–biotin complex was clearly very low since avidin inhibited the growth of microorganisms that require biotin concentrations of only 10^{-10} M to 10^{-11} M (Wright and Skeggs, 1944; Hertz, 1946). To obtain insight into the nature of this very firm binding, Fraenkel-Conrat *et al.*, (1952a,b) purified avidin more extensively and studied the effects of chemical modification on the activity. Although the activity was still only 70% that of later preparations, the close agreement of its amino acid composition with that subsequently determined suggests that most of the impurity was either avidin–biotin complex or some other form of inactivated avidin.

Little further interest was taken in avidin until the discovery of the coenzyme function of covalently bound biotin (Wakil *et al.*, 1958; Lynen *et al.*, 1959) made it clear that avidin could be a useful tool in the characterization of an important new class of enzymes. A more detailed study of avidin was commenced at this time in the author's laboratory (Green, 1963a,b,c; Melamed and Green, 1963).

Earlier reviews have dealt with avidin in relation to biotin (György, 1954) and biotinyl enzymes (Knappe, 1970; Moss and Lane, 1971). In this review the chemistry and binding properties will be discussed in more detail.

B. Occurrence and Function

Avidin activity has been observed in the eggs and oviducts of many species of birds and in the egg jelly of frogs, at a maximum concentration of about 0.05% of the total protein. The search for avidin in other tissues has not been extensive, but negative results have been reported for avian and amphibian intestinal mucosa and for mammalian oviduct (Hertz, 1946). The recent discovery of an avidinlike protein in culture filtrates of several species of *Streptomyces* (Chaiet *et al.*, 1963) was totally unexpected. It was observed during screening for antibiotics that these culture filtrates owed their activity against gram-negative bacteria to a synergistic action of high and of low molecular weight components. The antibacterial effects could be reversed by high concentrations of biotin in the medium, and it was shown that the high molecular weight component (MSD 235L, streptavidin) was a biotin-binding protein whereas the low molecular weight one (MSD 235 S, stravidin) inhibited biotin synthesis. The physical and chemical characteristics of the streptavidin proved to be remarkably similar to those of egg white avidin. The structure of stravidin has been determined by Baggaley *et al.* (1969).

The question of the biological function of biotin binding proteins is frequently raised and has not yet received a clear answer. A storage function is unlikely in view of the occurrence of these proteins in an essentially biotin-free state, although it must be admitted that if avidin were to exist with biotin already bound, the complex would not be detected by methods of assay for avidin. A general catalytic function is unlikely in view of the restricted distribution of the protein. Moreover, the catalytically active region of the biotin molecule is the part which interacts most strongly with the avidin so that it is unlikely to be available for catalysis. An antibacterial function is the most plausible one, but the evidence for it is only circumstantial. Egg white contains several antibacterial proteins (lysozyme, a flavin-binding protein, and conalbumin), and one more would not be surprising. The discovery of streptavidin is perhaps the most convincing support for this view in the light of the great versatility of streptomycetes in devising antibiotic systems.

II. Purification and Covalent Chemistry

A. *Purification*

The earliest isolation by Eakin *et al.* (1941) made use of selective solubilization of avidin by dilute salt, from alcohol-precipitated egg proteins. An extension of this method by Dhyse (1954) appears to be the source of most partially purified commercial avidin. The best products contained about 50% of active avidin (7 units/mg). [One unit of avidin binds 1 μg of biotin (4.1 nmoles). Methods of assay are discussed in Section VI,A.] Fraenkel-Conrat *et al.* (1952a) developed an alternative approach, used earlier for the purification of lysozyme (Alderton *et al.*, 1945), in which egg proteins were adsorbed on bentonite. Avidin was eluted with 1 *M* dipotassium phosphate and purified by ammonium sulfate fractionation to a specific activity of 8–10 units/mg. The product was obtained in three forms, a basic glycoprotein and complexes of this with an acidic glycoprotein (water soluble) and with low molecular weight fragments of DNA (water insoluble). Both of the complexes were dissociable by electrophoresis, and although evidence was produced for some specificity of interaction between the components, it seems most likely that the complexes were artifacts of the isolation procedure. They have not been observed during subsequent purifications. The nature of the low molecular weight DNA component of egg-white (approximately 1 mg per egg) does not appear to have been further investigated.

The unusually high isoelectric point of avidin and the availability of cellulose ion exchangers led logically to the development of an improved procedure based on adsorption of the basic proteins on CM-cellulose at high pH, followed by elution of avidin with ammonium carbonate. Further chromatography on CM-cellulose (Melamed and Green, 1963) gave a product which was later crystallized at pH 5 from ammonium sulfate (3 M) or potassium phosphate (3 M) (Green and Toms, 1970; Green and Joynson, 1970). This method is convenient for large-scale preparation. Affinity chromatographic methods using biotinyl cellulose (McCormick, 1965) or biocytinyl Sepharose (Cuatrecasas and Wilchek, 1968) have been described, but they have not yet been widely used.

B. *Homogeneity*

Avidin purified on carboxymethyl cellulose and then crystallized appears to contain two or three components when chromatographed on Amberlite CG-50 (Melamed and Green, 1963; Green and Toms, 1970). There were indications of differences of a few amino acid residues between the components but the variable amino acid has differed with different avidin preparations, and the analyses have not been repeated often enough to establish these small differences with certainty. There was some evidence that eggs derived from a single breed of hen gave a more homogeneous product (Melamed and Green, 1963). However, the amino acid sequence of avidin (Section II,D) was determined with protein purified from a commercial source of crude avidin (De Lange, 1970) probably derived from different breeds from those used by Melamed and Green, and the products from the two laboratories had almost identical amino acid compositions (see Table I). The sequence showed only one ambiguity; residue 34 was either threonine or isoleucine (De Lange and Huang,

1971). Since the variant peptides were recovered in equal amounts it is possible that each avidin molecule contains two of each type of chain, though the presence of heterogeneity suggests that the two chains are more likely to originate from different molecules. Small variations in carbohydrate composition have been observed (Huang and De Lange, 1971, and Table I), but further work is required to determine their extent.

C. General Properties and Stability

Avidin is a basic glycoprotein with its isoelectric point at pH 10 (Woolley and Longsworth, 1942). It is very soluble in water and salt solutions, and it is stable over a wide range of pH and temperature. It crystallizes from strong salt solutions between pH 5 and pH 7, but it has not yet been crystallized in the isoelectric region. In contrast, streptavidin has no carbohydrate, it has an acid isoelectric point, it is much less soluble in water and can be crystallized from water or 50% isopropanol (Chaiet *et al.*, 1963).

The early determinations of molecular weight (Woolley and Longsworth, 1942) and of composition (Fraenkel-Conrat *et al.*, 1952b) are surprisingly close to the results obtained with pure avidin, considering that the preparations used possessed only 50–70% of the maximum biotin binding capacity. It is likely that much of the impurity in these earlier preparations was avidin–biotin complex (Pai and Lichstein, 1964) or otherwise inactivated avidin. This might also explain why the crystals obtained by Pennington *et al.* (1942), which are similar in appearance to crystals of avidin, possessed a lower specific activity than the solution from which they were crystallized. The avidin–biotin complex is isomorphous with avidin (Green and Joynson, 1970) and could well crystallize more readily.

The biotin-binding capacities and amino acid compositions of avidin and streptavidin are compared in Table I. With the exception of the differences just mentioned there are marked similarities in composition. There are zero to two residues each of histidine, proline, cystine, and methionine, and the contents of tryptophan and threonine are very high, that of threonine uniquely so. Other properties of streptavidin, most of them similar to those of avidin, are summarized in Section VI,D.

The remarkable stability of avidin, and even more of the avidin–biotin complex, to heat and to proteolytic enzymes was noted very early. In a study of conditions required to release biotin from the complex by autoclaving it was observed that release was more rapid

TABLE I
Amino Acid Composition of Avidin and Streptavidin[a]

	Avidin residues/subunit	Streptavidin residues/subunit
Lysine	9	4
Histidine	1	2
Arginine	8	4
Aspartic acid	15 (14)	12
Threonine	20.5 (19)	19
Serine	9	10
Glutamic acid	10	9
Proline	2	2
Glycine	11	17
Alanine	5	17
Half-cystine	2	0
Valine	7	7
Methionine	2	0
Isoleucine	7.5 (8)	3
Leucine	7	8
Tyrosine	1	6
Phenylalanine	7	2
Tryptophan	4	8
Total residues	128	ca. 130
Amide	16	–
Mannose	4 (5)	0
Glucosamine	3	0
Subunit weight	15,600	14,000
Molecular weight	67,000[b]	60,000[c]
Biotin-bound/subunit 1%	0.97	0.95
E_{282} (1%)	15.4[d]	34[e] 47[c]
ϵ_{282} (subunit)	24,000	56,000

[a] The amino acid composition of avidin and the molecular weight of the subunit is based on the sequence (De Lange and Huang, 1971). The composition is identical with that given by Green and Toms (1970), except for figures included from the latter reference in parentheses. The results for streptavidin are from Chaiet and Wolf (1964) and Green (1968).

[b] See Table III.

[c] Chaiet *et al.* (1963).

[d] Green and Toms (1970).

[e] Green (1970).

in the absence of salts and was 88% complete after 10 minutes at 100°; however, autoclaving at 120° for 15 minutes was required for complete release in the presence of salts (Pai and Lichstein, 1964; Wei and Wright, 1964). A recent calorimetric study (Donovan and

Ross, 1973) has shown that avidin is inactivated in an endothermic transition at 85°C, whereas a similar transition with the avidin–biotin complex does not occur until a temperature of 132°C is reached (Section V,B). Loss of tertiary structure at 85°C has also been observed by measurement of the optical rotatory dispersion (ORD) spectrum (Pritchard *et al.*, 1966).

The avidin–biotin complex is resistant to proteolysis by the enzymes of the digestive tract and free avidin is not inactivated by trypsin or by Pronase. However, an early observation of György and Rose (1941) showed that biotin was released from the complex when it was administered intravenously or parenterally. This has been recently confirmed (Wei and Wright, 1964; Lee *et al.*, 1973), but nothing is known of the enzymes responsible. The breakdown is rather slow, and in short-term experiments Miller and Tausig (1964) showed that both avidin and streptavidin administered parenterally could protect mice against *Salmonella* infection in the presence of stravidin, an inhibitor of biotin synthesis.

The biotin-binding activity is not lost between pH values of 2 and 13 nor at concentrations of guanidinium chloride (GuHCl) below 3 *M*. Loss of activity outside these limits is accompanied by unfolding of the protein and dissociation into subunits, which is largely reversible (Green, 1963c, and Section V,A).

The instability of avidin under oxidizing conditions, particularly in strong light (György *et al.*, 1942), is probably the result of an unusually reactive tryptophan residue in the binding site (Section II,E).

D. Primary Structure

The complete sequence, established by De Lange and Huang (1971), is shown in Table II, aligned with that of hen egg white lysozyme. It had been suggested earlier (Green, 1968) that the two proteins might have arisen from a common ancestor on the rather tenuous evidence of similar amino acid composition and a weak binding of biotin by lysozyme. The alignment from the C-terminal end shows 15 identities (12%), which is little more than would be expected on a random basis. This observation that 88% of the residues are different is hardly convincing evidence even for a remote relationship. However, it is not very different from the 86% of differences observed between trypsinogen and the α-lytic protease of Sorangium (Dayhoff, 1972), where evidence for homology is made more convincing by the resemblance between the active site sequences. A further point against a chance resemblance is the ef-

TABLE II

Amino Acid Sequences of Avidin and Lysozyme from the Hen Egg White[a]

																		carbohydrate								
						83-S.S	5					10					15					20					25
Avidin			Ala	Arg	Lys	Cys	Ser	Leu	Thr	Gly	Lys	Trp	Thr	Asn	Asp	Leu	Gly	Ser	Asn	Met	Thr	Ile	Gly	Ala	Val	Asn	Ser
	Lys	Val	Phe	Gly	Arg	Cys	Glu	Leu	Ala	Ala	Ala	Met	Lys	Arg	His	Gly	Leu	Asp	Asn	Tyr	Arg	Gly	Tyr	Ser	Leu	Gly	Asn
						127-S.S	30					35					40					45					50
Avidin			Arg	Gly	Glu	Phe	Thr	Gly	Thr	Tyr	Ile Thr	Thr	Ala	Val	Thr	Ala	Thr	Ser	Asn	Glu	Ile	Lys	Glu	Ser	Pro	Leu	His
Lysozyme			Trp	Val	Cys	Ala	Ala	Lys	Phe	Glu	Ser	Asn	Phe	Asn	Thr	Glu	Ala	Thr	Asn	Arg	Asn	Thr	Asp	Gly	Ser	Thr	Asp
							55					60					65					70					75
Avidin			Gly	Thr	Glu	Asn	Thr	Ile	Asn	Lys	Arg	Thr	Gln	Pro	Thr	Phe	Gly	Phe	Thr	Val	Asn	Trp	Lys	Phe	Ser	Glu	Ser
Lysozyme			Tyr	Gly	Ile	Leu	Glu	Ile	Asn	Ser	Arg	Trp	Trp	Cys	Asn	Asp	Gly	Arg	Thr	Pro	Gly	Ser	Arg	Asn	Leu	Cys	Asp
							80					85					90					95					100
Avidin			Thr	Thr	Val	Phe	Thr	Gly	Gln	Cys	Phe	Ile	Asp	Arg	Asn	Gly	Lys	Glu	Val	Leu	Lys	Thr	Met	Trp	Leu	Leu	Arg
Lysozyme			Ile	Pro	Cys	Ser	Ala	Leu	Leu	Ser	Ser	Asp	Ile	Thr	Ala	Ser	Val	Asn	Cys	Ala	Lys	Lys	Ile	Val	Ser	Asp	Gly
							105					110					115					120					125
Avidin			Ser	Ser	Val	Asn	Asp	Ile	Gly	Asp	Asp	Trp	Lys	Ala	Thr	Arg	Val	Gly	Ile	Asn	Ile	Phe	Thr	Arg	Leu	Arg	Thr
Lysozyme			Asp	Glu	Met	Asn	Ala	Trp		Val	Ala	Trp	Arg	Asn	Arg	Cys	Lys	Gly	Thr	Asp	Val	Gln	Ala	Trp	Ile	Arg	Gly
Avidin			Gln	Lys	Glu																						
Lysozyme			Cys	Arg	Leu																						

[a] The lysozyme sequence (Canfield, 1963) has been aligned with that of avidin (De Lange and Huang, 1971) starting from the carboxyl terminus and allowing one deletion opposite residue 107 of avidin. The fifteen identities are enclosed in boxes. The tryptophan residues, the unique histidine residue, and the tyrosine residue of avidin are underlined.

fect of shifting the sequences one or two steps relative to each other in either direction. These shifts lead to a decrease in number of identities from 15 to 3, 4, 9, or 4 residues. A computer based comparison of the human lysozyme with avidin showed no significant resemblance (Dayhoff, 1972), but the two hen proteins have not been compared in this way.

Two notable features of the primary structure of avidin, both absent from streptavidin, are a single disulfide bond and an oligosaccharide linked via one of its acetylglucosamine residues to Asn (17). The structure of the carbohydrate is not known. It is probably heterogeneous (De Lange, 1970), and in this respect as well as in its overall composition it resembles the carbohydrate of ovalbumin, the structure of which has been determined (Montgomery, 1972).

E. Chemical Modification

Early experiments on the effect of chemical modification procedures on the activity of avidin (Fraenkel-Conrat *et al.*, 1952b) led to the conclusion that the activity was insensitive to extensive substitution of a variety of functional groups. Avidin was not inactivated by treatment with iodine at neutral pH, nor by acetylation of 60% of the amino groups, nor by esterification of 20% of the carboxyl groups. Significant ($>$ 70%) inactivation resulted from oxidation with H_2O_2 in the presence of Fe^{2+} and from treatment with formaldehyde in the presence of alanine, or with hydroxylamine at 50°C. The implications of these inactivations were not clear at the time but in retrospect most are consistent with modification of tryptophan residues, which subsequent work showed were vital for biotin binding.

The first evidence which directly implicated tryptophan came from biotin-induced difference spectra (Section III,B), and it was supported by the sensitivity of avidin to oxidation by *N*-bromosuccinimide (Green, 1962, 1963b). All four tryptophans were rapidly oxidized at pH 4.6 and biotin-binding activity was lost when an average of two had been destroyed. The avidin–biotin complex was almost completely resistant to oxidation by *N*-bromosuccinimide. The earlier results suggesting the presence of *N*-bromosuccinimide-resistant residues in avidin were shown later to have originated from the presence of avidin–biotin complex in the avidin preparation (Green and Ross, 1968). Other reagents which react with tryptophan in avidin, but not in the avidin–biotin complex are ozone (Mudd *et al.*, 1969) and periodate (Green, 1963b, 1974).

There are three interesting questions relating to the role of the tryptophans to which tentative answers are possible. These are: (1)

How many of the four tryptophans interact directly with bound biotin? (2) Is it possible to modify them selectively? (3) Is any one of them of particularly vital importance? Taking the last questions first, oxidation with *N*-bromosuccinimide showed that an average of one tryptophan could be destroyed with loss of only 25% of activity, but that all activity disappeared when a second residue was oxidized. Huang (1971) isolated tryptophan peptides after partial inactivation. His preliminary results showed that when half the activity was lost (about 1.3 tryptophans oxidized) about half of tryptophans 10 and 70 were destroyed, Trp 97 was intact, and the evidence on Trp 110 was inconclusive. From this it appears that either or both Trp 10 and Trp 70 are essential to the binding. More selective oxidation could be effected by sodium periodate (1 m*M*) at neutral pH (Green, 1963b, 1974). It produced a slow fall in E_{280} ($t_{1/2} = 35$ minutes) which stopped after oxidation of about one residue of tryptophan. Neutral periodate has been shown to oxidize indoles and tryptophan derivatives to formyl kynurenine derivatives (Dolby and Booth, 1966), but the rate is less than one-tenth as fast as that for the tryptophan in avidin. The spectrum of the oxidized avidin was consistent with oxidation of slightly more than one residue to formyl kynurenine. The product possessed 70% of the initial binding capacity and its affinity for biotin was decreased from 10^{-15} *M* to about 10^{-9} *M*. Its main feature of interest was the modified difference spectrum given with biotin, which showed peaks characteristic of *N*-formylkynurenine (Section IV,C). A product with similar spectroscopic properties has also been obtained by autoxidation of avidin during concentration of solutions by pervaporation (Green, 1963b), but the conditions are not readily reproducible.

To return to the questions posed above, the tryptophan oxidized by periodate does not appear to be of vital importance for biotin binding though the affinity is significantly reduced. In contrast, avidin in which Trp 10 and Trp 70 were oxidized was inactive, but it is not clear whether this was because the residues were different from that oxidized by periodate or because two residues were attacked rather than one. It may also be significant that oxindole, the major product of oxidation by *N*-bromosuccinimide, does not contribute to the difference spectrum of the partially inactivated product whereas formylkynurenine does contribute to that of the periodate-oxidized avidin (Green, 1963b, and Section IV,C). Although no compelling arguments can be made from this evidence, it seems likely that several tryptophans contribute to the biotin-binding site.

The only other functional group which has been implicated in the binding site of avidin is an amino group. It was observed that the initial rate of reaction of 1-fluoro-2,4-dinitrobenzene (FDNB) with avidin was faster than that with the avidin–biotin complex and that, if the reaction was stopped after one DNP had been introduced per subunit, almost no biotin binding activity remained (Green, 1975). This suggested that there was a unique amino group in the neighborhood of the binding site with which FDNB reacted preferentially. In contrast trinitrobenzene sulfonic acid and dansyl chloride (Green, 1963c) reacted without blocking this residue. The location of the lysine which is blocked by FDNB has not been determined. It may be one of the three lysine residues (9, 71, 111) which are adjacent to tryptophans in the sequence.

The single tyrosine residue (33) of avidin ionizes only at pH values above 12 and is probably buried (Green, 1963c). In the avidin–biotin complex it remained nonionized even in 0.5 *M* KOH. The single histidine, His (50), was alkylated by iodoacetamide only in the presence of 6 *M* guanidinium chloride (Huang, 1971). The single disulfide residue (4–62) could be reduced only after dissociation of the tetramer into subunits (Green, 1963c) and is also likely to be a buried residue. Preliminary attempts to regenerate activity from the reduced subunits (0.5 mg/ml) were unsuccessful (Green, 1963c).

It is unlikely that the carbohydrate plays any part in biotin binding. Most of the mannose residues would have been oxidized by the periodate treatment and the product still binds biotin strongly. Furthermore, streptavidin contains no carbohydrate.

III. Physical Properties

A. *Hydrodynamic Properties*

The hydrodynamic properties of avidin have not been systematically studied. Sedimentation velocity, sedimentation equilibrium, and diffusion coefficient measurements have been made, mostly at single concentrations, but there have been no reports of viscosity measurements. The available results are summarized in Table III together with the molecular weights determined by osmotic pressure, and the rotational relaxation times of dansyl avidin calculated from measurements of polarization of fluorescence. The results obtained by different methods or in different laboratories under different conditions agree well and suggest that the shape and conformation of avidin is not sensitive to changes in temperature, pH, and ionic strength. For example, the sedimentation and diffusion coefficients of avidin at pH 5.5 are the same as those of the avidin–biotin

TABLE III
Physical Properties of Avidin and Avidin–Biotin Complex

Parameter	Avidin: Neutral buffers	Avidin: Denaturing solvents	Avidin–biotin: Neutral buffers	Avidin–biotin: Denaturing solvents
$s_{20,w}$ (Svedberg units)	4.7[a] 4.55[d]	0.89 (6 *M* GuHCl)[c] 2.15 (0.1 *M* HCl)[e]		4.58 (0.1 *M* HCl)[d]
$D_{20,w}$ (cm²/sec × 10⁷)	5.98[d]	4.5 (6 *M* GuHCl)[c]		6.08 (0.1 *M* HCl)[d]
f/fo	1.33	3.7 (6 *M* GuHCl)[c]		
Molecular weight				
Sedimentation and diffusion	66,000[d]	16,400		66,000 (0.1 *M* HCl)[d]
Sedimentation equilibrium	68,000[d] 69,000[f]	18,700 (3 *M* GuHCl)[c]		
Osmotic pressure	66,000[b]	68,000 (8 *M* Urea)[b] 17,000 (3 *M* GuHCl)[c]		
Fluorescence polarization				
ρ_h, nsec (T varied)	56[c]	ca 25 (3 *M* GuHCl)[c]	50[c]	ca 50 (6 *M* GuHCl)[c]
ρ_h, nsec (isothermal)	100[d]		105[d]	

[a] Woolley and Longsworth (1942).
[b] Fraenkel-Conrat *et al.* (1952a).
[c] Green (1963c).
[d] Green (1964b).
[e] Green and Melamed (1966).
[f] De Lange (1970).

complex in 0.1 *M* HCl. The observation that crystals of avidin are isomorphous with those of the avidin–biotin complex (Green and Joynson, 1970) provides further evidence that there is little change in structure when biotin is bound. The results of molecular weight measurement are within 5–10% of that expected for a molecule containing four of the chemically defined subunits (62,400, Table I) and four biotin-binding sites (64,800). This is within the limits of error of the physical measurements. The molecule dissociates into four subunits in GuHCl (6 *M*) or HCl (0.1 *M*) without reduction of the disulfide bonds. The subunits do not reassociate until the GuHCl concentration is decreased below 3 *M* (Section V,A).

Measurements of polarization of fluorescence of dansyl avidin as a function of temperature gave a linear relation between $1/p$ and T/η

from which a rotational relaxation time, ρ_h, of 56 nsec could be calculated. This was approximately the same as ρ_0, the theoretical relaxation time for an anhydrous spherical protein of the same molecular weight. The ratio ρ_h/ρ_0 is usually nearer 2 than 1 for a rigid molecule. This result implies some freedom of rotation within the conjugate. Later measurements performed in sucrose at constant temperature (Fig. 1) (Green, 1964a) gave a relation between $1/p$ and T/η which departed from linearity in viscous solutions (about 40% sucrose), resembling the results of Wahl and Weber (1967) with labeled immunoglobulin. The value of ρ_h (100 nsec) calculated from the linear section of the curves was nearer the expected value for a rigid molecule, while the curvature at high viscosity implies a thermally activated rotation of the dansyl group relative to the protein, which was considerably greater in the avidin–biotin complex than in avidin. This would account for the lower apparent value of ρ_h obtained when temperature was varied. This greater freedom of rota-

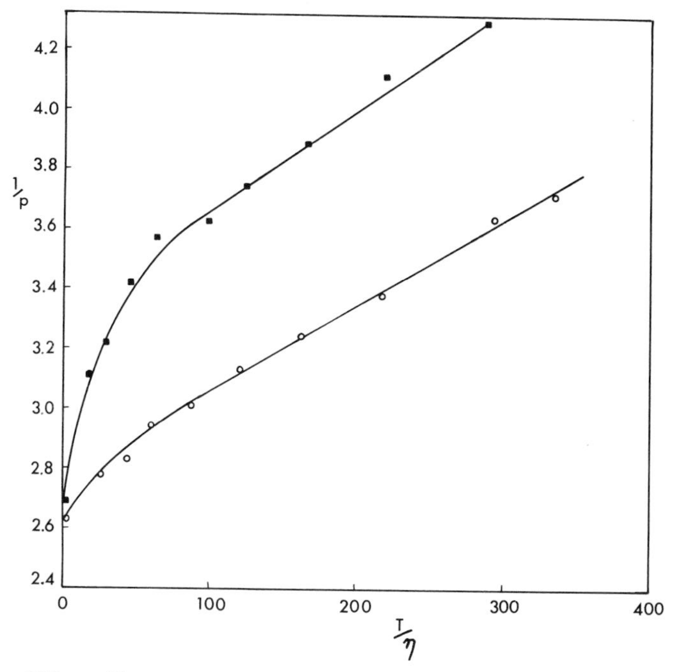

FIG. 1. Effect of biotin on the polarization of fluorescence of dansyl avidin. The polarization of fluorescence was measured at 25°C in the presence of increasing concentrations of sucrose, using avidin which carried one dansyl group per subunit, as described by Wahl and Weber (1967). ○, Dansyl avidin; ■, dansyl avidin + biotin.

tion induced by biotin could be correlated with a change in the dansyl fluorescence spectrum, which shifted from 508 nm to 514 nm and lost 40% of its intensity when biotin was bound. This suggests a more aqueous environment for the dansyl groups and implies that in avidin they are located near hydrophobic regions of the binding site and are displaced into solvent by biotin. Unlike the 2,4-dinitrophenyl groups discussed above (Section II,D), the dansyl groups do not block the binding site.

B. *Spectroscopic Properties*

1. *Ultraviolet Absorption*

The spectroscopic properties of avidin proved much more sensitive than the molecular kinetic properties to combination with biotin (Table IV), suggesting direct involvement of aromatic amino acid residues in the biotin-binding site (Green, 1962, 1963b). Since each subunit of avidin contains four tryptophan residues and only one tyrosine, the ultraviolet spectrum is dominated by the contribution of tryptophan. The sum of the contributions of tyrosine and tryptophan

TABLE IV

Spectroscopic Properties of Avidin

	Avidin[a]	Avidin–biotin[a]	Avidin In GuHCl 6.9 *M*	Avidin In GuHCl 3.4 *M*	Reference[b]
Ultraviolet					
ϵ_{282} (subunit)	24,000	24,000	23,000	–	1
ϵ_{232} (subunit)	93,000	118,000	74,000	–	1
$\Delta\epsilon_{294}$	0	1,800	−2,050	−2,400	1
$\Delta\epsilon_{232}$	0	25,000	−14,000	−19,000	1
Circular dichroism					
$(\epsilon_l - \epsilon_r)_{292}$	3.50	5.15	0.3	–	2
$(\epsilon_l - \epsilon_r)_{228}$	49.8	46.0	−2	–	2
Fluorescence					
Emission λ_{max}	338 nm	328 nm	350 nm	–	3
Intensity (tryptophan = 1.0 at 353 nm)	0.63	0.50	0.90	–	3

[a] Sodium phosphate, 0.05 *M*, pH 7.

[b] Unpublished measurements on crystalline avidin using a Cary 118 spectrophotometer, based on a subunit weight of 15,600; 2, Unpublished measurements on crystalline avidin using a Cary 61 circular dichroism spectropolarimeter. The values of $(\epsilon_l - \epsilon_r)$ were calculated per mole of tryptophan; 3, Green (1964a).

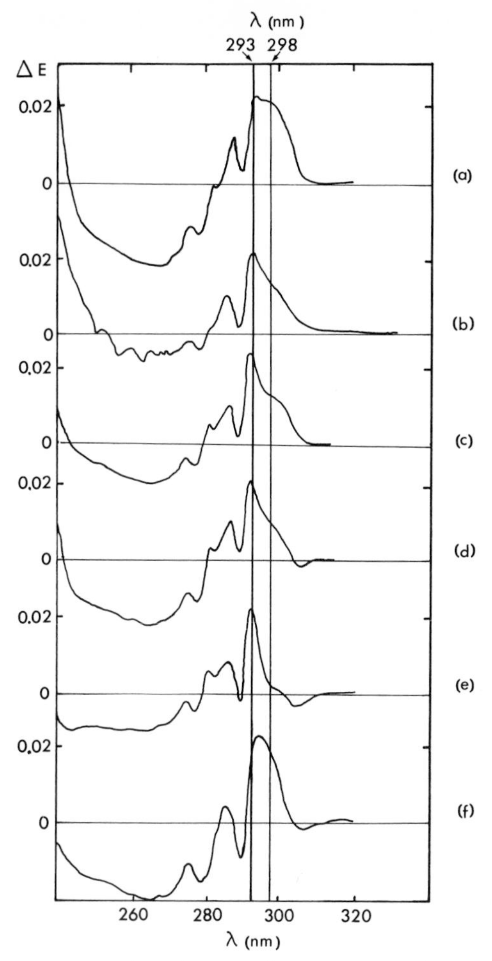

FIG. 2. Difference spectra of avidin complexes versus avidin. The avidin subunit concentration was 15.3 μM in 50 mM sodium phosphate pH 6.8 (a–e) or 50 mM sodium borate pH 9.2 (f). The ordinate scale of the first derivative spectrum is arbitrary. The measurements were made with a Cary 118 spectrophotometer. (a) Biotin; (b) first derivative of avidin spectrum; (c) biotin methyl ester; (d) desthiobiotin; (e) desthiobiotinol; (f) diamine from biotin.

accounts for 96% of the absorbance at 282 nm, allowing for a red shift of 2 nm relative to the free amino acids. When biotin binds to avidin there is a further red shift and a narrowing of the 280 nm band leading to the difference spectrum shown in Fig. 2, where it is compared with the difference spectra given by a number of analogs. The first derivative of the avidin spectrum is included to provide a standard showing the effects of a uniform red shift. Comparison of the biotin spectrum with the latter shows a much larger shoulder at 298 nm and a deeper trough at 260 nm than would be expected for a uniform shift. This result is consistent with a larger red shift of the ^1L_a bands of tryptophan (265 nm, 280 nm, and 297 nm) which lie on either side of the ^1L_b bands (at 284 nm and 291 nm) and are often more sensitive to solvent perturbation (Strickland *et al.*, 1969; Andrews and Forster, 1972). In this case, it appears possible to assign the extra perturbation of the ^1L_a bands in part to the negatively charged carboxyl group of biotin and in part to the sulfur atom. Removal of either of these features gives a much closer approach to the derivative spectrum, as can be seen from the spectra given by biotin methyl ester and by desthiobiotin. Removal of both negative charge and sulfur (desthiobiotinol) significantly sharpens the ^1L_b contribution. Whatever the theoretical origin of these features, they provide a useful characteristic of each biotin analog.

The difference peak near 233 nm is an order of magnitude larger than those in the 280 nm region (Table IV). It is much more uniform in shape, consistent with the simpler origins of the 219 nm band of tryptophan, shifted to 226 nm in the protein (Green, 1962, 1963b). The large difference extinction coefficient provides the basis of a spectrophotometric assay for avidin or biotin and of a method for determining dissociation constants of biotin analogs. It decreases as the resemblance of the analog to biotin becomes more remote (Table V). Evidence discussed below (Section IV) suggests that this is related to the number of different tryptophan residues that interact with the analog.

These effects can be understood qualitatively in terms of an exclusion of water from the immediate neighborhood of the tryptophan residues by the bound biotin, together with an additional perturbation from the negative charge affecting mainly ^1L_a bands. This effect of negative charge has also been seen with simple tryptophan derivatives (Ananthanarayanan and Bigelow, 1969).

The far ultraviolet absorption spectrum (Green and Melamed, 1966) is unaffected by combination with biotin, and the mean extinction of coefficient of the peptide chromophore is low ($\epsilon_{196} = 4800$), suggesting a high proportion of α-helix, when interpreted by the criteria of Rosenheck and Doty (1961). This conflicts with the conclusions from CD measurements that there is little or no helix present. A similar unexplained discrepancy exists between far UV and CD spectra of immunoglobulins (Gould *et al.*, 1964).

2. *Fluorescence*

The fluorescence emission spectrum of avidin (Green, 1964a) is typical of many proteins of high tryptophan content (Teale, 1960). The changes that follow combination with biotin or dissociation into subunits (Table IV) are consistent with the changes in exposure to the solvent deduced from the shifts in absorption spectrum.

3. *Circular Dichroism and Secondary Structure*

The CD and ORD spectra of avidin and streptavidin, like the absorption and fluorescence spectra, are dominated by the contribution of tryptophan (Green and Melamed, 1966). We shall consider only the CD spectra (Fig. 3 and Table IV) since they contain all the essential information. They are unusual in that they show two positive bands. The longer wavelength band extends from 250 nm to 300 nm and shows fine structure that can be matched closely by the CD spectrum of tryptophan measured at 77°K in methanol/glycerol mixtures (Strickland *et al.*, 1969). In avidin the ^1L_a bands predominate. In the complex with biotin the intensity of the ^1L_a band remains

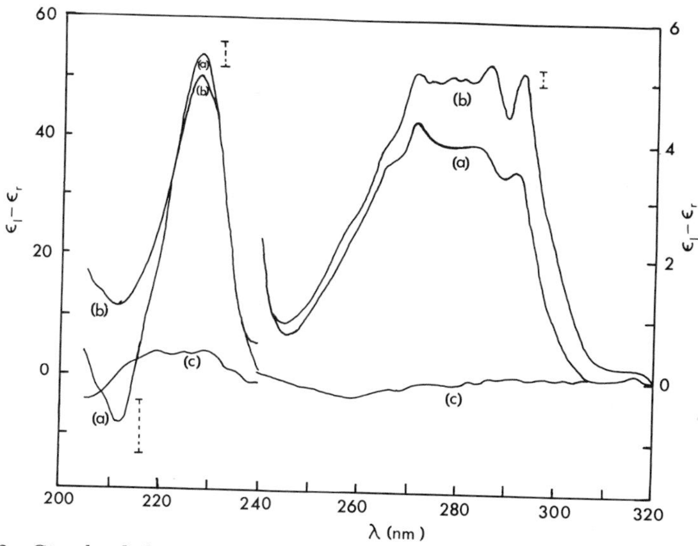

FIG. 3. Circular dichroism of avidin and avidin–biotin complex. Spectra were obtained with a Cary 61 CD spectropolarimeter [26 μM avidin subunits in 0.1 cm (left-hand scale) or 1-cm cuvettes (right-hand scale)]. (a) Avidin; (b) avidin–biotin complex; (c) base line. $(\epsilon_l - \epsilon_r)$ is expressed per mole of tryptophan. θ (°/cm²/decimole peptide bond) = 103 $(\epsilon_l - \epsilon_r)$.

about the same while that of the ^1L_b bands at 286 nm and 293 nm (2 nm red shift relative to free tryptophan) increase. The high intensity of the bands, four or five times that of the tryptophan derivatives at low temperature or of the aromatic bands of other proteins, implies a rigid asymmetric environment for the chromophores.

The shorter wavelength band (228 nm) showed no fine structure and only a small change of intensity when biotin was bound. Its position is almost exactly that predicted for the red-shifted 219 nm band of tryptophan (Green, 1962). It is unlikely that the peptide chromophore is contributing significantly to the CD in this region since at shorter wavelengths, where its effects should be most readily detectable, the dichroism is very low. In view of this it is difficult to draw any positive conclusions about tertiary structure, but it is probably safe to say that there is little or no α-helix in avidin. Chou and Fasman (1974) have recently proposed an empirical method for calculating the fractions of α-helix and β structure in a protein from the amino acid sequence. Application of their criteria to avidin leads to an estimate of zero α-helix and 50% of β-structure. This would be consistent with the very low CD of avidin below 215 nm, taking into account the averaged spectra calculated by Chen *et al.* (1972) for random coil and β-structures in a number of proteins. One remaining unknown factor is the contribution of the absorption bands of tryptophan in the region below 215 nm.

In GuHCl (6 *M*) there was almost no significant dichroism at wavelengths above 215 nm. The ORD spectrum showed that b_o was zero. In 0.1 *M* HCl the aromatic bands were still significant, showing that unfolding in this solvent was incomplete (Green and Melamed, 1966). This is also implied by the lower mobility of avidin subunits on gel filtration in 0.1 *M* HCl as compared with 6 *M* GuHCl (Green and Toms, 1972).

IV. Binding Properties

A. *Biotin*

The early work on avidin showed that the combination with biotin was very firm. Attempts to measure the dissociation constant by equilibrium dialysis with radioactive biotin (Launer and Fraenkel-Conrat, 1951) gave an upper limit of 10^{-10} *M*; the presence of radioactive impurities in the biotin made more exact interpretation difficult. In later work (Green, 1963a) the dissociation constant was calculated from the ratio of the rate constants for the forward and reverse reactions. The rate of dissociation was determined from the rate of exchange of bound ^{14}C-labeled biotin with an excess of unlabeled biotin, by separating free from bound with CM-cellulose. After a relatively rapid (18-hour) exchange of 5% of the biotin, the remainder of the reaction followed first-order kinetics for as long as it was measured (800 hours, Green and Toms, 1973) with a rate constant of 9×10^{-8} sec^{-1}. The rate was insensitive to ionic strength (N. M. Green, unpublished), but increased when the pH was below 3 (Table V). The rate constant measured by this method is that for the first dissociation step ($AB_4 \rightarrow AB_3 + B$). However, subsequent work (Section IV,E) has shown no interactions between the four binding sites, and it is likely that the calculated dissociation constant applies to all four sites.

The dissociation rate constant of biotin is so low that reequilibration of biotin after the initial binding can be ignored in any but very long-term experiments. This enables the forward rate to be measured by stopping the reaction with biotin-^{14}C by using an excess of unlabeled biotin. The rate constant obtained in this way, 7×10^7 M^{-1} sec^{-1}, was significantly less than that for a diffusion controlled reaction but similar to that observed for a number of enzyme–substrate and antibody–hapten combinations (Hammes, 1969). It should be possible to obtain more accurate values for the forward rate constant by stopped-flow spectrophotometry, but no measurements have been described.

It has been suggested that the stability of the avidin–biotin complex is decreased at very low ionic strength. The evidence is of two kinds. (1) The release of biotin on autoclaving the complex was more effective in the absence of salt (Wei and Wright, 1964; Pai and Lichstein, 1964); and (2) the combination of avidin with biotin appeared to be less complete at low ionic strength, when measured by dialysis or by separation of radioactive avidin–biotin complex by gel filtration (Wei and Wright, 1964). Many attempts have been made in the author's laboratory to show an increased dissociation constant in the absence of salt for complexes with biotin and several analogs, but no effects have been observed, either by following rates of exchange of biotin-^{14}C by measuring rates of dialysis or displacement of thiobiotin ($K = 10^{-13}$ M), or by spectrophotometric titration of more weakly bound analogs. Although the stability of the avidin–biotin complex to heat is less in the absence of salt, it is unlikely that the dissociation constant is changed. A possible source of misleading effects in gel filtration at low ionic strength is considered below (Section VI,B).

B. *Biotin Analogs*

1. *Measurement of Dissociation Constants*

Following the elucidation of the chemistry of biotin, the competition between it and a variety of analogs for avidin was studied by microbiological methods (Dittmer *et al.*, 1944; Wright and Skeggs, 1947; Wright *et al.*, 1951). In the more detailed studies of Wright, the concentration of analog required to halve the binding of biotin by an equivalent amount of avidin was measured (the inhibition ratio). It will be realized from the comments above that a true equilibrium situation did not exist in many of these experiments. Initially the high concentration of analog would almost saturate the avidin sites and equilibrium would be attained very slowly with firmly bound analogs. The biotin concentration would slowly decrease throughout the growth period of the *Lactobacillus arabinosus*, and the observed inhibition ratio would have only qualitative significance. Nevertheless, these early experiments established that the imidazolidone ring was much more important than the thiophan ring (Dittmer and du Vigneaud, 1944) and that modification of the carboxyl group had little effect on the binding to avidin (Wright *et al.*, 1951). Later work, showing that biotinyl enzymes are irreversibly inhibited by avidin, implies that covalent attachment of a macromolecule to the carboxyl group of biotin does not affect the binding.

The introduction of spectrophotometric methods for following the binding of biotin analogs (Green, 1963b) facilitated the quantitative comparison of a wide range of compounds (Table V). The dissociation constants of weakly bound analogs were calculated from the spectrophotometric titration curves. The lowest dissociation constant that could be measured directly was about 10^{-8} *M*. Approximate estimates for more stable complexes were made by measuring the rate at which the analog exchanged with biotin (Green, 1966), assuming that the rate of combination of the analogs with avidin was the same as that of biotin. The measurable half-times for exchange lay between 200 days for biotin and about 1 second for the *n*-hexylimidazolidones (compound 11; the compound numbers refer to numbers in column 1 of Table V) corresponding to dissociation constants between 10^{-15} and 10^{-8} *M*. For the latter compounds it was possible to compare the dissociation constants determined directly with those calculated from the rate constants for exchange. The agreement, within a factor of 5, was reasonable.

The dissociation constants (listed in Table V) cover a wide range from 10^{-15} *M* to 10^{-1} *M* and provide a detailed picture of the effects of

TABLE V
Difference Extinction Coefficients and Dissociation Constants of Avidin Complexes

			R_2	R_4	R_5	pH	$\frac{\Delta E_{max}}{E_{282}}$[b]	k_{-1}[c] (sec^{-1})	k_{-1}/k_1[c] (M)	K[c] (M)	$-\Delta G$[f] (kcal/mole)	$-\Delta H$[f] (kcal/mole) (6)	References[g]
1	Biotin and derivatives[a]	*d*-Biotin	O	COOH	–	2.0	0.90	2×10^{-5}	–	–	–	–	2,3
						3.0	–	9×10^{-7}	–	–	–	–	3
						5.0	–	9×10^{-8}	1.3×10^{-15}	–	20.4	21.5	1,2,3
						7.0	0.94	4×10^{-8}	–	–	–	–	3
						9.0	–	–	–	–	–	22.5	5
2		Biotin methyl ester	O	$COOCH_3$	–	7.0	0.74	–	–	–	–	–	4
3		Biotin sulfone	O	COOH	–	7.0	0.84	–	–	–	–	–	2
4		2′-Thiobiotin	S	COOH	–	9.0	1.09	3.6×10^{-5}	5×10^{-13}	–	16.9	17.8	6
5		2′-Iminobiotin	NH_2^+	COOH	–	9.0	0.76	–	–	5×10^{-8}	–	–	6
			NH	COOH	–	Calculated for free base[e]			–	3.5×10^{-11}	14.3	11.6	6
6		1′-*N*-Methoxycarbonylbiotin methyl ester	O	$COOCH_3$	–	4.6	0.43	–	–	7×10^{-6}	7.1	–	2
						6.8	0.52	–	–	4×10^{-7}	8.8	–	2
7		3′-*N*-Methoxycarbonylbiotin methyl ester	O	$COOCH_3$	–	4.6	0.34	–	–	1.5×10^{-8}	10.7	–	2
					–	6.8	0.45	–	–	ca 10^{-9}	12	–	2

(*Continued*)

TABLE V (*Continued*)

		R_2	R_4	R_5	pH	$\frac{\Delta E_{max}}{E_{282}}$[b]		k_{-1}[c] (sec^{-1})	k_{-1}/k_1[c] (M)	K[c] (M)	$-\Delta G^f$ (kcal/mole)	$-\Delta H^f$ (kcal/mole) (6)	References[g]
	Imidazolidone derivatives[a]												
8	*d*-Desthiobiotin	O	$(CH_2)_5COOH$	CH_3	4.0	0.46		7.6×10^{-5}	10×10^{-13}	–	16.5	–	4
					7.0	0.47		3.6×10^{-5}	5×10^{-13}	–	16.9	–	4
	dl-Desthiobiotin[d]	O	$(CH_2)_5COOH$	CH_3	4.0	0.46	*d*	6.3×10^{-5}	9×10^{-13}	–	16.5	–	4
					4.0	–	*l*	4.3×10^{-5}	6×10^{-11}	–	14.0	–	4
9	*dl*-Desthiobiotin methyl ester	O	$(CH_2)_5COOCH_3$	CH_3	7.0	0.39	*d*	2.9×10^{-4}	4×10^{-12}	–	15.6	–	4
					7.0	–	*l*	1×10^{-3}	1.5×10^{-11}	–	14.8	–	4
10	*dl*-Desthiobiotinol[d]	O	$(CH_2)_5CH_2OH$	CH_3	7.0	0.49	*d*	3.8×10^{-3}	5×10^{-11}	–	14.1	–	4
					7.0	–	*l*	7.7×10^{-2}	1×10^{-9}	–	12.3	–	4
11	D-4-*n*-Hexylimidazolidone	O	$(CH_2)_5CH_3$	H	7.0	0.38		0.24	3.5×10^{-9}	9×10^{-9}	11.1	–	4
	L-4-*n*-Hexylimidazolidone	O	$(CH_2)_5CH_3$	H	7.0	0.36		3.4	5×10^{-8}	2.3×10^{-7}	9.1	–	4
12	*dl*-4-*n*-Butylimidazolidone	O	$(CH_2)_3CH_3$	H	4.0	0.25		–	–	4×10^{-6}	7.4	–	4
					7.0	0.43		–	–	1.3×10^{-6}	8.1	–	4
					10.0	0.48		–	–	1.7×10^{-6}	7.9	–	4
13	*dl*-4-*n*-Propylimidazolidone	O	$(CH_2)_2CH_3$	H	7.0	0.24		–	–	5×10^{-6}	7.3	–	4
14	*dl*-4-Ethylimidazolidone	O	CH_2CH_3	H	7.0	0.22		–	–	8×10^{-6}	7.0	–	4
15	*dl*-4-Methylimidazolidone	O	CH_3	H	7.0	0.23		–	–	3.4×10^{-5}	6.1	–	4
16	Imidazolidone	O	H	H	6.8	0.26		–	–	5×10^{-4}	4.5	–	2
17	*dl*-4,5-Dimethylimidazolidone	O	CH_3	CH_3	7.0	0.17		–	–	2.3×10^{-5}	6.4	–	4
	meso-4,5-Dimethylimidazolidone	O	CH_3	CH_3	7.0	0.19		–	–	3×10^{-6}	7.5	–	4
18	*dl*-Norleucine hydantoin	O	$(CH_2)_3CH_3$	O	7.0	0.2		–	–	4×10^{-5}	6.0		4
19	D-4-*n*-Hexyl-2-thionoimidazolidine	S	$(CH_2)_5CH_3$	H	7.0	0.23		–	–	1.2×10^{-6}	8.1	–	4
20	D-4-*n*-Hexyl-2-iminoimidazolidine	NH^+	$(CH_2)_5CH_3$	H	11.8	0.26		–	–	4×10^{-6}	7.4	–	4
		NH^2	$(CH_2)_5CH_3$	H	Calculated for free base[e]				–	1×10^{-6}	8.2	–	4

	Oxazolidone derivatives[a]											
21	D-4-*n*-Hexyloxazolidone	O	$(CH_2)_5CH_3$	H	7.0	0.12	–	–	2.2×10^{-4}	5.0	–	4
22	D-5-*n*-Hexyloxazolidone	O	H	$(CH_2)_5CH_3$	7.0	0.26	–	–	4.4×10^{-6}	7.4	–	4
	Miscellaneous compounds											
23	Diamine from biotin [5-(3,4-diaminothiophan-2-yl]+ pentanoic acid	–	–	–	4.6	0.36	–	–	5×10^{-3}	2.7	–	2
					6.8	0.73	–	–	3×10^{-7}	9.0	–	2
					10.0	–	–	–	1.2×10^{-7}	9.5	4.7	6
24	Lipoic acid	–	–		6.8	0.39	–	–	7×10^{-7}	8.5	–	2
25	Tetrahydrofuran	–	–		7.0	0.28	–	–	0.9×10^{-2}	2.8	–	4
26	Urea	–	–		6.8	0.33	–	–	3.6×10^{-2}	2.2	–	2
27	Ethylene glycol	–	–		6.8	0.2	–	–	0.5	1	–	2
28	Hexyl alcohol	–	–		7.0	0.09	–	–	4.4×10^{-3}	3.2	–	4
29	Hexanoic acid	–	–		6.8	0.16	–	–	3×10^{-3}	4.9	–	2
						E_{500}/E_{282}						
30	4-Hydroxyazobenzene-2′-carboxylic acid	–	–		4.7	1.4	–	–	6×10^{-6}	7.2	–	7,8

[a] Note that C-3 and C-4 of biotin correspond to C-4 and C-5 of the imidazolidones. Since the latter were synthesized from α-amino acids, the α-CH of which became C-4 of the imidazolidone, they have been designated D or L according to their origin from the D or L series of amino acids. The absolute configuration of C-3 in biotin is the same as that of C-4 in the D-alkylimidazolidones. The relative configuration of the two oxazolidones is shown more clearly in Fig. 4. Further discussion of the stereochemistry and optical activity of some of these compounds is given by Green *et al.* (1970).

[b] The values of $\Delta E_{max}/E_{282}$ were determined with a Unicam SP. 700 spectrophotometer using avidin of specific activity 14. Redetermination of several of these values using pure avidin (specific activity 15) and a Cary 118 spectrophotometer gave values that were 10–15% higher (Table IV). The difference maxima were also shifted 1–2 nm to shorter wavelength, on account of the lower level of stray light from the double monochromator. The Unicam instrument gave wavelengths of maximum difference within 1 nm of 233 for all the compounds except 4, 18 (238 nm), 25, 27, 28, 29 (229 nm). The values of $\Delta E_{max}/E_{282}$ for urea and glycol have been corrected for a nonspecific red shift of the spectrum induced by the high concentration of reagent required to saturate the binding site. $\Delta E_{max}/E_{282}$ can be converted to a difference extinction coefficient per mole of analog bound by multiplying by 24,000.

[c] The values of k_{-1}/k_1 are based on the value of k_1 ($7 \times 10^7\ M^{-1}sec^{-1}$) determined for biotin at pH 5 (Green, 1963a). All measurements of rate and equilibrium constants were made at 25°C, except for those quoted from reference (2), which were at room temperature.

[d] The separate values of k_{-1} given for *d* and *l* isomers of *dl* desthiobiotin derivatives were calculated from the displacement rates of the *dl* mixtures, assuming in each case that the slower rate was that of the *d* isomer. The *dl* mixtures of the alkylimidazolidones gave linear Scatchard plots showing no significant difference in the affinity for *d* and *l* isomers.

[e] The value of K for the free base form was calculated from K measured at a lower pH, using the measured pK_a of the guanidinium group.

[f] ΔG and ΔH are the free energies and enthalpies of formation of the avidin complexes. They are all negative.

[g] 1, Green (1963a); 2, Green (1963b); 3, Green and Toms (1973); 4, Green and Toms (1975); 5, Suurkuusk and Wadsö (1972); 6, Green (1966); 7, Green (1965); 8, Green (1970).

stepwise modification of structure on the free energy of binding. The results obtained with close relatives of biotin confirmed the main conclusions of earlier workers concerning the relative importance of different parts of the biotin molecule. Biotin sulfone (compound 3), ester (compound 2) amides (Green, 1963b), and desthiobiotin (compound 8) all bind very firmly to avidin whereas the diamine (compound 23), in which the imidazolidone ring is broken, has a dissociation constant 10^7 times greater than that of biotin. Nevertheless, it was still bound quite firmly unless both amino groups were protonated. Other analogs (compounds 4 and 5) in which the imidazolone ring was less drastically modified were still bound strongly.

2. *Stereochemistry of Biotin*

Before considering the finer details of the effects of structure on binding, the stereochemistry of biotin and its derivatives will be outlined. The absolute configuration has been determined by X-ray crystallography (Trotter and Hamilton, 1966) and is shown in Table V. There are three asymmetric centers at C-2, C-3, and C-4. The two rings are fused in the cis configuration and the valeric acid side chain (C-2) is also cis in relation to the imidazolidone ring. There are four racemic pairs of stereoisomers (György, 1954). The binding of *dl*-epibiotin, epimeric at C-2, has not been studied. Wright and Skeggs (1947) showed that of the allo isomers (trans fusion of the rings) only the *dl*-epiallobiotin competed significantly with *d*-biotin (inhibition ratio, 6) and that *l*-biotin did not compete at all. The epiallo isomers resemble biotin in the relation of the valeric acid side chain at C-2 to N-1′ but differ in its relation to N-3′.

In the series of imidazolidones and oxazolidones shown in Table V, only the asymmetric center at C-3 remains. This becomes C-4 in the imidazolidones and when its configuration is D (see footnote *a* to Table V) it resembles that of *d*-biotin. The methyl group of desthiobiotin produces an extra asymmetric center at C-5, with the opposite configuration to that at C-4 (imidazolidone numbering). If the two alkyl chains were equivalent, desthiobiotin would be an internally compensated meso compound.

3. *Relation of Affinity to Structure*

The results obtained with the 4-alkyl imidazolidones (compounds 11–16) and their analogs are summarized in a diagram (Fig. 4). The D-hexyl derivative, stereochemically similar to *d*-biotin, binds most firmly with ΔG (−11 kcal/mole) 2 kcal greater than that of the L-

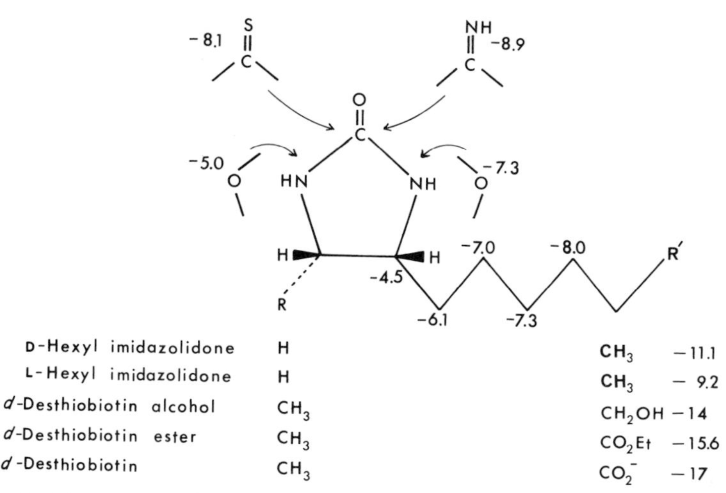

FIG. 4. Free energy of binding of hexyl imidazolidone and its analogs (Green and Toms, 1975). The numbers give the free energies of binding of a series of compounds regarded as derived from 4-*n*-hexylimidazolidone. The ring modifications were made one atom at a time using the D isomer. The series of shorter alkyl chains were all *dl* compounds.

isomer. When the alkyl chain was shortened, an approximately linear decrease in ΔG with decreasing chain length was observed implying that the whole of the alkyl chain was contributing to the binding energy. The average decrease of 0.7 kcal per methylene group was similar to that observed in hydrophobic interactions of other homologous series (Anderson *et al.*, 1965; Smith and Tanford, 1973).

Much larger free-energy changes followed modification of the ring structure, the most dramatic being the drop of 6 kcal when N-1′ was replaced by oxygen. This was considerably larger than that produced by a similar replacement at N-3′. A similar larger effect of modification of N-1′ versus N-3′ was seen when these positions in biotin were substituted with methoxycarbonyl groups (compounds 6 and 7). It seems likely that both NH groups are acting as H-bond donors to acceptor atoms on the protein and that the bond with N-1′ is particularly strong. The small difference between D and L isomers (except for the hexyl derivative) can be understood in terms of the "meso" structure of desthiobiotin. Interchange of C-4 and C-5 by rotation about an axis through the carbonyl group would convert an analog with the "wrong" (L) configuration at C-4 to one with the right configuration at C-5. One might predict on this basis that the

L-isomer when bound to avidin would be rotated 180° with respect to the bound D-isomer. Introduction of a methyl group in the 5 position and a carboxyl group at the end of the alkyl chain gives desthiobiotin, further increasing the free energy of binding.

It appears that almost every atom in the biotin molecule contributes something to the free energy of interaction with avidin. This is confirmed by the ability of avidin to bind analogs corresponding to quite small fragments of biotin such as urea, glycol (equivalent to C-3, C-4, N-1′, N-3′), tetrahydrofuran, and caproic acid. It is likely that each binds to a different part of the site since, for example, the spectral change produced by a mixture of imidazolidone and hexanoate is approximately the sum of the separate contributions (Green and Toms, 1975). It is also interesting that the free energy of binding of each of these two analogs was unaffected by the presence of an excess of the other. The sum of the free energies of binding of hexanoate (compound 29) and 4-methylimidazolidone (compound 15) is 11 kcal, considerably less than that of desthiobiotin (compound 8) (16.9 kcal), which is the equivalent single molecule. Thus although there is no evidence for cooperative interactions between different parts of the binding site, there is a considerable enhancement in the free energy of binding of biotin and its near analogs relative to that of the fragments.

This differs from the behavior of lactate dehydrogenase toward fragments of NAD (McPherson, 1970). The enzyme does not bind nicotinamide unless the other half of the coenzyme (AMP) is present, and under these conditions the sum of free energies of binding of the fragments (1.8 + 3.2 kcal) is close to that of the complete coenzyme (5.3 kcal).

Since binding of each fragment of biotin leads to red shifts of the tryptophan absorption spectrum and since each small modification of biotin decreases the red shift (Table V), it is likely that several tryptophans of each subunit interact directly with biotin. Further evidence for this comes from chemical modification experiments described in Section IV,C. Although there was a general trend toward a lower difference extinction coefficient as the free energy of binding decreased, the proportionality was not exact. For example, hydrolysis of the imidazolidone ring produced a large decrease in free energy of binding but only a small decrease in $\Delta\epsilon_{233}$ (compound 23), while removal of the sulfur from the thiophan ring had the converse effect (compounds 8, 9, 10). It is therefore difficult to draw any quantitative conclusions about the contribution of biotin–tryptophan interactions to the binding energy.

The wide range of compounds that will bind to avidin suggests a rather nonspecific binding site. But, apart from the dye (compound 30) and anilinonaphthalene sulfonate (Green, 1964b), they are all related to some part of the biotin molecule. On the other hand, a site of relatively high specificity is supported by studies on related compounds which did not bind significantly at the 1–10 m*M* level. These include the 20 natural amino acids, succinimide and imidazole-4-propionic acid (N. M. Green, unpublished experiments).

C. Periodate Oxidized Avidin

The effects of biotin on the spectrum of the formylkynureinine residue of oxidized avidin provided the opportunity of detecting differential effects of biotin analogs on at least three tryptophan residues. The difference spectrum produced by biotin (Fig. 5c) shows additional maxima at 242, 267, 276, and 345 nm, which are almost the same in position and relative magnitude as the peaks in the first derivative spectrum of α-*N*-acetyl-*N*-formylkynurenine, allowing for a red shift caused by the protein environment. Fortunately, they hardly overlap at all with the trypophan peaks so that differential effects are readily detectable. Examination of the spectra produced by analogs showed that esterification of the carboxyl group of biotin decreased the perturbation of the formylkynurenine residue without affecting the other tryptophans while removal of the sulfur atom showed the converse behavior. This implies that the oxidized tryptophan residue interacts in the neighborhood of the valeric acid side chain of biotin and that it is located away from the thiophan ring. One of the other tryptophans must be near the thiophan to account for the decreased $\Delta\epsilon_{233}$ when desthiobiotin was bound. The extent of the interaction of the valeric acid side chain with the formylkynurenine can be deduced from the different spectra given by propylimidazolidone, which produced a slight perturbation of the oxidized residue, and ethyl imidazolidone, which did not. Since ethyl imidazolidone gives a tryptophan difference spectrum it is likely that it interacts with a third tryptophan residue different from that which is perturbed by the thiophan ring.

D. Nature of the Binding Site

Although the strength of the avidin–biotin interaction suggests the possibility of a covalent attachment, most of the evidence favors noncovalent binding. The biotin can be released by 6 *M* GuHCl at pH 1.5 (Cuatrecases and Wilchek, 1968) or by autoclaving. The study of analog binding shows extensive noncovalent interaction with all

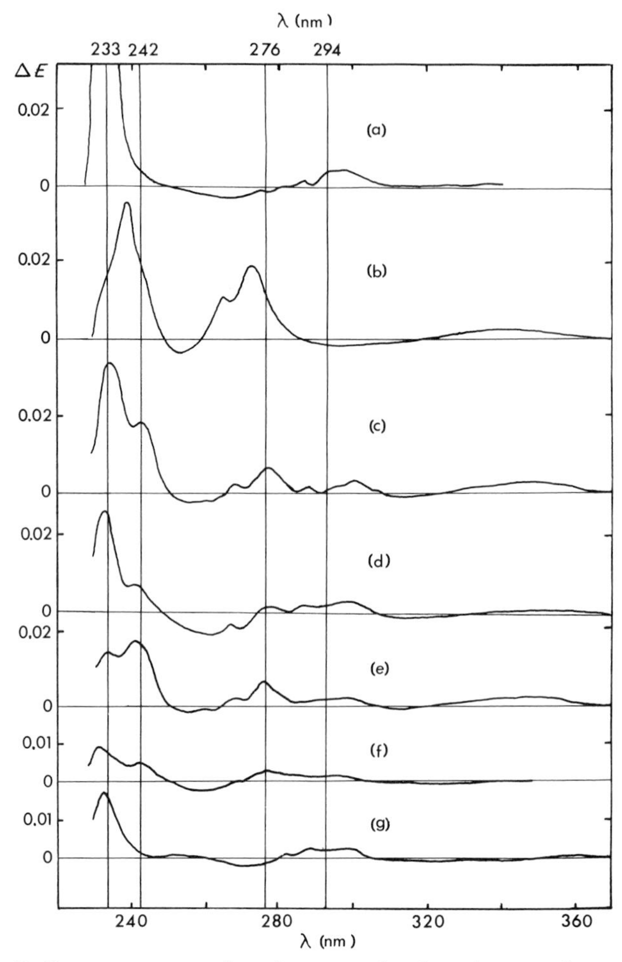

FIG. 5. Difference spectra of avidin or oxidized avidin complexes versus the ligand-free protein. The protein subunit concentration was 6.5 μM (0.4375 cm path) in 50 mM sodium phosphate pH 6.8 (Green, 1975). (a) Avidin–biotin; (b) first derivative of spectrum of α-*N*-acetyl-*N*-formylkynurenine; (c)–(g) Complexes of periodate oxidized avidin with a ligand, as follows: (c) biotin; (d) biotin methyl ester; (e) desthiobiotin; (f) propyl imidazolidone; (g) ethyl imidazolidone.

parts of the biotin molecule, and although the sum of the weak interactions is a significantly less than the free energy of binding of biotin the discrepancy is small enough to be explicable in terms of a slightly better fit of biotin as compared with the analogs. These results combined with those of chemical modification studies suggest that three or four tryptophan residues in each subunit interact directly with the biotin. In addition, there are groups which hydrogen bond to the ureido group and there is an amino group which reacts with FDNB to give a blocked binding site.

If this amino group were near to the tryptophan which is perturbed by the carboxyl group of biotin several other observations would fall into place. The susceptibility of this tryptophan to oxidation by periodate could be accounted for by binding of IO_4^- to NH_3^+. The observation that dinitrophenylation of this amino group prevents both the binding of biotin and the oxidation of the tryptophan by periodate (Green and Toms, 1975) would also be explicable. Although the amino group reacts readily with FDNB it does not react with the bulkier dansyl chloride and trinitrobenzene sulfonic acid reagents. This is consistent with a location near to the biotin carboxyl group, which studies with bisbiotinyl compounds (Section V,D) have shown to be some way beneath the van der Waals surface of the protein and hence subject to restricted access.

Since 70% of the biotin molecule is hydrophobic and much of it interacts with tryptophan, it was expected that the binding would be accompanied by an appreciable increase of entropy. However, calorimetric measurement of the enthalpy of binding (Green, 1966; Suurkuusk and Wadsö, 1972) led to the conclusion that there was no net entropic contribution to the free energy of binding. Nevertheless, the high negative ΔC_p (237 cal/°C/biotin) (Suurkuusk and Wadsö, 1972) confirmed a loss of hydrophobic surface accompanying binding. It was suggested that the high enthalpy of binding was a reflection of the formation of several strong hydrogen bonds to the imidazolidone ring. Since the entropy of formation of these bonds would be negative, the net entropy change might be very small. This interpretation is supported by the positive entropy of binding of the diamine (compound 23), in which the imidazolidone ring is broken. It would be possible to account for most of the large ΔH in these terms, only if the potential bonding atoms in avidin do not form hydrogen bonds to water when biotin is absent. It is difficult to see how this could happen without some local conformational change.

X-Ray crystallographic studies at present in progress could give a direct answer to this question. If the bound biotin analogs can be located with sufficient accuracy, it may also be possible to establish correlations between specific atomic interactions and the free energies of binding of analogs.

E. Evidence for Random Binding

The strength of the interaction between avidin and biotin is so great that it would not have been surprising to find that binding of biotin at one site influenced the interaction at neighboring sites. Nevertheless, several recent observations have confirmed the early conclusion (Green, 1964a) based on quenching of fluorescence, that the binding is a random process and that there is no detectable interaction between the sites: (1) Scatchard plots of the binding of analogs ($K > 10^{-8}\ M$) are linear (Green and Toms, 1975). (2) The binding constants of L-4-hexylimidazolidone and of 2-hydroxyazobenzene 4′-carboxylate were not altered when three quarters of the binding sites were presaturated with biotin (Green and Toms, 1975). (3) The heat of binding of biotin was not affected by partial presaturation of the binding sites (Green, 1966).

We will recapitulate the early evidence in more detail as it forms the basis of a general method for distinguishing cooperative from random binding that has recently been extended independently by Holbrook (1972) in a study of fluorescence quenching of dehydrogenases by NADH.

The method depends upon the ability of a dinitrophenylated biotin derivative to quench the fluorescence of subunits besides the one to which it is bound. This arises because R_0 [the distance between DNP and tryptophan at which quenching and emission have equal probability (30 Å)[1]] is comparable to the distance between binding sites (20–30 Å); even at a distance of 40 Å the probability of quenching is 16%. In consequence, the relation between fluorescence and fractional saturation of the sites will be nonlinear unless there is very strong positive or negative cooperativity that would eliminate species intermediate between A_4 and $A_4\ B_4$. The shape of the quenching curve can be calculated for different models of the binding process provided that the quantum yields for the different intermediate species can be assigned. This can be done either by a curve-fitting procedure (Holbrook, 1972) assuming $q_i/q_0 = (q_1/q_0)^i$ or by a semiempirical procedure (Green, 1964b, and Fig. 6) in which the fluorescence of randomly labeled DNP-avidin was measured as a function of the number of covalently bound DNP groups. The good agreement between the experimental points and the theoretical relationship for random binding provides strong evidence against any cooperative interactions.

[1] The value of 35 Å given by Green (1964a) was based on the assumption of too high a quantum yield for tryptophan.

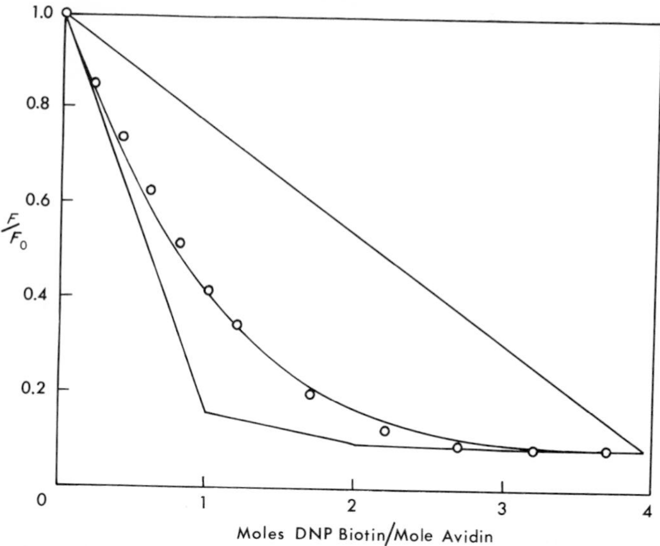

FIG. 6. Quenching of fluorescence of avidin by the DNP-hydrazide of biotin. The experimental points (Green, 1964b) are compared with theoretical curves recalculated for a four-subunit molecule from the relation $F/F_0 = \Sigma\ \alpha_i q_i$ where α_i is the fraction of avidin present as $A_4\ B_i$ and q_i is the relative quantum yield of the ith species. Knowing from the experimental points that $q_4 = 0.09$, it followed, from the fluorescence of avidin carrying covalently bound DNP (Green, 1964b), that $q_3 = 0.09$, $q_2 = 0.11$, $q_1 = 0.17$, $q_0 = 1$. Three curves were then calculated relating F/F_0 to x, the fractional saturation of the binding sites, using three different relationships between x and α_i: (1) sequential binding (left-hand line); (2) random binding (middle curve): α_i given by successive terms of the expansion of $[(1-x)+x]^4$; (3) cooperative binding (right-hand line): $\alpha_0 = (1-x)$, $\alpha_1 = \alpha_2 = \alpha_3 = 0$, $\alpha_4 = x$.

A similar analysis (Holbrook, 1972) of the quenching of the fluorescence of lactic dehydrogenase was consistent with random binding of NADPH by the four subunits. Here q_1/q_0 is considerably higher (0.59) than it is for the DNP-avidin system, since R_0 is smaller (25 Å) and the subunits are larger. In consequence the quenching curve is more nearly linear. By combination of appropriate curve-fitting procedures with accurate experimental results it was still possible to obtain information about the distribution of ligands among the binding sites.

V. SUBUNIT STRUCTURE

A. *Dissociation in Guanidinium Chloride*

Avidin resists unfolding in high concentrations of urea (Fraenkel-Conrat *et al.*, 1952b) or moderate concentrations of GuHCl (Green,

1963c). Above 3.5 *M* GuHCl the protein begins to unfold, as measured by loss of biotin binding activity, exposure of aromatic residues to the solvent and change of optical activity. At a concentration of 6 *M* GuHCl molecular weight measurement indicates dissociation to quarter-molecules. The rate of unfolding, measured by difference spectrophotometry, increased with the twelfth power of the concentration of GuHCl, and the kinetics could be accounted for as a sum of two exponential terms (recalculated from the published results). The loss of biotin-binding activity followed the same time course. These results could be accounted for in terms of a reversible unfolding of the protein with accumulation or reformation of a partially folded species, provided that the ultraviolet absorbance of the latter is the same as that of the fully unfolded form (Ikai *et al.*, 1973). If the extinction coefficients of the different unfolded forms were not the same, the kinetics of loss of activity would differ from the kinetics of the spectral change. However, it seems unlikely that this can in fact be the explanation of the complex kinetics since, as we shall see, the unfolded species are not in equilibrium with native avidin.

When the transition and its reversal were followed as a function of GuHCl concentration by difference spectroscopy (Fig. 7), the regain of tertiary structure and biotin binding activity did not occur until the concentration of GuHCl was lowered to 2 *M*. The simplest explanation of apparent hysteresis in systems of this type is that insufficient time has been allowed for the system to come to equilibrium (Steinhardt and Zaiser, 1953; Tanford, 1968). This is not the explanation here since in 4.5 *M* GuHCl no further change was observed between 18 hours and 60 hours.

Further information on irreversibility came from the addition of biotin to the system at various stages of the cycle. Over the whole of the dissociation branch, the binding of biotin was immediate and no slow changes were observed, even when the system was left for 48 hours. The amount bound was proportional to the fraction of native avidin left as measured by the spectral change. If any other species had been present which could bind biotin, complete reversal of the reaction would have been expected, since monomeric avidin–biotin complex reassociates to native tetramer even in 6 *M* GuHCl (Green and Toms, 1972). This unexpected irreversibility was not the consequence of a slow secondary change, since the same result was obtained if biotin was added during the course of denaturation in 6 *M* GuHCl. The biotin reacted immediately with the remaining native protein, but did not react at all with the freshly formed unfolded protein. Dilution of the reaction mixture to a concentration of 3 *M*

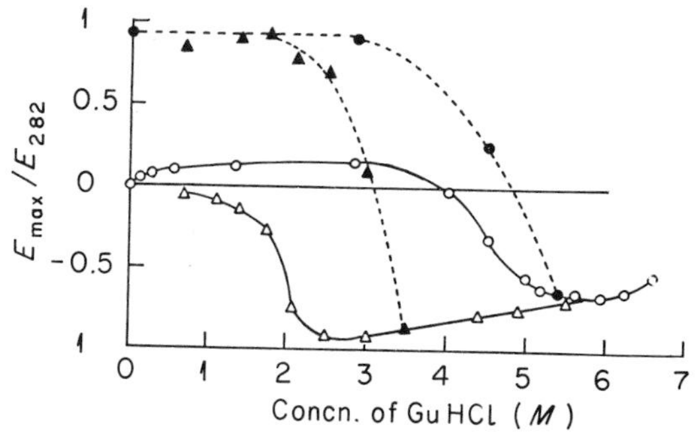

FIG. 7. Dissociation of avidin into subunits by guanidinium chloride (GuHCl). Avidin samples were left for 18 hours at the indicated concentration of GuHCl (in 50 m*M* sodium phosphate, pH 6.8), either before (○) or after (△) dissociation into subunits in 6.6 *M* GuHCl. The extent of dissociation was determined from the difference spectrum at 233 nm. One equivalent of biotin was added to each sample after the measurement, and the difference spectrum was redetermined (●, ▲).

GuHCl, without addition of biotin, also gave a frozen system containing native tetramer and unfolded monomer.

In contrast, the reassociation branch of the curve behaved in a more normal manner. The ability to recombine with biotin was regained when the guanidine concentration reached 3 *M*; a slow binding was then observed. At this concentration, in the absence of biotin, molecular weight measurements (Green, 1964b) showed no detectable tetramer. This was consistent with later results (Green and Toms, 1973) which showed that single subunits of avidin could bind biotin. In this region of the curve there appears to be a reversible equilibrium between unfolded and partly refolded subunits which can be displaced when the latter interact with biotin. When the refolding progresses to the stage where tetramers can re-form the reaction goes rapidly to completion even in the absence of biotin, which could account for the very sharp transition in this region. The kinetics of refolding were first order and the rate was inversely proportional to the twenty-fifth power of the concentration of GuHCl. A second denaturation of the refolded avidin followed the same course as the first.

The behavior described is in many respects similar to that of systems which show geniune hysteresis, such as agar gels (Rees, 1969) or polyglutamic acid (Jennings *et al.*, 1968) except that for these systems the temperature rather than the denaturant concentration was varied. It would be interesting to know whether avidin would show similar hysteresis with temperature at a constant GuHCl concentration in the region of 2 *M*.

The thermodynamics of these systems have been developed by Everett (Everett and Whitton, 1952; Everett, 1954, 1955), who pointed out that the transitions should be extremely sharp in homogeneous systems and that the finite width often observed is a consequence of the sequential melting of domains of different sizes. This is the one aspect of the behavior of avidin that is difficult to account for if we assume that it is a single molecular species. To elaborate this point, it seems most likely that at concentrations of GuHCl above 3 *M* the unfolded monomer represents the thermodynamically stable state. Nevertheless, the native tetramer is stable for periods of weeks in 3 *M* GuHCl and does not unfold at a significant rate until concentrations of 3.5–4 *M* are reached. It is difficult to understand why the unfolding does not go slowly to completion at this concentration, yet even at 4.5 *M* GuHCl it comes to a stop half way. It seems unlikely that the minor heterogeneity described by De Lange (1970) could account for the breadth of the transition, although it should be pointed out that heterogeneity could also explain the presence of two exponential terms in the unfolding kinetics. The refolding branch of the curve presents less of a problem since the transition is extremely sharp. Clearly more experimental work is required to resolve these contradictions, particularly on the effect of temperature on the system and on the formation of tertiary structure measured by CD.

Some time has been spent in this attempt to define the behavior of avidin in GuHCl since it is likely to be characteristic of the unfolding of other multisubunit proteins, few of which have been studied in detail. The refolding of aldolase after dissociation at low pH has been thoroughly examined by Teipel (1972). He showed a rapid regain of helical structure to give an inactive monomer–dimer mixture which slowly isomerized to a form which gave active tetramer. The second stage was accelerated by substrate. A similar sequence of events would be consistent with the more limited observations made with avidin, except that the inactive folded form of avidin is unlikely to be helical. The reversibility of the dissociation branch of the aldolase curve has not been studied, but preliminary results of Deal *et al.* (1963) using urea suggest that it may show irreversibility similar to that of avidin.

The dissociation of apoferritin at acid pH shows marked hysteresis, which has been followed by difference spectroscopy and by sedimentation velocity (Crichton and Bryce, 1973). Avidin also shows hysteresis at low pH (Green, 1963c).

B. *Stabilization by Biotin*

The effect of adding biotin during the process of unfolding has been considered in the preceding section. Although biotin could not induce refolding in 6 *M* GuHCl, it completely prevented unfolding (Section III,A) even in 8 *M* GuHCl or in 0.1 *M* HCl. A combination of 6 *M* GuHCl with low pH (1.5) does dissociate the complex and has been used to recover avidin from biocytinyl-Sepharose columns (Cuatrecasas and Wilchek, 1968).

1. *Dissociation of Partially Saturated Avidin–Biotin Complexes*

Further information about the pathway of unfolding of the tetrameric avidin and the nature of the stabilization by biotin has been obtained from measurements on avidin molecules partially saturated with biotin. Since subunits combine at random with biotin such preparations will be mixtures in which the number of occupied sites per molecule is binomially distributed. Occupied subunits present in a tetramer were resistant to 6 *M* GuHCl or to *N*-bromosuccinimide. The question was asked whether such subunits were equally resistant when neighboring subunits were unoccupied and, if so, whether they had any stabilizing influence on their neighbors. The answer to the first question was "yes" and to the second, "almost none," as will appear from the following section.

When avidin is treated with *N*-bromosuccinimide, oxidation of the tryptophan residues is followed by dissociation of the tetramer into subunits which partially aggregate (Green and Ross, 1968). The presence of biotin completely protected the subunit to which it was bound, but it did not affect the reactivity of neighboring subunits. The oxidized subunits dissociated, while the protected ones recombined to form tetrameric avidin–biotin complex. The effects of 6 *M* GuHCl were identical, except that the subunits were not oxidized (Green and Toms, 1972). The rate of unfolding of the vacant subunits in GuHCl was decreased by a factor of two in molecules which contained two or three occupied subunits, a small effect when considered in relation to the free energy of binding of biotin.

These observations have several further implications. Since all four tryptophan residues of the vacant subunits were oxidized and all those of the occupied subunits were protected, it follows that each binding site was almost certainly situated within a subunit rather than at an interface between subunits. This conclusion was also supported by the small effect of biotin bound to one subunit on the stability of the neighboring subunits. This last observation also suggested that GuHCl initiated the unfolding in the region of the biotin binding site rather than at the intersubunit bonds. If the latter pathway were important, a greater effect of biotin on the unfolding of neighboring subunits would have been expected. Both sets of experiments imply that single subunits with bound biotin are stable, at least for short periods, at concentrations of GuHCl up to 6 *M*, since they must be intermediates in the overall reaction

$$A_4B \rightarrow 3A + AB$$
$$4AB \rightarrow A_4B_4$$

2. *Thermal Stability*

The thermodynamics of the dissociation of avidin has not yet been studied since the appropriate conditions for reversibility have not been established. Calorimetric measurements of the irreversible heat denaturation are of some interest, however, in relation to the stabilizing effect of biotin. By differential scanning calorimetry Donovan and Ross (1973) showed that avidin underwent a cooperative thermal transition at 85°C with a ΔH of 76 kcal per subunit. This is within the range normally found for denaturation of proteins. The corresponding transition in the avidin–biotin complex took place at 132°C under pressure and ΔH was increased, by almost 200 kcal, to 270 kcal per subunit. Some of this can be accounted for from the heat of binding of biotin, which could be as high as 46 kcal per subunit at 132°C (ΔC_p is 240 cal/°C per subunit, Suurkuusk and Wadsö, 1972). Donovan and Ross pointed out that most of the remaining difference could also be an effect of the high transition temperature and the large measured ΔC_p of 3.0 kcal/°C per subunit.

C. *Binding of Biotin by Single Subunits*

Single subunits have been obtained free in solution only in the presence of GuHCl or at acid pH. To study their properties under normal conditions it was necessary to have them coupled to a Sepharose matrix to prevent them from reassociating (Green and Toms, 1973). Low concentrations of tetramer were coupled using (1) small amounts of cyanogen bromide, to minimize the proportion of doubly linked molecules and (2) low concentrations of tetramer, to avoid interactions between the covalently coupled subunits after the 75% of noncovalently bound subunits had been removed with 6 *M* GuHCl. In spite of these precautions the flexibility of the Sepharose matrix permitted some interaction between covalently linked subunits. In consequence the preparation showed three classes of binding site in approximately equal proportions. The strongest was indistinguishable from tetrameric avidin while the weakest ($K = 10^{-7}$ M) was assigned to single subunits. The proportion of the latter was doubled when the Sepharose was stabilized by cross-linking with divinylsulfone. All classes of covalently bound subunits were able to reform tetramers when subunits in 3 M GuHCl were added and the system was diluted. It can be concluded that the Sepharose matrix did not interfere with refolding of the peptide chain and that the formation of active subunits was not dependent on interactions with other subunits. Subunit interaction, however, was required for firm binding of biotin.

The question of whether there exists a finite concentration of single subunits in equilibrium with the tetramer has not been answered. The most sensitive method for detecting such an equilibrium is to look for formation of hybrids after long incubation. Preliminary experiments in the author's laboratory have used mixtures of trinitrophenyl avidin (TNP_4A_4) and unlabeled avidin. The fluorescence of the mixture should decrease by half when a normal subunit is exchanged for a TNP subunit. Less than 5% change in fluorescence was observed over periods of several days showing that the dissociation constant for formation of monomers or dimers from tetramers must be less than 10^{-12} M. [If a rate constant of 10^6 M^{-1} sec^{-1} is assumed for the association reaction between macromolecular species an approximate dissociation constant (K) can be calculated from the rate of subunit exchange (= rate of dissociation), $K = 10^{-6}k$.]

Further experiments along these lines in the presence of GuHCl could provide information on the species which are in equilibrium with each other in the different states illustrated in Fig. 7.

D. Bifunctional Biotin Derivatives and Subunit Structure

Although examination of avidin in the electron microscope did not reveal any subunits (Fig. 8A), it was possible to deduce the symmetry of their arrangement from the morphology of polymers produced when avidin combined with bisbiotinyldiamines (Green *et al.*, 1971). These polymers were formed only when the chain joining the carboxyl groups of biotin was more than 15 Å long. When the chain length was increased to 18 Å the second biotin was bound essentially irreversibly. With intermediate chain lengths it was possible to depolymerize the products with an excess of reagent or with the dye 4-hydroxyazobenzene 2′-carboxylate. The polymers (Fig.

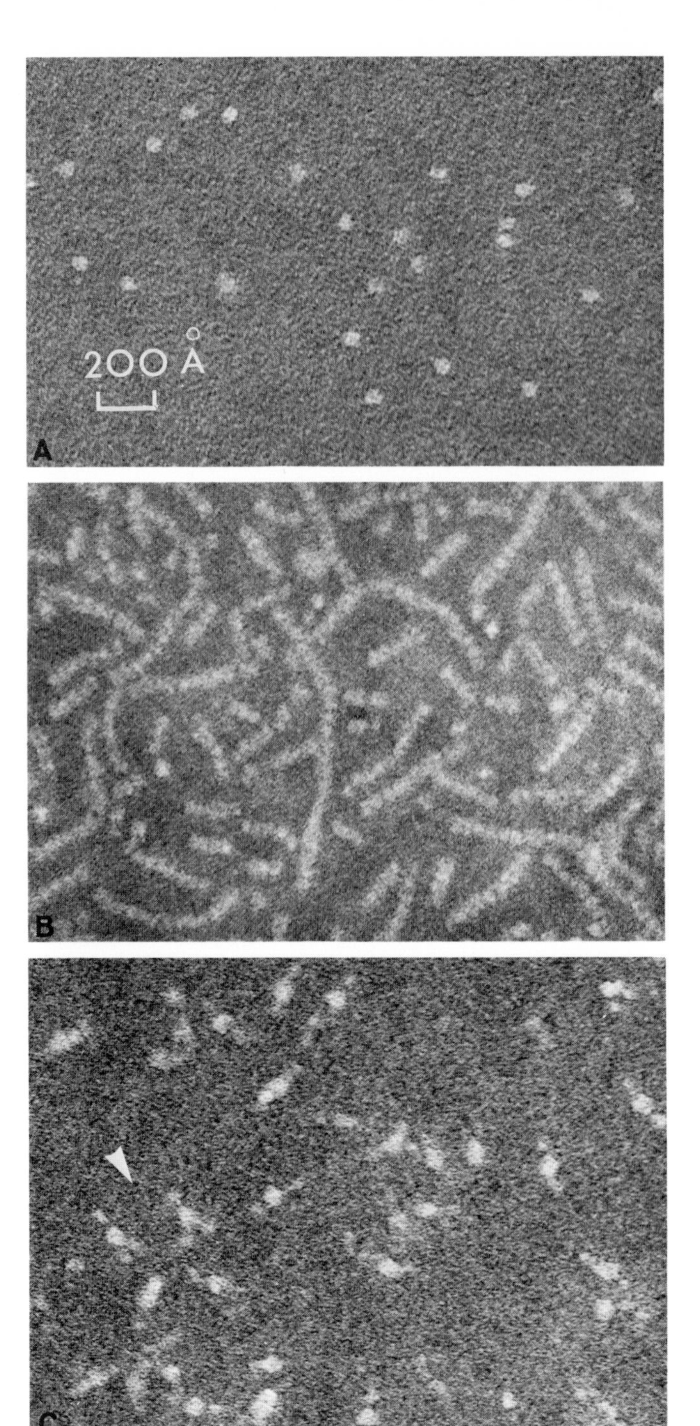
200 Å
A
B
C

8b) were linear, unbranched, and as wide as a single avidin molecule (55 Å). The repeat distance of 41 Å suggested that each molecule was a cylinder or an oblate ellipsoid 55 Å × 41 Å, consistent with the size of the unit cell of the crystal (62 Å × 107 Å × 43 Å) which contained two molecules (Green and Joynson, 1970). A tetramer with 2-fold symmetry would give linear polymers when linked by chains of minimal length, whereas one with 4-fold symmetry would not. Since spectrophotometric titration showed that all the sites were saturated, each molecule was doubly linked to its neighbor and the binding sites were arranged in pairs on the 55 Å faces as indicated in the diagram (Fig. 9). When linking chains longer than 23 Å were used the length of the polymers was markedly decreased, suggesting termination by intramolecular binding of the bifunctional ligand.

In the stable polymers the carboxyl group of the bound biotin lies 8–9 Å beneath the van der Waals surface of the molecule. Even so it must be accessible to the solvent since large molecules can be coupled to it without disturbing the binding. This suggests that there may be a shallow depression in the surface as illustrated in Fig. 9. This would account for the ability of a 23 Å chain to bridge neighboring sites without bringing the sites improbably close together.

When avidin was combined with four molecules of 1-biotinamido-12-dinitrophenylaminododecane only two of the four DNP groups were titratable with Fab fragments of anti-DNP antibody and with few exceptions only two Fab fragments could be seen bound to each avidin in electron micrographs of the product (Fig. 8C) (Green, 1972). This confirms the 2-fold symmetry of the avidin molecule and shows that the members of each pair of sites are too close together to accommodate 2 Fab fragments (diameter 40 Å) without considerable strain. The occasional avidin carrying three Fab fragments suggests that weak binding of extra Fabs is possible.

An extension of this experimental approach which may prove useful is to employ avidin as a bifunctional reagent for linking biotin labeled macromolecules together. The biotin could be linked directly to a lysine residue of a protein or combined with a firmly bound specific ligand. The products may be useful for electron

---

FIG. 8. Electron micrographs of avidin. (A) Avidin; (B) avidin polymerized with bisbiotinamidodocane; (C) avidin labeled with Fab fragments of anti-DNP antibody using 1-dinitrophenylamino-12-biotinamidodocane. The molecules with three Fab fragments (arrow) represented 2 or 3% of the total population. ×425,000. From Green (1972).

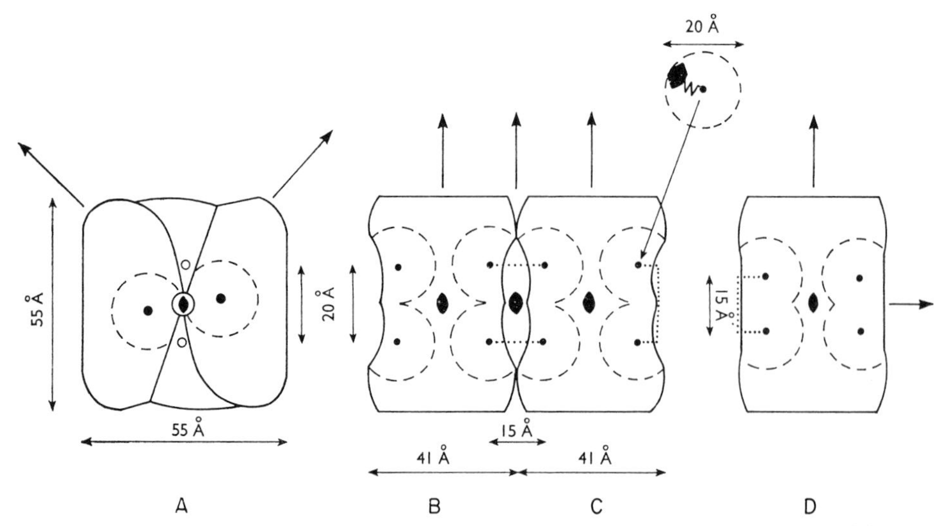

FIG. 9. Structure of avidin polymers with bisbiotinyldiamines. The location of the biotin carboxyls (●) relative to the surface is defined by the minimal chain length of diamine required for stable polymers (18 Å). Their intramolecular separation is defined by the chain length (23 Å) leading to termination of the polymers and by the depth of the depression in which the sites are situated. (A) Polymer chain viewed end on. The sites on the lower face of the terminal molecule are shown as open circles. The rotation of one molecule relative to the next is determined by the angle between pairs of sites. In the alternative side views, (B) and (C), this angle has been arbitrarily set at 0° to simplify illustration. The 20 Å separation of sites in (B) requires a deeper depression than the 15 Å separation in (C). Arrows and elongated solid ellipses represent 2-fold symmetry axes. From Green *et al.* (1971).

microscopy, for studies of energy transfer in fluorescent systems or for immunological studies of linked antigens. Provided that the macromolecule is more than 40 Å in diameter, no more than one should combine on each face of the avidin molecule.

VI. MISCELLANEOUS TOPICS

A. *Methods for Estimation of Avidin*

The early methods for estimation of avidin were based on microbiological assay of biotin using mainly yeast (Hertz, 1943) or *L. arabinosus* (Wright and Skeggs, 1944) as the test organism. They are discussed in detail elsewhere (Snell, 1950; György, 1954). They are extremely sensitive, being capable of measuring biotin concentrations of 10^{-11} *M*, but are rather time consuming. A number of more convenient methods are now available based either on the use of radioactive biotin or on spectral changes that accompany biotin binding.

Since avidin is strongly adsorbed by carboxymethyl cellulose (Green, 1963a) or by bentonite (Fraenkel-Conrat *et al.*, 1952a), the amount of bound radioactive biotin is easily determined. The adsorption by bentonite is more efficient and the sensitivity of the method is limited only by the specific activity of the biotin (Korenman and O'Malley, 1967) and can be of the same order as that of the microbiological assay. The estimation of radioactive avidin–biotin complex by gel filtration (Wei and Wright, 1964) should be used with caution since avidin is adsorbed by Sephadex, particularly at low salt concentrations. The effect should be small in 0.2 *M* ammonium carbonate, as used by Wei and Wright, but it can be significant in 0.1 *M* NaCl at pH 7 (N. M. Green, unpublished observations).

The spectrophotometric method is more convenient but much less sensitive (20 μg of avidin per milliliter, 10^{-6} *M*). It is based on the large spectral change in the dye 4-hydroxyazobenzene-2′-carboxylic acid when it is bound to avidin (Green, 1965). A new absorption band appears ($\epsilon_{500} = 34{,}000$) characteristic of the quinonoid form of the dye, and this disappears when the dye is displaced by biotin. The amount of avidin may be calculated directly from the increased absorbance at 500 nm or the dye may be used as an indicator in a spectrophotometric titration with biotin. Since the extinction coefficient and the dissociation constant of the avidin–dye complex remain constant between pH 4 and pH 8 and over a wide range of salt concentrations, the method is very convenient (Green, 1970). It can be used to measure protein bound biotin, preferably after partial digestion with pronase. Satisfactory results have been obtained both with purified proteins (Scrutton and Mildvan, 1968; Gerwin *et al.*, 1969) and with chloroplast preparations (Kannangara and Stumpf, 1973). The only misleading system (Wellner *et al.*, 1968; Huston and Cohen, 1969) also gave misleading results when microbiological procedures were used. Further details of specificity and experimental procedure for this and other methods of assay for avidin and biotin can be found elsewhere (McCormick and Wright, 1970).

B. Avidin and Biotinyl Enzymes

Avidin has been used in the study of several different aspects of biotinyl enzymes. In the first place it has provided a criterion for the involvement of biotin in enzymic reactions. This is particularly useful since biotin is always covalently bound and is difficult to detect by other means. In detailed studies of the molecular characteristics of biotinyl enzymes avidin has been used to measure changes in the availability of biotin in the presence of various ligands and conversely to measure the binding of these ligands by the enzyme. It has also been used to locate the biotinyl residue in electron micrographs of complex enzymes (Green *et al.*, 1972). It may yet prove possible to employ it or some modified version of lower affinity for enzyme purification.

The biotinyl enzymes themselves have been comprehensively reviewed by Moss and Lane (1971), and only those points specifically related to avidin will be considered here. Before 1950, studies of biotin-deficient organisms had implicated biotin more or less directly in a number of metabolic pathways. Some of the first direct evidence for its involvement in a specific enzyme came from a study of the effect of avidin in blocking incorporation of CO_2 into oxaloacetate (Wessman and Werkman, 1950). However, it was not until 1960 that the enzyme responsible, pyruvate carboxylase, was purified and shown to contain biotin (Utter and Keech, 1960). Meanwhile two other biotinyl enzymes which carboxylated acetyl-CoA and β-methylcrotonyl-CoA had been characterized (Wakil *et al.*, 1958; Lynen *et al.*, 1959) and had been shown to be inhibited by avidin.

1. *Kinetics of Inhibition*

Since avidin has four binding sites and most biotinyl enzymes possess at least four subunits, complex kinetics changing with molar ratio of the reactants are to be expected. Although the limited studies that have been made confirm this expectation, it has still been possible to draw some interesting conclusions from the results. When only a small excess of pure avidin was used an initial fast stage of inhibition followed by a slow second stage was observed for both pyruvate carboxylase (Scrutton and Mildvan, 1968) and transcarboxylase (N. M. Green, unpublished experiments). With a larger excess of avidin the inhibition of pyruvate carboxylase was first order in time but of order 1.5 with respect to avidin concentration (Scrutton and Utter, 1965). These effects probably reflect the steric barriers to reactions within and between polymeric complexes. Approximate values of rate constants (Table VI) show considerable differences between different enzymes and some interesting effects of ligands on the availability of biotin to avidin. When acetyl-CoA carboxylase is activated by isocitrate it forms polymers that do not react with avidin (Ryder *et al.*, 1967). Conversely, avidin prevents activation and polymer formation if added before isocitrate. Since it does not reverse the polymerization, it can be used to follow the

TABLE VI
Rate of Inhibition of Biotinyl Enzymes by Avidin

Enzyme	Ligand added	Approximate rate constant[a] ($M^{-1}sec^{-1}$)	Reference
Methylmalonyl-CoA pyruvate transcarboxylase	–	10^6	N. M. Green (unpublished)
Propionyl-CoA carboxylase	–	5×10^4 [b]	Kaziro *et al.* (1960)
Acetyl-CoA carboxylase	–	10^4 to 10^5	Moss and Lane (1972)
Pyruvate carboxylase	Citrate	0	
	–	500	Scrutton and Utter (1965)
	ATP	0	Scrutton and Utter (1965)
	Acetyl-CoA	2500	Scrutton and Utter (1967)

[a] Except for propionyl-CoA carboxylase the second-order rate constants were calculated from first-order rate constants measured in the presence of excess avidin.

[b] The rate of inhibition was measured in the presence of one equivalent of avidin. A previously published figure (Green, 1963a) calculated from the same results was overestimated by a factor of 60.

kinetics of polymerization and depolymerization (Moss and Lane, 1972).

The availability of the biotin of pyruvate carboxylase to avidin is less than that of biotin in other enzymes, but it is increased by the allosteric activator, acetyl-CoA. Other ligands (oxalate, pyruvate, oxaloacetate) also have small effects on the rate which have been used to determine their dissociation constants (Mildvan *et al.*, 1966).

The resistance of the activated form of acetyl-CoA carboxylase to inhibition by avidin provides the only example of such behavior of a biotinyl enzyme. It shows that resistance to avidin does not by itself eliminate the possibility that an enzyme reaction may involve biotin. There was one apparent example of the converse behavior. The synthesis of δ-aminolevulinate involved a decarboxylation step and the synthetase enzyme was inhibited by impure avidin. However, the inhibition was not prevented by biotin, and more highly purified avidin had no effect (Gibson *et al.*, 1962).

2. *Reversibility of Inhibition*

In all biotinyl enzymes the biotin is covalently linked to a protein lysine and in almost all it acts as a carrier of CO_2 between substrates on two different active sites each catalyzing half of the overall reaction. The biotin has to be free to move through appreciable distances (at least 10–20 Å), so it is not surprising that it is available to avidin and that the inhibition is usually irreversible. [If "irreversible" means less than 10% reactivation in 1 hour, this implies a dissociation rate constant of $< 3 \times 10^{-5}$ sec^{-1} and a dissociation constant of $< 3 \times 10^{-11}$ *M*. (cf. Section V,C).]

There are some reports of a partial reversal of avidin inhibition (Halenz *et al.*, 1962; Scrutton and Mildvan, 1968) and even of complete reversal of one of the half-reactions catalyzed by propionyl-CoA carboxylase (Halenz *et al.*, 1962; Friedman and Stern, 1961). Restoration of a half-reaction could be an effect of noncovalently bound biotin which might bind and be carboxylated in the absence of competition from the covalently bound biotin, now held by the avidin. Such a mechanism would be consistent with the high concentration of biotin (1 m*M*) required to restore activity. The partial reactivation of pyruvate carboxylase has been studied in some detail (Scrutton and Mildvan, 1968) and has a different origin. It was observed only when minimal concentrations of avidin were used to inhibit the enzyme. Under these conditions the reaction products contained a mixture of species of high molecular weight, in which the tetrameric enzyme was linked together by avidin molecules. The system could be depolymerized by an excess of avidin or by biotin. The latter restored up to 50% of the original activity while excess avidin rendered the inhibition irreversible. It appeared that when avidin was involved in cross-bridging of enzyme molecules it was unable to bind firmly, a situation reminiscent of the behavior of avidin polymers made with the shorter chain bisbiotinyl compounds (Section V,D). These results confirm that the biotin residues of pyruvate carboxylase are less accessible to avidin than they are in most other enzymes.

3. *Carboxylated Enzymes*

Avidin is able to combine with N-carboxylated derivatives of biotin (Table V) with a reduced affinity. It also inactivates the carboxylated form of propionyl-CoA carboxylase (Kaziro and Ochoa, 1961), but no extensive experiments with other enzymes have been reported. The inhibition may in part arise from the avidin increasing the rate of decarboxylation of the carboxybiotinyl enzyme. Edwards and Lane (quoted by Moss and Lane, 1971) have found that the rate of decarboxylation of acetyl-CoA carboxylase is increased several hundred times by avidin. It is likely that this enhanced rate is a consequence of a higher p*K* of the carboxyl group in the hydrophobic environment of the biotin-binding site. The protonated form of *N*-carboxyimidazolidone decarboxylates several thousand times faster than the anion (Caplow and Yager, 1967), so an increased p*K* would increase the rate of decarboxylation at a given pH.

4. *Isolation of Biotinyl Enzymes*

The use of avidin affinity columns to purify biotinyl enzymes remains an unfulfilled dream; no method for releasing the enzyme in a native state has been found. The combination of 6 *M* GuHCl and pH 1.5 has been used to release biotinyl peptides from avidin-Sepharose columns, but the recovery was only 24% (Bodanszky and Bodanszky, 1970). The biotin-containing subunit of acetyl-CoA carboxylase has also been obtained in unspecified yield by this method (Moss and Lane, 1971; Landman and Dakshinamurti, 1973).

Several attempts have been made in the author's laboratory to use either avidin subunits (Green and Toms, 1973) or periodate-oxidized avidin (N. M. Green, unpublished) coupled to Sepharose 4B to isolate biotinyl enzymes. Although the affinity of these derivatives for biotin was lower than that of avidin ($K = 10^{-7}$ to 10^{-9} M), it was still too high to allow significant release of enzyme, possibly because of multiple binding of several biotin residues on each molecule. There is still hope that further selective modification of the binding site may give derivatives with sufficiently low affinity to be useful for enzyme purification.

It is worth noting that avidin can be used as a specific adsorbent without covalent coupling since its high isoelectric point leads to strong adsorption on CM-cellulose (Green, 1963a).

The significant affinity of avidin for lipoate (Table V) suggests that avidin columns might be of use in the isolation of lipoyl enzymes or peptides.

C. *Biosynthesis of Avidin*

Certain aspects of avidin biosynthesis have attracted much attention recently. It had long been known that avidin synthesis was specifically stimulated by progesterone (Hertz *et al.*, 1943) and this led O'Malley and his colleagues to a detailed study summarized by O'Malley *et al.* (1969). The main conclusions were that avidin is synthesized in the goblet cells of the epithelium of the oviduct whereas the other major proteins of egg white are synthesized in the underlying tubular gland cells and that this accounts for the differential effects of estrogen and progesterone on the two classes of protein (Kohler *et al.*, 1968). Progesterone probably acts by control of transcription of the avidin mRNA and recent work has been concentrated on this aspect of the process. The mRNA has been partially purified (Rosenfeld *et al.*, 1972) and its synthesis in response to progesterone has been followed (Chan *et al.*, 1973). Introduction of the partially purified mRNA into the estrogen primed oviduct stimulated synthesis of avidin in the absence of progesterone (Tuohimaa *et al.*, 1972).

D. Streptavidin

No detailed characterization of this interesting relative of avidin has been published since the papers describing its discovery and isolation (Chaiet *et al.*, 1963; Chaiet and Wolf, 1964; Tausig and Wolf, 1964). In spite of considerable differences in composition (Table 1) it is in most other respects remarkably similar to avidin from eggs.

It has a high tryptophan content, which gives rise to positive bands in the CD spectrum at 280 and 230 nm (Green and Melamed, 1966) that are characteristic of avidin and hitherto unique to these proteins. The tryptophan residues are less susceptible to oxidation by *N*-bromosuccinimide than are those of avidin. Biotin produces the same characteristic different spectrum when it binds and it protects the tryptophan against oxidation by *N*-bromosuccinimide. One or two of the tryptophan residues are susceptible to oxidation by periodate and biotin binds to the product giving difference peaks characteristic of formylkynurenine (N. M. Green, unpublished experiments). The heat of binding of biotin (23 kcal/mole) is similar to that shown by avidin (Green, 1966).

The molecular weight of streptavidin is about 60,000 and 1 mole of biotin is bound per 14,000 g (17–18 μg/mg streptavidin, Chaiet and Wolf, 1964). It is even more resistant than avidin to dissociation into subunits by guanidinium chloride.

No detailed study has been made of its binding of biotin analogs, but Lichstein and Birnbaum (1965) used microbiological methods to show that substitution in the carboxyl group does not affect the binding, but that neither the diamine (compound 23) nor desthiobiotin compete significantly with biotin, so that both rings must be intact for strong binding. Like avidin, it gives a complex with four moles of the dye 2-hydroxyazobenzene 4′-carboxylate ($K = 10^{-4}$ *M*), which has the same extinction coefficient as the avidin complex (Green, 1970).

REFERENCES

Alderton, G., Lewis, J. C., and Fevold, H. L. (1945). *Science* **101**, 151.
Ananthanarayanan, V. S., and Bigelow, C. C. (1969). *Biochemistry* **8**, 3717.

Anderson, B. M., Reynolds, M. L., and Anderson, C. D. (1965). *Biochim. Biophys. Acta* **99**, 46.
Andrews, L. J., and Forster, L. S. (1972). *Biochemistry* **11**, 1875.
Baggaley, K. H., Blessington, B., Falshaw, C. P., Ollis, W. D., Chaiet, L., and Wolf, F. J. (1969). *Chem. Commun.* **3**, 101.
Boas, M. (1927). *Biochem. J.* **21**, 712.
Bodanszky, A., and Bodanszky, M. (1970). *Experientia* **26**, 327.
Canfield, R. (1963). *J. Biol. Chem.* **238**, 2698.
Caplow, M., and Yager, M. (1967). *J. Amer. Chem.* Soc. **89**, 4513.
Chaiet, L., and Wolf, F. J. (1964). *Arch. Biochem. Biophys.* **106**, 1.
Chaiet, L., Miller, R. W., Tausig, F., and Wolf, F. J. (1963). *Antimicrob. Ag. Chemother.* **3**, 28.
Chan, L., Means, A. R., and O'Malley, B. W. (1973). *Proc. Nat. Acad. Sci. U. S.* **70**, 1870.
Chen, Y. H., Yang, J. T., and Martinez, H. M. (1972). *Biochemistry* **11**, 4120.
Chou, P. Y., and Fasman, G. D. (1974). *Biochemistry* **13**, 222.
Crichton, R. R., and Bryce, C. F. H. (1973). *Biochem. J.* **133**, 289.
Cuatrecasas, P., and Wilchek, M. (1968). *Biochem. Biophys. Res. Commun.* **33**, 235.
Dayhoff, M. O. (1972). *In* "Atlas of Protein Sequence and Structure," Vol. 5, p. 109. Nat. Biomed. Res. Found., Washington, D. C.
Deal, W. C., Rutter, W. J., and Van Holde, K. E. (1963). *Biochemistry* **2**, 246.
De Lange, R. J. (1970). *J. Biol. Chem.* **245**, 907.
De Lange, R. J., and Huang, T.-S. (1971). *J. Biol. Chem.* **246**, 698.
Dhyse, F. G. (1954). *Proc. Soc. Exp. Biol. Med.* **85**, 515.
Dittmer, K., and du Vigneaud, V. (1944). *Science* **100**, 129.
Dittmer, K., du Vigneaud, V., György, P., and Rose, C. S. (1944). *Arch. Biochem.* **4**, 229.
Dolby, L. J., and Booth, D. C. (1966). *J. Amer. Chem. Soc.* **88**, 1049.
Donovan, J. W., and Ross, K. D. (1973). *Biochemistry* **12**, 512.
Eakin, R. E., Snell, E. E., and Williams, R. J. (1941). *J. Biol. Chem.* **140**, 535.
Everett, D. H. (1954). *Trans. Faraday Soc.* **50**, 1077.
Everett, D. H. (1955). *Trans. Faraday Soc.* **51**, 1551.
Everett, D. H., and Whitton, W. I. (1952). *Trans. Faraday Soc.* **48**, 749.
Fraenkel-Conrat, H., Snell, N. S., and Ducay, E. D. (1952a). *Arch. Biochem. Biophys.* **39**, 80.
Fraenkel-Conrat, H., Snell, N. S., and Ducay, E. D. (1952b). *Arch. Biochem. Biophys.* **39**, 97.
Friedman, D. L., and Stern, J. R. (1961). *Biochem. Biophys. Res. Commun.* **4**, 266.
Gerwin, B. I., Jacobson, B. E., and Wood, H. G. (1969). *Proc. Nat. Acad. Sci. U. S.* **64**, 1315.
Gibson, K. D., Neuberger, A., and Tait, G. H. (1962). *Biochem. J.* **83**, 539.
Gould, H. J., Gill, T. J., and Doty, P. (1964). *J. Biol. Chem.* **239**, 2842.
Green, N. M. (1962). *Biochim. Biophys. Acta* **59**, 244.
Green, N. M. (1963a). *Biochem. J.* **89**, 585.
Green, N. M. (1963b). *Biochem. J.* **89**, 599.
Green, N. M. (1963c). *Biochem. J.* **89**, 609.
Green, N. M. (1964a). *Biochem. J.* **90**, 564.
Green, N. M. (1964b). *Biochem. J.* **92**, 16C.
Green, N. M. (1965). *Biochem. J.* **94**, 23C.
Green, N. M. (1966). *Biochem. J.* **101**, 774.
Green, N. M. (1968). *Nature (London)* **217**, 254.

Green, N. M. (1970). *In* "Methods in Enzymology" (D. B. McCormick and L. D. Wright, eds.), Vol. 18A, p. 418. Academic Press, New York.
Green, N. M. (1972). *In* "Mosbach Colloquium" (R. Jaenicke and E. Helmreich, eds.), No. 23, p. 183. Springer-Verlag, Berlin and New York.
Green, N. M. (1975). In preparation.
Green, N. M., and Joynson, M. A. (1970). *Biochem. J.* **118**, 71.
Green, N. M., and Melamed, M. D. (1966). *Biochem. J.* **100**, 614.
Green, N. M., and Ross, M. E. (1968). *Biochem. J.* **110**, 59.
Green, N. M., and Toms, E. J. (1970). *Biochem. J.* **118**, 67.
Green, N. M., and Toms, E. J. (1972). *Biochem. J.* **130**, 707.
Green, N. M., and Toms, E. J. (1973). *Biochem. J.* **133**, 687.
Green, N. M., and Toms, E. J. (1975). In preparation.
Green, N. M., Mose, W. P., and Scopes, P. M. (1970). *J. Chem. Soc., C* pp. 1330–1333.
Green, N. M., Konieczny, L., Toms, E. J., and Valentine, R. C. (1971). *Biochem. J.* **125**, 781.
Green, N. M., Valentine, R. C., Wrigley, N. G., Ahmad, F., Jacobson, B. E., and Wood, H. G. (1972). *J. Biol. Chem.* **247**, 6284.
György, P. (1954). *In* "The Vitamins" (W. H. Sebrell, Jr. and R. S. Harris, eds.), Vol. 1, p. 527. Academic Press, New York.
György, P., and Rose, C. S. (1941). *Science* **94**, 261.
György, P., Rose, C. S., Hofmann, K., Melville, D. B., and du Vigneaud, V. (1940). *Science* **92**, 609.
György, P., Rose, C. S., and Tomarelli, R. (1942). *J. Biol. Chem.* **144**, 169.
Halenz, D. R., Feng, J., Hegre, C. S., and Lane, M. D. (1962). *J. Biol. Chem.* **237**, 2140.
Hammes, G. G. (1969). *Advan. Protein Chem.* **23**, 1.
Hertz, R. (1943). *Proc. Soc. Exp. Biol. Med.* **52**, 15.
Hertz, R. (1946). *Physiol. Rev.* **26**, 479.
Hertz, R., Fraps, R. M., and Sebrell, W. H. (1943). *Proc. Soc. Exp. Biol. Med.* **52**, 142.
Holbrook, J. J. (1972). *Biochem. J.* **128**, 921.
Huang, T.-S. (1971). Ph. D. Thesis, University of California, Los Angeles.
Huang, T.-S., and De Lange, R. J. (1971). *J. Biol. Chem.* **246**, 686.
Huston, R. B., and Cohen, P. P. (1969). *Biochemistry* **8**, 2658.
Ikai, A., Fish, W. W., and Tanford, C. (1973). *J. Mol. Biol.* **73**, 165.
Jennings, B. R., Spach, G., and Schuster, T. M. (1968). *Biopolymers* **6**, 635.
Kannangara, C. G., and Stumpf, P, K. (1973). *Arch. Biochem. Biophys.* **155**, 391.
Kaziro, Y., and Ochoa, S. (1961). *J. Biol. Chem.* **236**, 3131.
Kaziro, Y., Leone, E., and Ochoa, S. (1960). *Proc. Nat. Acad. Sci. U. S.* **46**, 1319.
Knappe, J. (1970). *Annu. Rev. Biochem.* **39**, 757.
Kohler, P. O., Grimley, P. M., and O'Malley, B. W. (1968). *Science* **160**, 86.
Korenman, S. G., and O'Malley, B. W. (1967). *Biochim. Biophys. Acta* **140**, 174.
Landman, A. D., and Dakshinamurti, K. (1973). *Anal. Biochem.* **56**, 191.
Launer, H. F., and Fraenkel-Conrat, H. (1951). *J. Biol. Chem.* **193**, 125.
Lee, H. M., Wright, L. D., and McCormick, D. B. (1973). *Proc. Soc. Exp. Biol. Med.* **142**, 439.
Lichstein, H. C., and Birnbaum, J. (1965). *Biochem. Biophys. Res. Commun.* **20**, 41.
Lynen, F., Knappe, J., Lorch, E., Jutting, G., and Ringelmann, E. (1959). *Angew. Chem.* **71**, 481.
McCormick, D. B. (1965). *Anal. Biochem.* **13**, 194.
McCormick, D. B., and Wright, L. D. (1970). *In* "Methods in Enzymology" (D. B. McCormick and L. D. Wright, eds.), Vol. 18A, p. 379. Academic Press, New York.

McPherson, A. (1970). *J. Mol. Biol.* **51**, 39.
Melamed, M. D., and Green, N. M. (1963). *Biochem. J.* **89**, 591.
Mildvan, A. S., Scrutton, M. C., and Utter, M. F. (1966). *J. Biol. Chem.* **241**, 3488.
Miller, A. K., and Tausig, F. (1964). *Biochem. Biophys. Res. Commun.* **14**, 210.
Montgomery, R. (1972). *In* "Glycoproteins" (A. Gottschalk, ed.), BBA Library, Vol. 5A, p. 519. Elsevier, Amsterdam.
Moss, J., and Lane, M. D. (1971). *Advan. Enzymol.* **35**, 321.
Moss, J., and Lane, M. D. (1972). *J. Biol. Chem.* **247**, 4944.
Mudd, J. B., Leavitt, R., Ongun, A., and McManus, T. T. (1969). *Atmos. Environ.* **3**, 669.
O'Malley, B. W., McGuire, W. L., Kohler, P. O., and Korenman, S. G. (1969). *Recent Prog. Horm. Res.* **25**, 105.
Pai, C. H., and Lichstein, H. C. (1964). *Proc. Soc. Exp. Biol. Med.* **116**, 197.
Pennington, D., Snell, E. E., and Eakin, R. E. (1942). *J. Amer. Chem. Soc.* **64**, 469.
Pritchard, A. B., McCormick, D. B., and Wright, L. D. (1966). *Biochem. Biophys. Res. Commun.* **25**, 524.
Rees, D. A. (1969). *Advan. Carbohyd. Chem. Biochem.* **24**, 267.
Rosenfeld, G. C., Comstock, J. P., Means, A. R., and O'Malley, B. W. (1972). *Biochem. Biophys. Res. Commun.* **47**, 387.
Rosenheck, K., and Doty, P. (1961). *Proc. Nat. Acad. Sci. U. S.* **47**, 1775.
Ryder, E., Gregolin, C., Chang, H., and Lane, M. D. (1967). *Proc. Nat. Acad. Sci. U. S.* **57**, 1455.
Scrutton, M. C., and Mildvan, A. S. (1968). *Biochemistry* **7**, 1490.
Scrutton, M. C., and Utter, M. F. (1965). *J. Biol. Chem.* **240**, 3714.
Scrutton, M. C., and Utter, M. F. (1967). *J. Biol. Chem.* **242**, 1723.
Smith, R., and Tanford, C. (1973). *Proc. Nat. Acad. Sci. U. S.* **70**, 289.
Snell, E. E. (1950). *In* "Vitamin Methods" (P. György, ed.), Vol. 1, p. 327. Academic Press, New York.
Steinhardt, J., and Zaiser, E. M. (1953). *J. Amer. Chem. Soc.* **75**, 1599.
Strickland, E. H., Horwitz, J., and Billups, C. (1969). *Biochemistry* **8**, 3205.
Suurkuusk, J., and Wadsö, I. (1972). *Eur. J. Biochem.* **28**, 438.
Tanford, C. (1968). *Advan. Protein Chem.* **23**, 218.
Tanford, C. (1970). *Advan. Protein Chem.* **24**, 1.
Tausig, F., and Wolf, F. J. (1964). *Biochem. Biophys. Res. Commun.* **14**, 205.
Teale, F. W. J. (1960). *Biochem. J.* **76**, 381.
Teipel, J. W. (1972). *Biochemistry* **11**, 4100.
Trotter, J., and Hamilton, J. A. (1966). *Biochemistry* **5**, 713.
Tuohimaa, P., Segal, S. J., and Koide, S. (1972). *Proc. Nat. Acad. Sci. U. S.* **69**, 2814.
Utter, M. F., and Keech, D. B. (1960). *J. Biol. Chem.* **235**, PC17.
Wahl, P., and Weber, G. (1967). *J. Mol. Biol.* **30**, 371.
Wakil, S. J., Titchener, E. B., and Gibson, D. M. (1958). *Biochim. Biophys. Acta* **29**, 225.
Wei, R. D., and Wright, L. D. (1964). *Proc. Soc. Exp. Biol. Med.* **117**, 341.
Wei, R. D., Kou, D. H., and Hoo, S. L. (1971). *Experientia* **27**, 366.
Wellner, V. P., Santos, J. I., and Meister, A. (1968). *Biochemistry* **7**, 2848.
Wessman, G. E., and Werkman, C. H. (1950). *Arch. Biochem.* **26**, 214.
Woolley, D. W., and Longsworth, L. G. (1942). *J. Biol. Chem.* **142**, 285.
Wright, L. D., and Skeggs, H. R. (1944). *Proc. Soc. Exp. Biol. Med.* **56**, 95.
Wright, L. D., and Skeggs, H. R. (1947). *Arch. Biochem.* **12**, 27.
Wright, L. D., Skeggs, H. R., and Cresson, E. L. (1951). *J. Amer. Chem. Soc.* **73**, 4144.

CARBONYL–AMINE REACTIONS IN PROTEIN CHEMISTRY

By ROBERT E. FEENEY,[1] GUNTER BLANKENHORN,[2]
and HENRY B. F. DIXON[3]

[1] Department of Food Science and Technology, University of California, Davis, California.

[2] During the time that this review was written, on leave at the University of California, Davis, from Fachbereich Biologie, Universität Konstanz, Konstanz, West Germany, on a grant from the Deutsche Forschungs Gemeinschaft BL 12912.

[3] Department of Biochemistry, University of Cambridge, Cambridge, England.

I. Introduction

Reaction of a carbonyl group with an amino group is one of the commonest natural reactions of proteins. It is the key reaction in many enzymic and other biological processes, such as vision. In addition, it occurs in natural deteriorative processes of tissues, fluids, and preparations of biological materials, particularly of different types of concentrated food products or those of low moisture content. The carbonyl–amine reaction is also a valuable tool for determining or modifying amino groups of proteins. For example, for nearly a half century formaldehyde has been used in Sörensen's formol titration of amino groups and for the chemical modification of microbial toxins to give toxoids. The widespread participation of the carbonyl–amine reaction in protein chemistry, however, has been recognized only during the last decade.

In this review, we have attempted to organize and correlate the different types of carbonyl–amine reactions found in proteins with their organic chemistry. One of the objectives has also been to show how one type of chemical reaction participates in so many dissimilar biological processes. Although the article is prepared primarily for biochemists, we have hoped that it will also be of value to organic chemists in providing an insight into the reactions that are unique or novel because they are directed to serve a specific biological function.

The breadth of the subject has necessitated selection of particular examples from extensive materials. Omissions are the fault of the authors.

II. The Chemistry of Carbonyl Groups, of Amino Groups, and of the Carbonyl–Amine Reaction

The chemistry of the carbonyl group and the amino group has been reviewed in great detail (Patai, 1966, 1968, 1970). We have therefore selected a few examples illustrating the important principles as far as they seemed to be relevant for this review.

A. *Amino Groups*

Perhaps the most characteristic chemical property of amines is their ability to act as nucleophiles because they possess a lone pair of electrons on the nitrogen atom. They also can act as bases by accepting protons from a variety of acids, such as water in aqueous solution. The base strengths of saturated aliphatic amines vary only within one order of magnitude (Table I). Their pK_a values in

TABLE I
Typical Amines and Their Properties[a]

Amine	Name	B.p. (°C)	M.p. (°C)	Water solubility (g/100 ml)	K_B in water[b]
NH_3	Ammonia	−33	−77.7	90	1.8×10^{-5}
CH_3NH_2	Methylamine	−6.5	−92.5	1156	4.4×10^{-4}
$CH_3CH_2NH_2$	Ethylamine	16.6	−80.6	∞	5.6×10^{-4}
$(CH_3)_3CNH_2$	*t*-Butylamine	46	−67.5	∞	2.8×10^{-4}
$(CH_3CH_2)_2NH$	Diethylamine	55.5	−50	v. sol.[c]	9.6×10^{-4}
$(CH_3CH_2)_3N$	Triethylamine	89.5	−115	1.5	4.4×10^{-4}
$(CH_3CH_2CH_2CH_2)_3N$	Tri-*n*-butylamine	214		sl. sol.	
(cyclic structure) NH	Piperidine	106	−9	∞	1.6×10^{-3}
(cyclic structure) N	Pyridine	115	−42	∞	1.7×10^{-9}
(cyclohexyl)—NH_2	Cyclohexylamine	134		sl. sol.	4.4×10^{-4}
(phenyl)—NH_2	Aniline	184.4	−6.2	3.4	3.8×10^{-10}
$H_2NCH_2CH_2NH_2$	Ethylenediamine	116	8.5	sol.	8.5×10^{-5}

[a] From Roberts and Caserio "Basic Principles of Organic Chemistry," copyright 1964 by W. A. Benjamin, Inc., Menlo Park, California.
[b] Usually at 20–25°C.
[c] Abbreviations: sol., soluble; v., very; sl., slightly.

aqueous solution are usually between 9 and 10, whereas aromatic amines are much weaker bases with pK_a values around 5.

The properties of amines depend very much on the degree of substitution on the nitrogen. Primary and secondary amines (compounds I and II), for example, form hydrogen bonds of the N—H---N type in contrast to tertiary amines (III), which can form this type of hydrogen bond only if protonated by external acids.

$R—NH_2$	R_2NH	R_3N	$R_4N^+X^-$
Primary amine	Secondary amine	Tertiary amine	Quarternary ammonium salt
(I)	(II)	(III)	(IV)

The reactivity of secondary and tertiary amines as compared to primary amines depends on steric and electronic effects of the amine substituents. Substituents that increase electron density at nitrogen will increase the amine's nucleophilicity whereas substituents that decrease electron density at nitrogen will decrease its nucleophilicity. Bulky substituents at the amine will reduce its nucleophilic power because attack at the electrophile is hindered. After reaction with an electrophilic center, primary and secondary amines can lose a proton and hence become a poor leaving group. This makes the reaction of R-CO-Cl with a primary amine virtually irreversible, whereas its reaction with a tertiary amine is not.

As a result of hydrogen bonding between amine and water of the type —N····H—O—H, low-molecular-weight amines are very water soluble. A characteristic property of the infrared spectra of primary and secondary amines is an absorption at 3500 to 3300 cm^{-1} which appears as a doublet for primary amines.

B. Carbonyl Groups

1. Physicochemical Properties

The carbon–oxygen double bond is both strong and reactive. Its bond energy (179 kcal) is more than that of two carbon–oxygen single bonds (2 × 85.5 kcal) in contrast to the carbon–carbon double bond (145.8 kcal), which is weaker than two carbon–carbon single bonds (2 × 82.6 kcal). The reactivity of the carbonyl bond is primarily due to the difference in electronegativity between carbon and oxygen, which leads to a significant contribution of the dipolar resonance form, oxygen being negative and carbon being positive, [Eq. (1)]. On approach of a nucleophile to the carbon atom the polarity of the carbonyl bond will further increase since the electrons of the π-bond interacting with the nucleophile will be shifted further toward oxygen.

$$>C{=}O \longleftrightarrow >C^{+}{-}O^{-} \sim >C^{\delta+}{-}O^{\delta-} \quad (1)$$

The polarity of the carbonyl group is documented by many of its properties in aldehydes and ketones. Boiling points for the lower members of a homologous series are 50–80°C higher than those of the corresponding hydrocarbons. Most aldehydes and ketones have dipole moments of approximately 2.7 D, which correspond to 40–50% ionic character for the carbonyl bond. The characteristic infrared stretching frequencies for the carbonyl groups of aldehydes and ketones generally are observed in the range from 1705 to 1740 cm^{-1}. In accord with the polarity of the carbon–oxygen double bond, the absorption intensities are much greater than for carbon–carbon double bonds.

2. *Chemical Properties and Reactivity*

The important reactions of carbonyl groups usually involve additions to the carbon–oxygen double bond. In some cases the reaction stops as soon as the two groups have been added, but in many others subsequent reactions take place. Most of them are of two types:

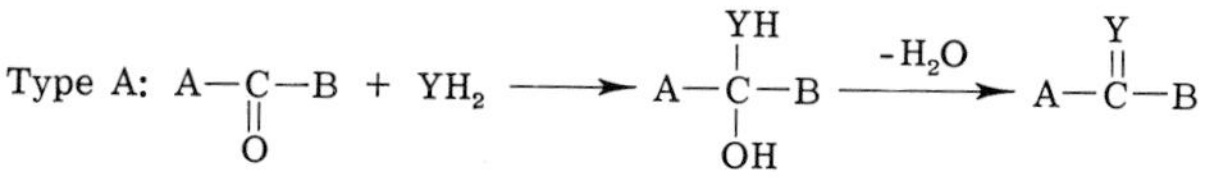

$$\text{Type A: } A{-}\underset{\underset{O}{\|}}{C}{-}B + YH_2 \longrightarrow A{-}\underset{\underset{OH}{|}}{\overset{\overset{YH}{|}}{C}}{-}B \xrightarrow{-H_2O} A{-}\overset{\overset{Y}{\|}}{C}{-}B \tag{2}$$

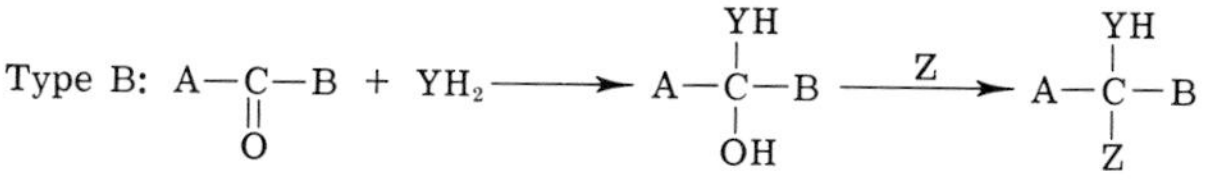

$$\text{Type B: } A{-}\underset{\underset{O}{\|}}{C}{-}B + YH_2 \longrightarrow A{-}\underset{\underset{OH}{|}}{\overset{\overset{YH}{|}}{C}}{-}B \xrightarrow{Z} A{-}\underset{\underset{Z}{|}}{\overset{\overset{YH}{|}}{C}}{-}B \tag{3}$$

In type A, the adduct loses water and the net result of the reaction is the substitution of C=Y for C=O [Eq. (2)]. In type B, there is rapid substitution, the hydroxy group of the tetrahedral intermediate being replaced by another group Z, the Z very often being another YH moiety [Eq. (3)].

The orientation of asymmetrical additions to the carbon–oxygen double bond is determined by its strongly polar character. Negative attacking species always add at the carbon, and positive ones add at the oxygen.

A nucleophile will approach the carbonyl bond perpendicular to the plane of the bond, so that it starts to conjugate with the pi orbital (V). Therefore, depending on the side of approach, two stereoisomeric products are possible. An enzyme attack of the nucleophile will often be limited to only one side.

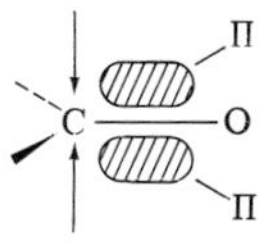

C —— O, Π, Π

(V)

Many of the addition reactions to the carbon–oxygen double bond are subject to both acid and base catalysis. Base can catalyze the reaction by converting a reagent of the form YH_2 to the more powerful nucleophile YH^-. The equilibrium concentration of YH^- as a fraction of the total reagent concentration depends only on the pH of the solution, so that bases increase this by increasing the concentration of OH^- ions, the "specific base" in water, and such catalysis is called "specific base catalysis." Even at a fixed pH, however, an increase in the concentration of other bases may catalyze the reaction by removing a proton from YH as it reacts, even though they are too weak to remove it in a prior equilibrium. Such a reaction is said to be "general-base catalyzed." Acids may catalyze the reaction by protonating the oxygen atom of the carbonyl compound, making its carbon atom a much stronger electrophile. Acid catalysis may be "specific acid catalysis," depending only on the pH of the solution, because the concentration of any reactive species depends on the concentration of the specific acid H^+, or it may be "general acid catalysis," general in that other acid species can protonate the carbonyl group as it reacts. Indeed an acid group may form a hydrogen bond with the carbonyl group in a prior equilibrium, even though it is too weak to protonate it until reaction (VI):

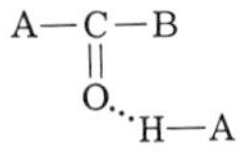

$$\begin{array}{l} \mathrm{A{-}\underset{\overset{\|}{O}}{C}{-}B} \\ \qquad \cdots \mathrm{H{-}A} \end{array}$$

(VI)

This also results in a decreased electron density at the carbonyl carbon. Similar catalysis may be found with metal ions, such as Cu^{2+}, which may act as Lewis acids.

Substituent effects on the reactivity of the addition reaction may be summarized as follows: If A and/or B are electron-donating groups, rates are decreased. Electron-withdrawing substituents increase rates. Steric factors are also quite important: decreasing reactivity can be expected with increasing bulkiness of substituents.

3. *Hydrations*

The reaction of water is a typical nucleophilic addition. Thus, aldehydes and ketones form hydrates in aqueous solution [Eq. (4)]. The position of the equilibrium depends very much

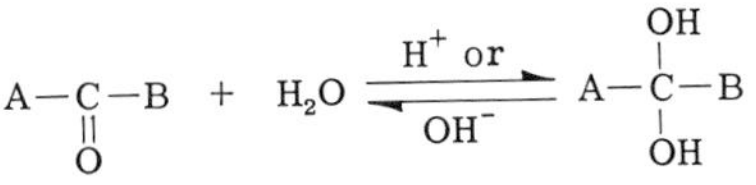

$$\mathrm{A{-}\underset{\overset{\|}{O}}{C}{-}B} + \mathrm{H_2O} \underset{OH^-}{\overset{H^+ \text{ or}}{\rightleftharpoons}} \mathrm{A{-}\underset{OH}{\overset{OH}{C}}{-}B} \tag{4}$$

on the structure of the hydrate. Thus, formaldehyde in water at 20°C exists 99.99% in the hydrated form, while acetaldehyde is hydrated

58%; for acetone the hydrate concentration is negligible (Bell and Clunie, 1952).

Hydration of carbonyl groups is subject to both general and specific acid-base catalysis.

C. Reactions Involving Carbonyl and Amino Groups

1. The Carbonyl–Amine Reaction

A wide variety of substances with $—NH_2$ groups condense with carbonyl compounds. Excellent reviews on the chemistry of this reaction are available (Patai, 1968; Jencks, 1969). Both types of addition reactions, described in Section II,B,2 occur. Strongly basic amines like aliphatic amines or hydroxylamine react according to type A [Eq. (5)]:

$$\mathrm{A{-}\underset{\underset{O}{\|}}{C}{-}B} + \mathrm{R{-}NH_2} \rightleftharpoons \mathrm{A{-}\underset{\underset{OH}{|}}{\overset{\overset{NHR}{|}}{C}}{-}B} \underset{}{\overset{-H_2O}{\rightleftharpoons}} \mathrm{A{-}\overset{\overset{N-R}{\|}}{C}{-}B} \qquad (5)$$

This condensation of primary amines with aldehydes and ketones to give imines was first discovered by Schiff (1900). The overall equilibrium greatly favors hydrolysis in aqueous solution for aliphatic aldehydes. With aromatic aldehydes, the equilibrium is shifted in favor of Schiff base formation. It is important to note that increasing the nucleophilic strength of the amine will increase the rate of the carbonyl–amine reaction but will have almost no effect on the position of the equilibrium.

Aliphatic aldehydes that possess a hydrogen atom on the carbon adjacent to the carbonyl group do not generally yield Schiff bases, because the imines formed initially undergo condensation to dimeric or polymeric materials (Section II,C,3). The properties of the $>C{=}N{-}R$ bond and the $>C{=}O$ bond are similar, so it is not surprising that certain imines show chemical behavior similar to that of carbonyl compounds. Weak bases, such as amides, urea or melamine, or secondary amines, can undergo reaction according to type B [Eq. (6)].

$$\mathrm{A{-}\underset{\underset{O}{\|}}{C}{-}B} + \mathrm{R{-}NH_2} \rightleftharpoons \mathrm{A{-}\underset{\underset{OH}{|}}{\overset{\overset{NHR}{|}}{C}}{-}B} \overset{RNH_2}{\rightleftharpoons} \mathrm{A{-}\underset{\underset{NHR}{|}}{\overset{\overset{NHR}{|}}{C}}{-}B} + \mathrm{H_2O} \qquad (6)$$

The tetrahedral intermediate formed in the initial addition reaction does not undergo intramolecular dehydration, but it reacts with an-

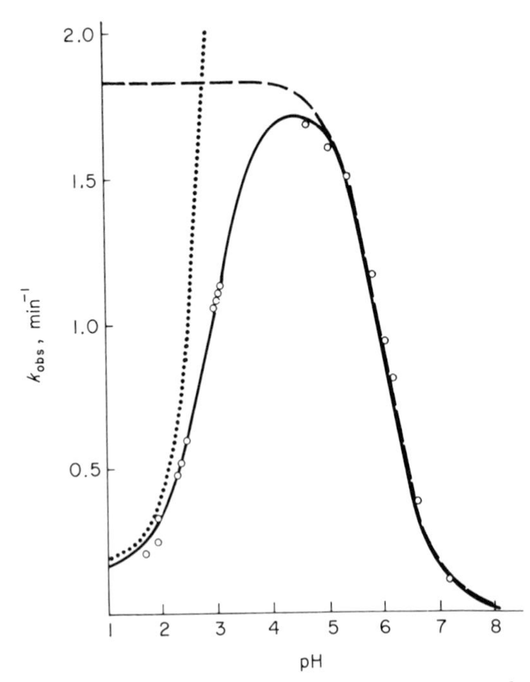

FIG. 1. Effect of pH on the pseudo first-order rate constant for the reaction of 5×10^{-4} M acetone with 0.0167 M total hydroxylamine, showing the change in rate-determining step with changing pH. Dotted line: rate of attack of free hydroxylamine on acetone. Dashed line: rate under conditions in which acid-catalyzed dehydration is rate determining. Solid line: calculated, from the steady-state. From Jencks (1969).

other molecule of amine to form alkylidenediamines or *gem*-diamines.

The rate of the carbonyl–amine reaction usually shows a characteristic pH dependence that results in a bell-shaped curve (Fig. 1). At neutral pH, amine addition to the carbonyl compound is fast and the loss of water from the tetrahedral intermediate is rate determining. With decreasing pH, the rate of acid-catalyzed dehydration continues to increase, but, as the free amine becomes protonated, the equilibrium concentration of the addition compound decreases. At pH values well below the pK of the amine, these two effects offset each other and the calculated rate becomes independent of pH. Since the observed rate does not reach a plateau with decreasing pH but decreases after going through a maximum, another step in the reaction sequence must have become rate determining: the rate of acid-catalyzed dehydration becomes so fast that the rate of formation of the addition intermediate from carbonyl compound and free amine can no longer keep up with the rate of its dehydration. The attack of free amine on the carbonyl group becomes rate determining. The rate of this reaction is proportional to the concentration of amine present as the base which, at low pH values, is directly proportional to the hydroxide ion activity. The rate of amine attack at the carbonyl group therefore increases with increasing pH as shown in Fig. 1. For strongly basic amines, such as aliphatic amines and hydroxylamine, the change in the rate-determining step generally occurs between pH 2 and pH 5. Below this pH the attack and loss of water is fast and the attack and loss of free amine is rate determining [Eq. (7)].

$$R{-}NH_2 \quad {>}C{=}O \underset{\text{in acid}}{\overset{\text{slow}}{\rightleftharpoons}} R{-}\overset{+}{N}H_2{-}\overset{|}{\underset{|}{C}}{-}O^- \tag{7}$$

Above this pH, the attack and loss of free amine is fast and the attack and loss of water or hydroxide ion is rate determining [Eq. (8)].

$$B \quad H{-}\underset{H}{O} \quad {>}C{=}\overset{+}{N}{<}^{H}_{R} \underset{\text{in base}}{\overset{\text{slow}}{\rightleftharpoons}} BH + \underset{H}{O}{-}\overset{|}{\underset{|}{C}}{-}N{<}^{H}_{R} \overset{\text{fast}}{\rightleftharpoons} {}^-O{-}\overset{|}{\underset{|}{C}}{-}\overset{+}{N}H_2{-}R \tag{8}$$

The protonated imine is the reactive species. Water attack on the protonated imine is subject to general base catalysis, the expelling of hydroxide ion from the intermediate is generally acid catalyzed. The rate constant for the attack of hydroxide ion on the protonated amine exceeds the value on the Brönsted plot for general base catalysis (Koehler *et al.*, 1964). This indicates that the reaction is a direct nucleophilic attack of hydroxide ion rather than an attack of water which is catalyzed by hydroxide ion [Eq. (9)].

$$HO^{\ominus} \quad {>}C{=}N^+ \rightleftharpoons HO{-}\overset{|}{\underset{|}{C}}{-}\ddot{N}{<} \tag{9}$$

Reaction of weakly basic amines, such as semicarbazide and anilines with the carbonyl group, proceeds by additional pathways, caused by the electron withdrawing effects of substituents on the amine. The change in the rate-determining step occurs at the same or slightly higher pH as in the reactions of the more basic amines. The addition and loss of water follows the same mechanism as with more basic amines. The smaller driving force of the less basic nitrogen atom results in a higher degree of acid catalysis for the dehydration step. As the amine becomes more acidic, base catalysis for water attack and expulsion becomes significant, as in reactions of hydroxylamine, semicarbazide, and anilines with carbonyl derivatives.

Amines with electron-withdrawing substituents, such as amides, urea, and thiourea, attack carbonyl compounds through a specific base-catalyzed mechanism, the anion R—NH$^-$ becoming the active species.

A variety of other compounds, such as alcohols, mercaptans, and nitriles, add in a similar manner to carbonyl compounds as water or amines. Detailed discussions of these reactions may be found elsewhere (Patai, 1966, 1968, 1970).

2. *Chemical Properties of Schiff Bases*

In this chapter, the chemical properties of Schiff bases, mainly from pyridoxal and related compounds with amino acids, will be discussed because they are biologically by far the most important ones. They have been reviewed recently (Bruice and Benkovic, 1966; Jencks, 1969; Snell and Di Mari, 1970). The chemical properties of Schiff bases are similar to those of the carbonyl group. Since the imino group with a p*K* around 7 is more basic than the carbonyl group, it is protonated to a much higher extent at neutral pH. For this reason, the carbon atom in the protonated imine group is more electrophilic than in the unprotonated carbonyl group.

a. Transimination. Schiff bases react with amines predominantly by attack on the protonated imine. This conversion of one imine into another has been termed transimination [Eq. (10)]. It is an

$$\mathrm{R{-}NH_2} \quad {>}\mathrm{C{=}\overset{+}{N}}{<} \rightleftharpoons \mathrm{R{-}\overset{H}{\underset{H}{N^{+}}}{-}\overset{|}{\underset{|}{C}}{-}N}{<} \rightleftharpoons \mathrm{R{-}\underset{H}{N}{-}\overset{|}{\underset{|}{C}}{-}\overset{H}{\underset{|}{N^{+}}}{-}} \rightleftharpoons \mathrm{\overset{R}{\underset{H}{>}}\overset{+}{N}{=}C}{<} \quad \mathrm{\overset{H}{\underset{|}{N}}{-}} \tag{10}$$

example of the susceptibility of the protonated imine to undergo nucleophilic addition reactions.

b. Reactions Involving Schiff Bases of Pyridoxal. In Schiff bases of pyridoxal and its derivatives with amino acids, the considerable displacement of electrons at the α-carbon of the amino acid residue toward the nitrogen of the imine is achieved by the conjugation of the imine function with an efficient electron sink, the pyridine ring. In addition, the pyridine nitrogen under most reaction conditions is partly protonated, thereby withdrawing electrons from the imine bond. Additional activation is achieved by the presence of chelating metal ions, such as Cu^{2+}, which increase the polarization of the imine bond, thus weakening the three bonds at the α-carbon atom of the amino acid residue. Therefore three types of reactions can result depending on which of the three bonds at the α-carbon atom will be broken. Abstraction of a proton from the α-carbon leads to a resonance-stabilized ion that can be looked at either as a carbanion of the amino acid or a pyridoxamine Schiff base [Eq. (11)] (Jencks, 1969). Readdition of a proton to the amino acid residue leads to racemization. If the proton is added to the carbonyl carbon of pyridoxal, the Schiff base of pyridoxamine and an α-keto acid is formed that will hydrolyze subsequently [Eq. (11)]. Reversion of this reaction sequence with a different α-keto acid results in regeneration of pyridoxal and formation of a new amino acid. The net result of this reaction sequence in two directions is the transamination of α-amino and α-keto acid [Eq. (12)] (Jencks, 1969).

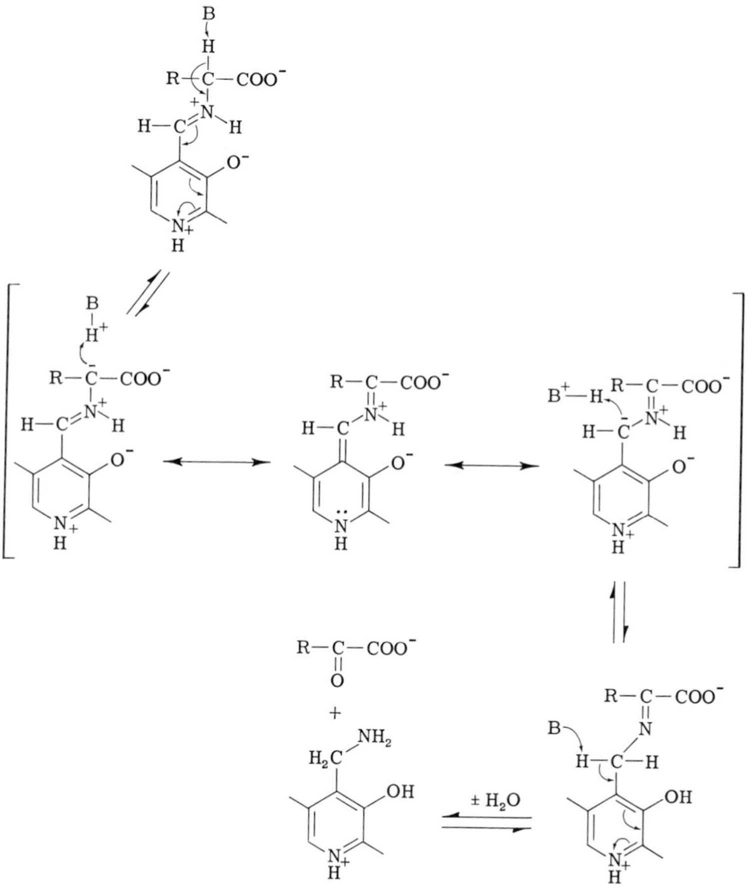

(11)

$$R_1-\underset{^+NH_3}{\overset{H}{C}}-COO^- + R_2-\underset{O}{\overset{\|}{C}}-COO^- \rightleftharpoons R_1-\underset{O}{\overset{\|}{C}}-COO^- + R_2-\underset{^+NH_3}{\overset{H}{C}}-COO^- \quad (12)$$

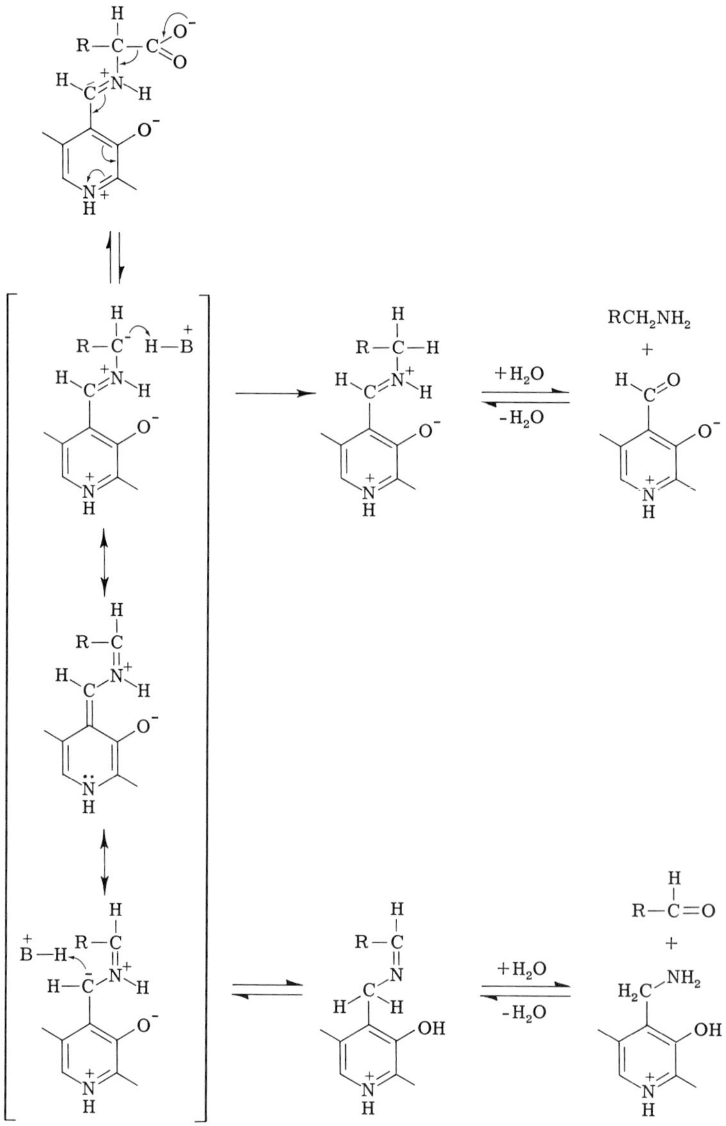

(13)

Electron withdrawal from the carboxylate group of the Schiff base between pyridoxal and amino acid results in decarboxylation. If the anion that is left behind adds a proton to the resulting amine carbon of the Schiff base, a simple decarboxylation product is obtained. If the anion adds a proton to the pyridoxal carbon, pyridoxamine and an aldehyde will be formed after hydrolysis [Eq. (13)] (Jencks, 1969).

Schiff bases of pyridoxal and appropriate amino acids, such as serine, may undergo electron removal from the side chain. Initially, formaldehyde is eliminated, leaving behind an anionic intermediate. Addition of a proton to this intermediate at the α-carbon leads to the formation of glycine and pyridoxal as the end products [Eq. (14)] (Jencks, 1969).

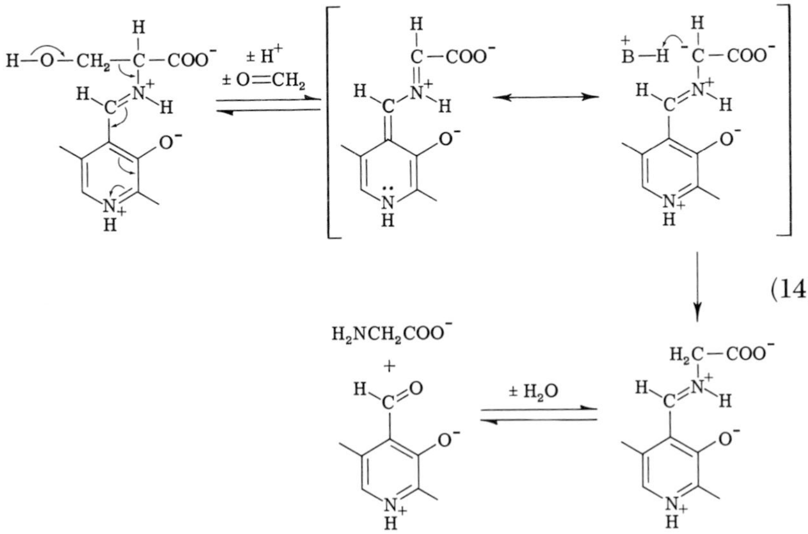

H–O–CH_2–C(H)–COO^- ; H–C=N^+–H ; O$^-$; N^+–H

$\pm H^+$

$\pm O{=}CH_2$

[H–C(–COO^-)=N^+–H ; H–C ; O$^-$; N: H ⟷ B^+–H $^-$C(H)–COO^- ; H–C=N^+–H ; O$^-$; N^+ H]

(14)

$H_2NCH_2COO^-$ + H–C=O ; O$^-$; N^+ H

$\pm H_2O$

H_2C–COO^- ; H–C=N^+–H ; O$^-$; N^+ H

3. *Aldol Condensation*

An important property of carbonyl compounds is that the loss of a proton from an adjacent carbon atom to form an enol or enolate is promoted by carbonyl groups [Eq. (15)].

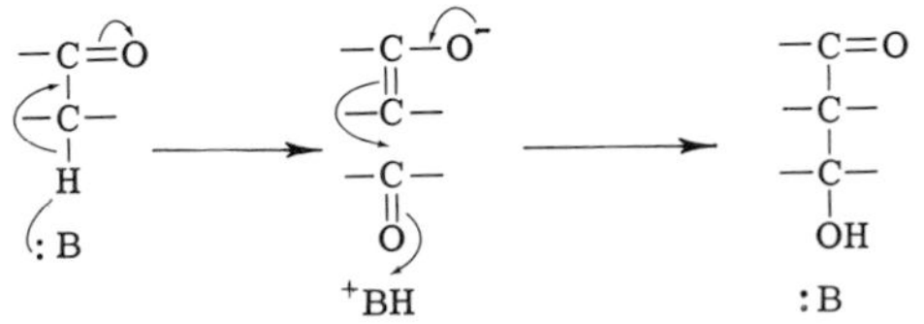

$$
\begin{array}{ccccc}
-C{=}O & & -C{-}O^- & & -C{=}O \\
-C- & \longrightarrow & \| & \longrightarrow & -C- \\
H & & -C- & & -C- \\
:B & & -C- & & OH \\
 & & O & & :B \\
 & & {}^+BH & &
\end{array}
\qquad (15)
$$

In the second step a new carbon–carbon bond with the acceptor carbonyl group is formed to yield an aldol adduct [Eq. (15)]. Amines can catalyze aldol condensations [Eq. (16)]. The amine combines with the carbonyl compound to form a protonated imine, whose positively charged nitrogen atom attracts electrons and thus facilitates loss of a carbon-bound proton and thus the formation of an intermediate enamine. The electron pair on its nitrogen atom provides the driving force for carbon–carbon bond formation through attack at the carbonyl group.

$$H^+ + -C(=O)-CH- + R-NH_2 \underset{+H_2O}{\overset{-H_2O}{\rightleftharpoons}} -C(-CH-)=\overset{+}{N}H-R \;\; (:B) \rightleftharpoons -C(=C-)-\ddot{N}H-R \;\; (-C=O,\; {}^{+}BH)$$

$$\rightleftharpoons -C(-C(OH)-)=\overset{+}{N}H-R \;\; (B:) \underset{-H_2O}{\overset{+H_2O}{\rightleftharpoons}} H^+ + R-NH_2 + -C(=O)-C-C(OH)- \qquad (16)$$

Proton transfer from and to the carbon atom that reacts as the carbanion or carbanion-like species is achieved by keto–enol tautomerization but is facilitated by amines through the energetically more favorable imine–enamine tautomerization [Eq. (17)].

$$-CH-C=O \underset{+HNR_2}{\rightleftharpoons} -CH-C=N^+R_2 \overset{\pm H^+}{\rightleftharpoons} \bar{C}=C\!\cdots\!N^+R_2 \qquad (17)$$

Aldol condensations, with and without amine catalysis, are biologically important. Amine catalysis has been shown to occur in enzymic aldolase reactions like that of fructose-1,6-diphosphate aldolase from animal muscle (see Section IV,A,4). Aldol condensations occur in the formation of certain types of cross-links in connective tissues like collagen and elastin (see Section IV,C,2).

4. *Electron Transfer to Form Radical Species*

Kon and Szent-Györgyi (1973) suggested that electron transfer occurs between amines and aldehydes, so that radical species are formed. This reputedly occurs under conditions in which the reactants are combined as imines. Experimental confirmation of such a suggestion would be desirable.

III. Reaction of Carbonyl Compounds with Amino Acids and Proteins

A. *Reactions with Formaldehyde*

Formaldehyde, the simplest of all aldehydes, has unique reactivities and has been used extensively as a reagent in protein chemistry. It is one of the more familiar reagents to protein chemists as a consequence of its early use by Sörensen (1908) for the formol titration of amino groups, its extensive use in industry for tanning, and in pharmacy for the manufacturing of toxoids. A comprehensive review of the reactions of formaldehyde with both amino acids and proteins by French and Edsall (1945) described in detail information available at that time. Bowes (1948) has summarized the literature with particular reference to formaldehyde tanning before 1945. A series of monographs on the general chemistry and reactions of formaldehyde is also available (Walker, 1964). Because it has been used extensively by industry, there is also a large patent literature on formaldehyde reactions.

Formaldehyde is commercially available in 37–40% aqueous solutions, named formalin, which contains small amounts of methanol. In these solutions, formaldehyde primarily exists as a series of low-molecular-weight polymers of the type $H(OCH_2)_nOH$. Paraformaldehyde is a stable solid composed of high-molecular-weight polymers of the same type. Heating the latter is a method used to generate pure gaseous formaldehyde. Solutions of formaldehyde may also contain small amounts of formic acid, a contaminant that is important in formol titration, wherein it may contribute appreciably to the end point of a titration. Various types of polymers of formaldehyde dissociate to the monomer in dilute aqueous solution. As pointed out already (Section II,B,3), the predominant species in aqueous solution is the formaldehyde hydrate.

The fact that formaldehyde reacts with amino groups of amino acids was early recognized. However, the exact mechanism of this reaction has long been controversial (Bowes, 1948). Schiff (1900, 1901) first suggested that the methylene derivative was formed.

$$R{-}NH_2 + H{-}\underset{OH}{\overset{H}{C}}{-}OH \rightleftharpoons R{-}\underset{H}{N}{-}\underset{H}{\overset{H}{C}}{-}OH + H_2O \not\rightleftharpoons R{-}N{=}C\genfrac{}{}{0pt}{}{H}{H} + H_2O \quad (18)$$

$$R{-}\underset{H}{N}{-}\underset{H}{\overset{H}{C}}{-}OH \overset{CH_2O}{\rightleftharpoons} R{-}\underset{CH_2OH}{N}{-}CH_2OH$$

However, the presence of significant concentrations of this Schiff base is now not generally accepted. The reaction is rather considered to proceed to the dihydroxymethyl derivative by the addition of another molecule of formaldehyde to the hydroxymethyl initially formed (French and Edsall, 1945; Kallen and Jencks, 1966). This second step seems to be faster than loss of water from the intermediate [Eq. (18)]. For simple amino acids such as glycine, the reaction can proceed further to form a variety of cyclic products. One of these, trimethylenetriamine (VII), was postulated by French and Edsall (1945).

$$3\,RNH_2 + 3\,CH_2O \rightleftharpoons \text{R–N}\begin{matrix} \diagup H_2C{-}N{-}R \diagdown \\ \diagdown H_2C{-}N{-}R \diagup \end{matrix} CH_2 + 3\,H_2O$$

(VII)

Formaldehyde reacts not only with primary amino groups in proteins, but also with sulfhydryl groups. Since sulfhydryl groups are better nucleophiles than amino groups, they react more rapidly. However, only one molecule of formaldehyde is capable of reacting with one sulfhydryl (Lewin, 1956; Barnett and Jencks, 1967). It has been generally assumed in the earlier literature (French and Edsall, 1945) that the imidazole group of histidine does not react with formaldehyde in formol titrations. Kallen and Jencks (1966), however, showed that this assumption might very well be incorrect. Their results strongly suggest that a monohydroxymethyl adduct is formed.

Amide groups of glutamine and asparagine side chains will also react with formaldehyde. For asparagine, a well-defined crystalline product was described by Schiff as early as 1900.

B. Glutaraldehyde

Although the reactions of formaldehyde and pyridoxal derivatives with amino groups are better known, those of various other carbonyl compounds have also been examined for their use in protein chemistry. Some of these have been directed at forming insolubilized preparations for commercial uses, which will be discussed in more detail in Section VI,A.

The bifunctional carbonyl reagent glutaraldehyde in recent years has become important for the modification of proteins and ribosomes. Aqueous solutions of glutaraldehyde have been found to consist of free glutaraldehyde (VIII), the cyclic hemihydrate (IX), and its oligomers (X) (Korn *et al.*, 1972). The oligomers are rapidly converted to (VIII) and (IX) upon dilution.

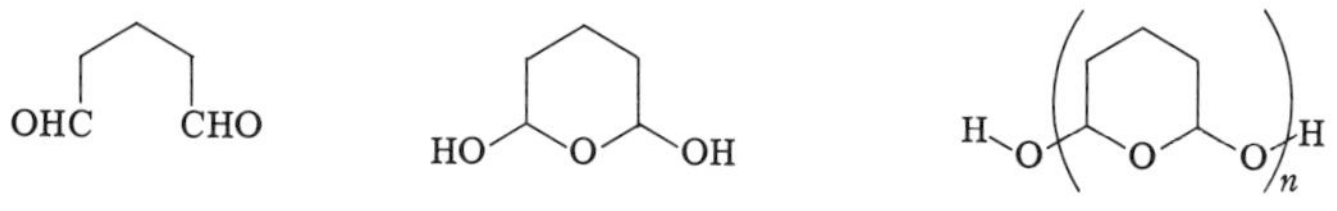

OHC CHO (VIII)   HO O OH (IX)   H O O O H, n (X)

Reaction with lysine residues may involve aldol condensation in addition to simple imine formation, since it is not freely reversible (see Richards and Knowles, 1968). Quiocho and Richards (1966) reported that glutaraldehyde insolubilized crystals of carboxypeptidase, with only a slight loss of catalytic activity. In addition to lysine, residues of cysteine, histidine, and tyrosine also react to some extent (Habeeb and Hiramoto, 1968). Addition of reducing agents before the addition of aldehyde apparently increased specificity toward the lysyl amino groups (Ottesen and Svensson, 1971b). In this respect, the reaction is analogous to that of formaldehyde (Section III,F).

The dissociation of free ribosomes into subunits at lowered concentrations of Mg^{2+}-ions or by hydrostatic pressure, as in ultracentrifugation, can be prevented if they are treated with glutaraldehyde (Subramanian, 1972). Dissociation is effectively prevented over a wide range of ribosome and glutaraldehyde concentrations in the buffers commonly used for studying ribosome function; no preincubation of ribosomes with glutaraldehyde is required. Glutaraldehyde probably cross-links neighboring ribosome proteins or ribosome bound cofactors and ribosomal proteins.

Glutaraldehyde modification of ribosomes also has been used as a tool for mapping their topography. In two-dimensional acrylamide gel electrophoresis protein patterns of glutaraldehyde-treated ribosomes, certain spots are "missing" compared to the patterns given by untreated preparations (Kahan and Kaltschmidt, 1972). These "missing spots" are assumed to represent individual proteins that have been modified in the reaction.

The modified ribosomal proteins most likely are no longer seen in their proper spot because in the process of cross-linking both size and charge of these proteins are changed. The size of some proteins is increased by cross-linking with adjacent proteins and the resulting modified amino groups bear no charge under the experimental conditions of electrophoresis; the original amino groups, however, were protonated.

Since treated 30 S ribosomes retained some functional activity it can be assumed that the modification did not disrupt the tertiary or quaternary structure of the ribosome. From this it is concluded that the modified proteins, exposed to the reagent, are external in the native structure of the ribosome. Unmodified proteins either do not contain primary amino groups or they are not exposed to the reagent and therefore must be internal. The number of individual ribosomal proteins modified by glutaraldehyde increases with increasing reagent concentration indicating their different reactivities toward glutaraldehyde.

C. Malonaldehyde

Since malonaldehyde is one of the main carbonyl compounds derived from oxidation of polyunsaturated lipids in biological materials, its reaction with amino acids (Chio and Tappel, 1969a), ribonuclease (Chio and Tappel, 1969b), and DNA (Reiss *et al.*, 1972) were investigated.

The products formed have been attributed to cross-linking of amino acids by malonaldehyde forming the amino-iminopropene structure (XI).

$$R{-}NH{-}CH{=}CH{-}CH{=}N{-}R'$$

(XI)

These products have a characteristic maximum fluorescence peak at 465 nm when excited at 390 nm.

Pancreatic ribonuclease was irreversibly inactivated by malonaldehyde. The degree of inactivation was proportional to the formation of conjugated imine as judged by the appearance of fluorescence at 470 nm.

When DNA was treated with malonaldehyde, fluorescent products were formed with a fluorescence peak at 460 nm when excited at 390 nm. Model studies with adenine and guanine have shown that similar spectral species were formed in their reaction with malonaldehyde. These results suggest that the amino groups of adjacent bases in the DNA structure have been cross-linked to form the conjugated amino-iminopropene structure (XI).

Since the formation of fluorescent products correlates with the loss of template activity of DNA in the malonaldehyde reaction, it is probable that malonaldehyde alters the spatial structure of DNA. Proteins that have been allowed to react with malonaldehyde besides ribonuclease include bovine serum albumin (Kwon and Brown, 1965; Crawford *et al.*, 1967) and myosin (Buttkus, 1967). Bifunctional protein reagents were reviewed by Wold (1967).

D. Salicaldehyde

Salicaldehyde is one of the few aromatic aldehydes that has been used for the modification of amino groups in proteins (Williams and Jacobs, 1966, 1968; Mühlrad *et al.*, 1970). In cytochrome *c* all 19 lysine ε-amino groups react with salicaldehyde to form an insoluble product that has no activity in the NADH–cytochrome *c* reductase system from heart (Williams and Jacobs, 1968). The modification reaction is readily reversed upon dialysis, but the product is only 20% active. When the dialyzed reaction product is momentarily exposed to pH 3 or pH 11 buffer, however, full activity is restored. Most likely this behavior is a result of oligomer formation of cytochrome *c*, which has been shown to occur in the course of the modification reaction. Monomers are reformed after treatment with acid or alkali.

Mühlrad *et al.* (1970) studied the nature and specificity of the reaction between salicaldehyde and myosin. It was found that the reaction was specific for the lysyl residues of myosin. From the 400 lysyl residues present in myosin of molecular weight (MW) 500,000, approximately 10 have a high affinity for salicaldehyde ($K_{ass} = 1.8 \times 10^5\ M^{-1}$); for 130 lysyl residues the association constant is $2.2 \times 10^3\ M^{-1}$ and the majority of the lysyl residues can be modified only after denaturation. Salicaldehyde modification of myosin has been found to be fully reversible, the extent of modification increased with increasing pH.

The blocking of amino groups with salicaldehyde had almost no effect on the Ca^{2+}-activated ATPase activity of myosin whereas the Mg^{2+}-inhibited ATPase of myosin was activated, activation being maximal with incorporation of 25–70 moles of salicaldehyde per mole of myosin. When the reaction was carried out in the presence of ATP, less salicaldehyde was incorporated. This effect was not observed if only inorganic phosphate was present.

5-Nitrosalicaldehyde condenses with an ε-lysyl amino group at the active site of acetoacetate decarboxylase and reduction of the resulting Schiff base with $NaBH_4$ introduces a 2-hydroxy-5-nitrobenzylamino residue as a "reporter group" into the protein (Frey *et al.*, 1971; Kokesh and Westheimer, 1971). The p*K* of the nitrophenyl residue is 2.4 as compared to a value of 5.9 in the model compound, *N*-methyl-2-hydroxy-5-nitrobenzylamine. In addition, a p*K* value of 6.0 could be assigned as that of the ε-ammonium ion of lysine at the active site of the enzyme, which corresponds to an increase in acidity of >4 units relative to the model compound. These changes of acidity upon binding must be ascribed to the electrostatic effect of nearby positive charges. A possible functional role for the decreased basicity of the essential amino group is that a high fraction of it should remain unprotonated at physiological pH. Thus it could react more rapidly with a carbonyl group than one of more normal basicity which would be overwhelmingly protonated (Kokesh and Westheimer, 1971).

E. Pyridoxal 5′-Phosphate

Pyridoxal 5′-phosphate not only plays a very important role as a cofactor in many enzymic reactions, but it can also be used for the chemical modification of amino and thiol groups in proteins. Pyridoxal 5′-phosphate is much more reactive than pyridoxal, because the aldehyde group of the latter is largely masked, like that of sugars, by intramolecular hemiacetal formation (Heyl *et al.*, 1951; Metzler and Snell, 1955).

1. Rabbit Muscle Aldolase

Rabbit muscle aldolase loses activity when exposed to low concentrations of pyridoxal 5′-phosphate (Shapiro *et al.*, 1968). The inactivation is prevented specifically by substrates and substrate analogs, and reversed by dilution, addition of amino thiols, and addition of excess substrate. The attachment was made irreversible by reducing the imines formed with borohydride, and six lysine ε-amino groups proved to be modified. These did not include the one that forms a Schiff base with the substrate. Hence Shapiro *et al.* concluded that pyridoxal 5′-phosphate forms an imine with lysine residues at one of the phosphate-binding sites of the enzyme.

2. Glyceraldehyde-3-phosphate Dehydrogenase

When glyceraldehyde-3-phosphate dehydrogenase is incubated with pyridoxal 5′-phosphate, up to 10–11 ε-aminolysyl residues per molecule of enzyme become modified, with increasing inhibition of the enzyme (Ronchi *et al.*, 1969; Forcina *et al.*, 1971). The inhibition can be fully reversed by the addition of substrate, lysine or valine, or by dialysis. For the enzyme from rabbit muscle it was demonstrated that lysine residues 191 and 212 are most reactive toward pyridoxal 5′-phosphate. Other lysyl residues appeared to have limited reactivity as indicated by the incorporation of 4 moles of pyridoxal 5′-phosphate per subunit of the enzyme.

3. *6-Phosphogluconic Dehydrogenase*

6-Phosphogluconic dehydrogenase behaved very similarly to glyceraldehyde-3-phosphate dehydrogenase on incubation with pyridoxal 5′-phosphate (Rippa and Pontremoli, 1969). The inhibition of the enzymic activity was due to the formation of a Schiff base between a single ϵ-aminolysine group at the active center of the enzyme and pyridoxal 5′-phosphate. The inhibition was reversed by addition of substrate or amino acids, by dilution, or by dialysis. No inactivation was observed with pyridoxal, pyridoxine, or pyridoxamine phosphate, although in the case of pyridoxal a Schiff base with the enzyme was formed.

4. *Glutamate Dehydrogenase*

When glutamate dehydrogenase was incubated with pyridoxal 5′-phosphate and subsequently reduced with $NaBH_4$, the lysine residue in position 126 was substituted (Smith *et al.*, 1970; Anderson *et al.*, 1966; Piszkiewicz *et al.*, 1970). Both the glutamate and alanine dehydrogenase activity were lost simultaneously after incubation but could be fully restored by dialysis of the inactivated enzyme against water when it had not been treated with $NaBH_4$. The association–dissociation equilibrium of the six identical subunits was disturbed by the modification; only monomers were present in the modified inactivated enzyme.

In a recent investigation, Hucho *et al.* (1973) showed that during irradiation of the enzyme–pyridoxal 5′-phosphate complex, pyridoxal 5′-phosphate slowly becomes irreversibly incorporated. This irreversible incorporation of pyridoxal 5′-phosphate was interpreted as a photoactivated addition of an imidazole group to the double bond of a Schiff base formed between pyridoxal 5′-phosphate and a lysyl ϵ-amino group. After irradiation it was still possible to modify lysine-126 according to Anderson *et al.* (1966). This indicates that the ϵ-amino group of lysine 126 in glutamate dehydrogenase is not affected by the photoactivated pyridoxal 5′-phosphate incorporation.

5. *Ribonuclease A*

When ribonuclease A is treated with excess pyridoxal 5′-phosphate, only two of the ten lysine residues present in the molecule undergo Schiff base formation with retention of 20% of the catalytic activity (Raetz and Auld, 1972). Modification by pyridoxal 5′-phosphate is inhibited in the presence of phosphate ions. Reduction with $NaBH_4$ leads to irreversible incorporation of pyridoxal 5′-phosphate (Means and Feeney, 1971). The modified amino acid residues of the enzyme could be identified as the spatially near lysine-7 and lysine-41 after tryptic–chymotryptic digests. Although ribonuclease modified only at lysine-7 retains considerable catalytic activity, the lysine-41 modified derivative is largely inactive. The residual activity of modified enzyme was therefore ascribed to incompleteness of modification of lysine-41. This interpretation is substantiated by the fact that the change in circular dichroism (CD) spectra is consistent with the reaction of a single lysine residue per molecule, suggesting that modification of lysine-7 prevents reaction of lysine-41 and vice versa. The close proximity of these two amino acids in the three-dimensional structure of ribonuclease A has been demonstrated by cross-linking with the bifunctional reagent 1,5-difluoro-2,4-dinitrobenzene (Marfey *et al.*, 1965).

F. Glyoxal

When glutamate dehydrogenase is incubated with excess glyoxal, the catalytic activity with respect to the substrate 2-oxoglutarate is inhibited competitively, but the enzyme retains largely its hydrodynamic and allosteric properties (Deppert *et al.*, 1973). The reaction is fully reversible upon dilution, but the Schiff base intermediate can be trapped by borohydride reduction. One amino group per polypeptide chain was shown to react with glyoxal.

From the analysis of K_m and (V) in the range pH 7.8 to pH 9.2 the authors concluded that the amino group reacting with glyoxal has a pK value of 8.2 and is involved in binding the substrate. In further investigations Rasched *et al.* (1974) identified this amino acid residue as lysine-27. Lysine residue 126, which reacts with pyridoxal phosphate, is not modified by glyoxal but also participates in binding the substrate. These results lead to the conclusion that lysine residues 27 and 126 may be close together in the tertiary structure of the enzyme.

G. Reductive Alkylation of Amino Groups

Reductive alkylation of protein-amino groups by carbonyl compounds was first discovered as a special interaction between certain enzymes and their substrates or coenzymes, such as pyridoxal 5′-phosphate (Fischer, 1964). These reactions will be discussed in Section IV,A. The general applicability of the reduction of carbonyl-amine adducts to modify specifically protein-amino groups was apparently not recognized until some time later (Means and Feeney, 1968).

The method of reductive alkylation is based on the observation that, in protein solutions treated with low concentrations of simple aliphatic aldehydes or ketones and small amounts of sodium borohydride, amino groups are converted in high yield to the corresponding mono- or dialkylamino derivatives. The composition of the reaction products depends on both temperature and pH (Fig. 2). Low pH and high temperature favor the breakdown of borohydride to hydrogen gas and borate. These conditions are therefore not favorable. Optimal conditions for modifying specifically amino groups are at pH 9, or slightly above, and 0°. Specificity for amino groups is favored by adding the reducing agent before the addition of the carbonyl compound. Lack of appreciation of this detail may well have been the reason why this simple method was not adopted earlier.

With formaldehyde as the carbonyl component, the reaction proceeds rapidly converting lysine to ε-*N,N*-dimethyllysine residues as the principal products. Monomethyllysine is formed initially, but its conversion into dimethyllysine is very fast. The monomethyllysine predominates only when formaldehyde is added in much less than molar amounts. With other aldehydes and ketones, however, reaction with a second carbonyl molecule is retarded, mainly for steric reasons, so that the mono alkyllysyl residue is predominantly obtained. Mixed ε-methylalkyllysines such as ε-*N*-methyl-*N*-isopropyllysine can be prepared by adding the higher substituted

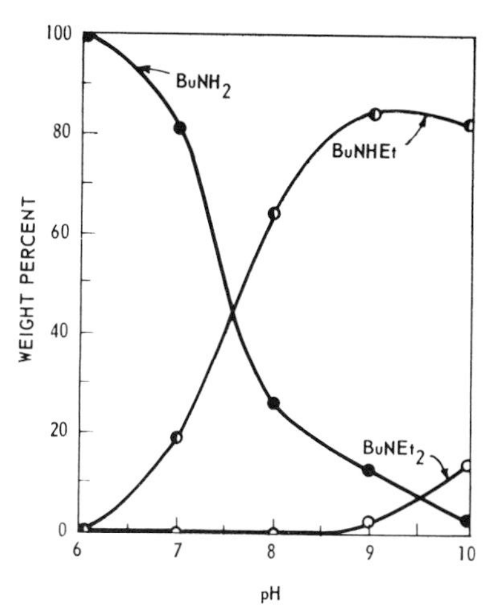

FIG. 2. Reductive ethylation of butylamine with acetaldehyde. Effect of pH upon the composition of reaction mixtures. From Means and Feeney (1968).

carbonyl compound before formaldehyde. From studies of model systems, reaction sequences as shown in Eq. (19) (Means and Feeney, 1968) have been proposed for the reductive alkylation.

$$\begin{array}{c} RNH_2 \\ H^+ \updownarrow \\ RNH_3^+ \end{array} + R'CHO \rightleftharpoons RN{=}CHR' \xrightarrow{[H]}$$

$$\begin{array}{c} RNHCH_2R' \\ \updownarrow H^+ \\ R\overset{+}{N}H_2CH_2R' \end{array} \underset{}{\overset{R'CHO}{\rightleftharpoons}} \begin{array}{c} R\overset{+}{N}{=}CHR' \\ | \\ CH_2 \\ | \\ R' \end{array} \xrightarrow{[H]} RN(CH_2R')_2 \qquad (19)$$

The optimal pH was near pH 10 or above but, because most proteins are unstable at such a high pH, pH 9 was recommended for a general procedure. A pH as low as 7 can be used for proteins sensitive to higher pH, but the yield of modified product is decreased. Increasing the concentration of formaldehyde above stoichiometric amounts leads to significant competition of formaldehyde for the reducing agent. This results in a concomitant lowering of the yield of alkylated products as the availability of reducing agent becomes limiting. This competition appears to be made particularly effective by a greater displacement of the amine aldehyde–Schiff's base equilibrium in favor of the unreactive dihydroxymethyl compound (Eq. 18). Reducing agent should therefore always be in excess. Any excess of carbonyl compound not required for reductive alkylation will then be reduced to the corresponding alcohol, and this may prevent side reactions. Because of the relatively small changes in basicity and the relatively small space occupied by the methyl groups, ε-*N*,*N*-dimethyllysine residues more closely resemble unmodified lysines than do most other modified lysine residues. As a consequence, methylation causes very little disruption of protein conformations and few changes in other general properties. It is

TABLE II
Properties of Reductively Alkylated Ribonuclease[a]

	Lysine	Monoalkyllysine	Dialkyllysine	b_0	a_0	λ_c	s_{20} (S)
Dimethyl	0.7	0.3	8.4	−150	−350	242	1.8
Isopropyl	2.5	7.8	0	−209	−190	274	1.9
Native	10	0	0	−133	−465	236	1.8

[a] From Means and Feeney (1968).

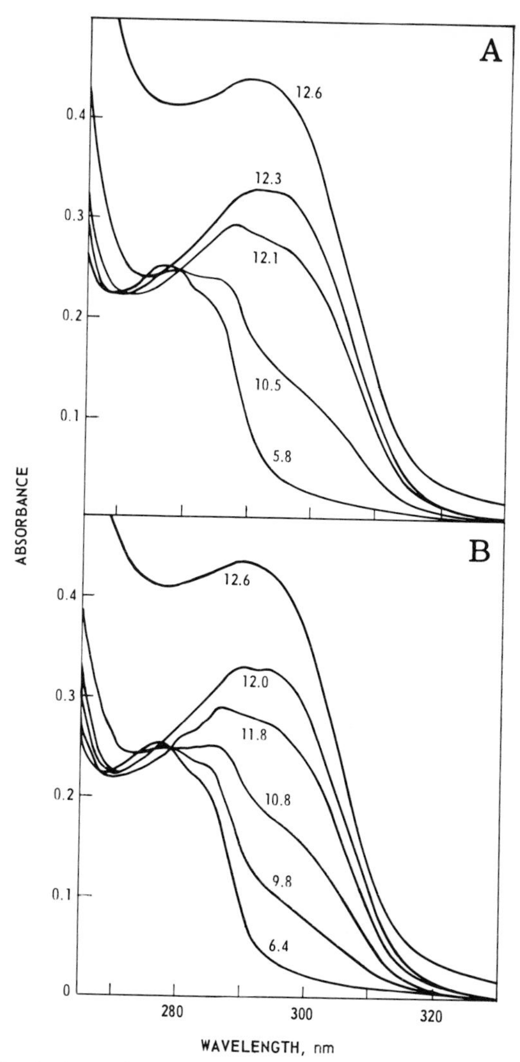

FIG. 3. Spectrophotometric titrations of RNase and reductively methylated RNase showing the loss of the 278-nm isosbestic points above (A) pH 12.2 for the native enzyme and (B) pH 11.8 for the methylated enzyme. From Means and Feeney (1968).

therefore one of the best specific methods for detecting functions in which a primary amino group participates.

Means and Feeney (1968) showed that reductive methylation of approximately 80% of the lysyl ϵ-NH_2 groups in ribonuclease only slightly changed its general properties (Table II, Fig. 3). Analogous

TABLE III
Reductive Alkylation of Selected Proteins[a]

Protein	Distribution of products		
	Dialkyllysine (%)	Monoalkyllysine (%)	Lysine (%)
Chymotrypsin			
Methyl	98	1	<1
Chymotrypsinogen			
Methyl	98	1	<1
Human serum transferrin (iron form)			
Methyl	78	6	16
Isopropyl	0	29	71
Lysozyme			
Methyl	98	<1	<1
Cyclopentyl	0	32	68
Turkey ovomucoid			
Methyl	60	14	26
Isopropyl	0	22	78

[a] From Means and Feeney (1968).

results were obtained with polylysine and calf thymus histone, which were reductively methylated (Paik and Kim, 1972). Reductive alkylation of several selected proteins (Table III) shows some difference in the susceptibility of different amino groups to alkylation by formaldehyde as well as by other carbonyl compounds. Ottesen and Svensson (1971a,b) applied reductive methylation with formaldehyde to bovine serum albumin and to subtilisin novo and reported it to be highly specific for amino groups and to be particularly useful for differentiating between those amino groups exposed to solvent and those that are masked or buried. Modification of groups other than amino groups may occur only under extensive treatment with reagent (Ottesen and Svensson, 1971a,b).

Reductive alkylation has also been used to introduce radioactive label into proteins (Rice and Means, 1971; Ottesen and Svensson, 1971b). Reductive methylation with [^{14}C]formaldehyde is efficient and economical, and it may cause less change to the physicochemical properties of proteins than alternative methods such as iodination; heterogeneity is not usually exhibited by treated proteins in separations that depend on charge, such as electrophoresis or ion-exchange chromatography.

Another simple application of reductive methylation is the development of an assay for the activity of proteolytic enzymes. Proteolysis of *N,N*-dimethyl-substituted proteins can be determined by direct measurement of the bonds hydrolyzed. This is possible because the use of an appropriate amino group reagent such as trinitrobenzenesulfonic acid allows the direct determination of the appearance of new terminal amino groups. The low blank values obtainable with methylated proteins greatly increase the sensitivity and accuracy in comparison with unmodified proteins (Lin *et al.*, 1969). With these substrates it is possible to calculate directly the number of bonds cleaved.

A further elaboration of this concept has led to comparatively specific assays for carboxypeptidase A and carboxypeptidase B (Lin *et al.*, 1969). In these assays, the protein substrates are prepared by treating them with either α-chymotrypsin or trypsin and then reductively methylating the amino groups of the resulting mixture of peptides. Reductively methylated chymotrypsin-treated proteins consist of a mixture of peptides containing hydrophobic COOH-terminal amino acids. These are readily hydrolyzed by carboxypeptidase A. Reductively methylated trypsin-treated proteins consist of a mixture of peptides with lysyl or arginyl COOH terminals. The arginine residue is readily hydrolyzed by carboxypeptidase B. With the use of new fluorescent reagents such as fluorescamine (Udenfriend *et al.*, 1972) for the detection of primary amines, assays for the activity of proteolytic enzymes using methylated proteins should be 10^4 to 10^6 times more sensitive than any other procedures using proteins as substrates.

Reaction of aldehydes or ketones with ammonia, primary amines, and secondary amines, and subsequent reduction with sodium cyanohydridoborate ($NaBH_3CN$) at about pH 7, produces primary, secondary, or tertiary amines (Borch *et al.*, 1971). Reaction of substituted pyruvic acids with ammonia in the presence of BH_3CN^- was reported to be an excellent method for the synthesis of amino acids and, in particular, for inexpensive isotopic labeling with ^{15}N by the use of $^{15}NH_3$. The basis of the selectivity of BH_3CN^- appears to be that it is too weak a hydride donor to reduce carbonyl compounds at an appreciable rate; it can, however, reduce the protonated form of the imine. Its comparatively low reactivity allows its use under relatively acidic conditions where imines will be largely protonated.

Either $NaBH_4$ or $NaBH_3CN$ has been used to immobilize enzymes by reductive alkylation of their amino groups with aldehyde derivatives of polymers (G. P. Royer, personal communication, 1974).

E. N. Shaw and H. B. F. Dixon (personal communication from H. B. F. Dixon) have obtained reductive carboxymethylation of α- and ϵ-amino groups of peptides by adding glyoxylate and reducing the intermediate imine with BH_3CN^-.

H. Modification and Removal of NH_2-Terminal Residues of Proteins by Transamination

The α-amino group can be specifically removed and replaced with a carbonyl group by transamination (Dixon and Fields, 1972). The modified NH_2-terminal residue may subsequently be selectively removed so that the protein molecule is shorter by one amino acid. Both of these steps involve a carbonyl–amine reaction for most amino acids. Ideally, it would thus be possible to use the sequential procedures for modifying the NH_2-terminal residue, removing the entire group to expose the adjacent amino acid, and then successively repeating the entire process. Techniques available so far,

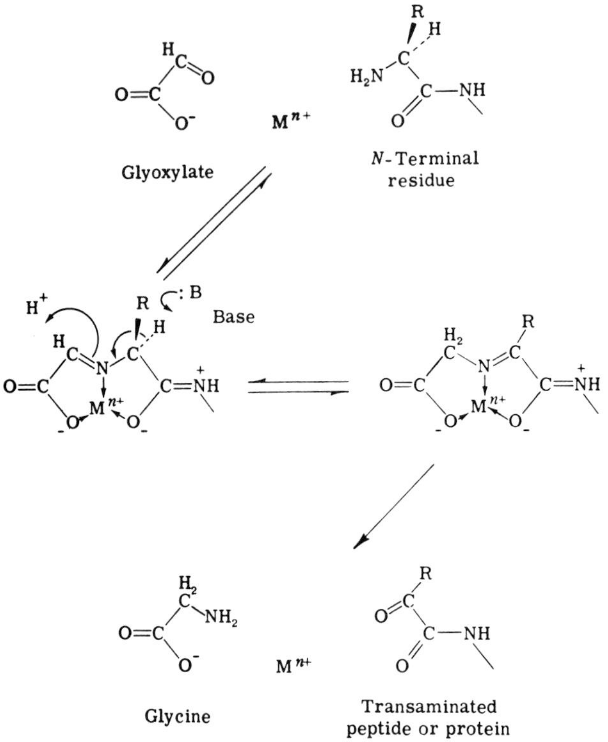

FIG. 4. Modification of NH_2-terminal residues by transamination. From Dixon and Fields (1972).

however, especially for the scission reaction, involve conditions that are too harsh to avoid denaturation in many proteins.

Essential components for the transamination reaction are (Fig. 4): (a) an acceptor functional group to bind the terminal amino group of the protein as a Schiff base; this is usually the carbonyl group of glycoxylate or a pyridoxal derivative; (b) a heavy metal cation like Cu^{2+} or Ni^{2+}; (c) a high concentration of base, usually acetate ion or pyridine.

In the coordination complex of the Schiff base, the proper orientation between the carbonyl acceptor function and the peptide for the transamination reaction is achieved. The specificity of this method

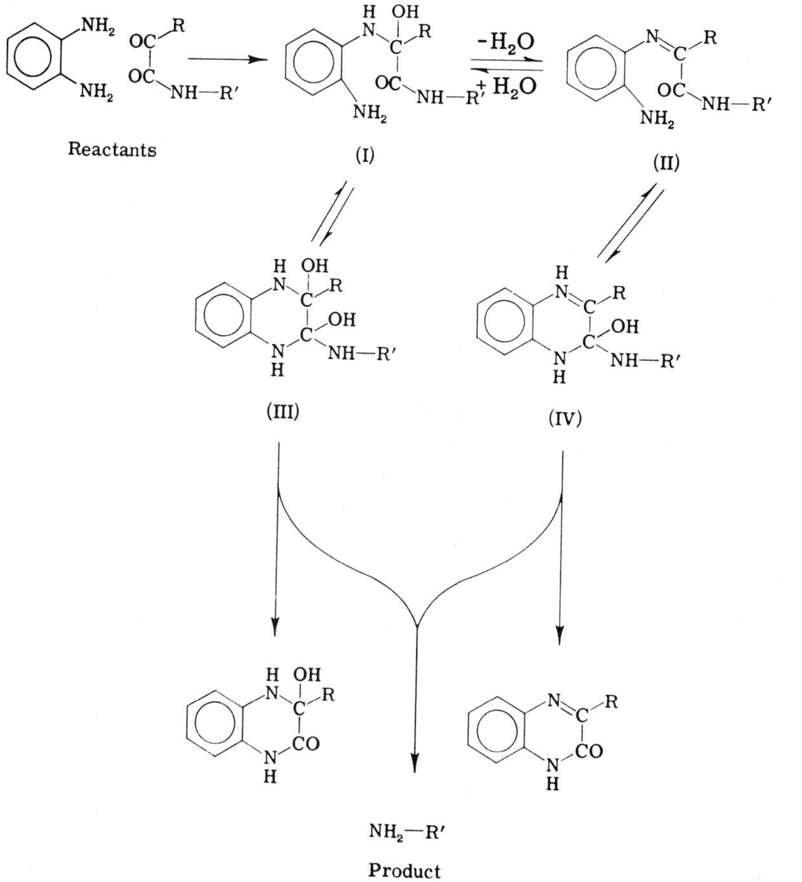

FIG. 5. Scissions of α-ketoacyl residues in transamination. From Dixon and Fields (1972).

for α-amino groups arises from the fact that the peptide carbonyl group is coordinated to the metal ion in the active complex. Thereby the α-proton is labilized sufficiently for the transamination to take place. Since a new carbonyl group is formed in the protein replacing a primary amino group, methods for determining either carbonyl groups or changes in charge can be applied for quantitating the reaction. The scission of the α-ketoacyl residue is achieved by reacting the electrophilic carbon of the keto group with a reagent that will specifically attack and break the adjacent peptide bond (Fig. 5).

The reagent has to contain two nucleophilic groups that are separated by one or two atoms. One nucleophilic group attacks the carbonyl carbon, and places thereby the other correctly for attack on the peptide carbon forming a 5- or 6-membered ring; *o*-phenylenediamine proved to be the most satisfactory reagent so far. The cyclic intermediate in the final step of the scission reaction is stabilized by expulsion of the new peptide which contains one amino acid less than the original peptide (or protein) introduced to the transamination reaction. Many side reactions can occur in the course of this process, but they seem to be reversible since the yield of the desired product has been greater than 95% in the cases tried.

An alternative route exists for converting *N*-terminal serine and threonine residues into 2-oxoacyl groups, i.e., their oxidation by peroxidate to yield glyoxyloyl residues. These groups, unlike the ketoacyl groups of transaminated proteins, can themselves be transaminated under mild conditions, so that terminal serine and threonine can be converted into glycine (Dixon and Fields, 1972). In the case of the hormone corticotropin most of the biological activity was lost upon oxidation, and restored when transamination restored the presence of a terminal amino group (Dixon and Weitkamp, 1962).

IV. The Carbonyl–Amine Reaction in Biological Processes

Carbonyl–amine reactions play important roles in very different biological processes. They include catalytic reactions in enzymes, cross-linking in structural proteins like collagen and elastin and the visual process.

A. *Schiff Bases in Enzyme Catalysis*

The properties of a Schiff base make it suitable for its participation in enzymic reactions. All steps of its formation from carbonyl and amine compounds via the tetrahedral amino alcohol are reversible in aqueous solution. The reactivity of the compounds involved allows their participation in an extraordinary variety of apparently different enzymically catalyzed processes. Detailed descriptions of amino acid decarboxylases (Boeker and Snell, 1972) and Schiff base intermediates in enzyme catalysis (Snell and Di Mari, 1970) have recently been presented. We have therefore attempted to limit this discussion to those subjects that appear to be suited best for explaining the activity of fundamental catalytic intermediates. Furthermore, we have attempted to point out the common principles that relate the carbonyl–amine reaction in enzyme catalysis to its role in other biological processes.

According to Snell and Di Mari (1970), two categories of enzymes that operate via Schiff base mechanisms may be defined. In category I, the carbonyl group is present in the enzyme as part of a prosthetic group or coenzyme, the substrate being amines of various types. This category includes the pyridoxal phosphate-dependent enzymes and a more recently discovered series of enzymes that contain covalently bound pyruvate as the prosthetic group. Enzymes of category II lack a carbonyl group but contain as part of the active site an amino group that reacts with a carbonyl containing substrate. This category includes certain β-keto acid decarboxylases as well as aldolases.

1. Interaction of Apoenzymes with Pyridoxal 5′-Phosphate

Apart from its chemical function in the catalytic process, pyridoxal 5′-phosphate appears to be essential in many proteins in maintaining the active conformation. The contribution of products from the carbonyl–amine reaction to the structural stability of a protein represents a property that is not only common in pyridoxal 5′-phosphate-dependent enzymes but also in visual pigments like rhodopsin and structural proteins like collagen and elastin.

Studies with pyridoxal 5′-phosphate analogs have shown that every substituent of the pyridoxal 5′-phosphate molecule contributes to its binding by the apoenzyme. The presence of the 5′-phosphate group is particularly important because it represents a potential ionic binding site. Thus the affinity of apoaspartate aminotransferase for pyridoxal 5′-phosphate is more than 10^3 times that for free pyridoxal (Wada and Snell, 1962). From such studies it can be concluded that binding of pyridoxal 5′-phosphate results from a mixture of ionic and other forces, such as hydrophobic interactions. The imine bond contributes to binding, since it is easier to dissociate pyridoxamine phosphate than pyridoxal phosphate from the apoenzyme of a transaminase; clearly the bond is not essential for binding, since the cofactor does not dissociate when the imine bond is broken during catalysis.

Most pyridoxal 5′-phosphate enzymes contain two or more subunits whose dissociation is generally favored by dissociation of the cofactor. Reassociation of subunits is favored by addition of pyridoxal 5′-phosphate. However, reactivation and reassociation of subunits are separate phenomena, since pyridoxal 5′-phosphate analogs that permit reassociation of subunits do not necessarily regenerate enzymic activity. Inactivation of one subunit does not affect the activity of another. Association with pyridoxal 5′-phosphate frequently results in a readily measurable conformational change in the enzyme. The holoenzyme forms of most pyridoxal 5′-phosphate enzymes are substantially more resistant toward denaturation than the corresponding apoenzymes.

In this context the function of pyridoxal 5′-phosphate in phosphorylases is particularly significant. Pyridoxal 5′-phosphate is apparently not directly involved in the catalytic step in glycogen phosphorylase—indeed reduction of the imine it forms with an amino group of the enzyme leaves the enzyme active; nevertheless its removal, even under mild and fully reversible conditions, results in a total loss of catalytic activity and decreased stability of the enzyme (Shaltiel *et al.*, 1966; Hedrick *et al.*, 1966). The pyridoxal 5′-phosphate site of glycogen phosphorylase not only strongly interacts with the catalytic site of the enzyme but also with the known regulatory sites, mainly, the site that accommodates the allosteric activator AMP or the site that contains the serine which becomes phosphorylated upon conversion of phosphorylase *b* to *a* (Hedrick *et al.*, 1969).

From these facts it becomes apparent that the function of pyridoxal 5′-phosphate may not be merely that of a structural building block. It is possible, for example, that pyridoxal 5′-phosphate is part of the catalytic site of the enzyme or that it occupies another site which may be involved in the recognition of the substrate or in the transfer of a regulatory signal from or to phosphorylase (Bresler and Firsov, 1968; Shaltiel *et al.*, 1969).

Fischer (1964) and his associates proposed that at acidic or alkaline pH pyridoxal 5′-phosphate is bound to the protein through a Schiff base linkage. This suggestion was supported by the finding that at pH 4–5 $NaBH_4$ reduction results in irreversible attachment of pyridoxal 5′-phosphate at a lysyl residue of the protein. However, this reduction does not occur at neutral pH.

Recent studies by Cortijo and Shaltiel (1972) on $NaBH_4$-reduced glycogen phosphorylase suggest that at pH 7, where the enzyme exhibits maximal catalytic activity, the pyridoxamine 5′-phosphate residue is "buried" in a hydrophobic environment. Upon gradual lowering of the pH of the medium to pH 4.8, there are structural changes in the enzyme which are reflected in an increase of the molar absorption around 330 nm and the quantum yield of the cofactor fluorescence. Since these transitions could be simulated with free pyridoxamine 5′-phosphate by increasing the water content of a dimethylformamide–water mixture from 20% to 80%, acidification of the medium might cause structural changes in the enzyme which expose the cofactor site. This conclusion could account for the lack of reduction of the hydrogen-bonded Schiff base in phosphorylase at neutral pH. $NaBH_4$ was found to decompose in water much faster than it could react with pyridoxal 5′-phosphate in this hydrophobic pocket, where the pyridoxal 5′-phosphate may also be sterically inaccessible (Cortijo and Shaltiel, 1972).

The differences in microenvironment in the region of the internal Schiff base between the formyl group of pyridoxal 5′-phosphate and the ϵ-amino group of a lysine residue on the protein are reflected in their absorption spectra. They vary between 410 and 435 nm and shift substantially to near 360 nm with increasing pH in some enzymes but not in others. Studies with model compounds (Heinert and Martell, 1963) have assigned the absorbance maximum near 415 nm to the protonated Schiff base, where the hydroxy group is deprotonated. A second maximum at 330 nm can be assigned to a structure with an intact hydroxy group. The variation in position of the absorbance maximum near 415 nm from one enzyme to another is not understood but presumably reflects subtle differences in the microenvironment around the cofactor site supplied by different apoenzymes. A detailed discussion of this phenomenom will be given in Section IV,B.

2. *Mechanisms of Enzymic Reactions*

The binding of the aldehyde group of pyridoxal phosphate to an amino group of lysine as an imine evidently lowers its free energy (since binding occurs). This binding is therefore unfavorable to substrate binding, in the sense that the dissociation constant must be higher than if the aldehyde had been free. This is not a serious drawback, since it can be overcome by other, favorable, binding factors, so it can be advantageous if it assists catalysis (compare the straining of substrates on binding). In this case, although the *equilibrium* of binding is less favorable, its *rate* is enhanced, possibly because of the more symmetrical transition states involved than for reaction with a free aldehyde (Cordes and Jencks, 1962).

After formation of the Schiff base between enzyme and substrate, the subsequent events during catalysis appear to proceed by mechanisms essentially similar to those of the related nonenzymic reactions (see Section II,B,5). Their general outlines were laid down by Braunstein and Shemyakin (1953) and Metzler *et al.* (1954). Each of the reactions is a consequence of cleavage of one of the three bonds to that substrate carbon adjacent to the imine nitrogen. Unlike nonenzymic reactions, however, the enzymic reactions, with the exception of certain amine oxidases, are not catalyzed by di- or trivalent metal ions. The functional role of the metal ion is being played much more effectively by the apoenzyme, which may enhance the rates of the nonenzymic reactions by factors as high as 10^7 or more (Snell and Di Mari, 1970). Part of the role of the metal in the nonenzymic reactions was to hold substrate and pyridoxal derivative together; the binding of each to the apoenzyme can fulfill this part. Another part was to provide electron attraction; this can be provided by protonation on the imine nitrogen, and the electron-withdrawing effects of the ring (Eq. 11).

As an example we shall discuss the mechanism suggested for enzymic transamination (Snell, 1962) (Fig. 6). The pyridoxal 5′-phosphate form of the enzyme possesses specific affinity for its substrate and undergoes transimination with amino acid to yield the enzyme–substrate complex, liberating the ϵ-amino group of a lysine residue. It may be this same lysine amino group that acts as a general base and removes the proton from the 2-carbon of the amino acid, and adds it to the 4′-carbon of the coenzyme, thus catalyzing the prototropic shift from aldimine to ketimine (see also Ivanov and Karpeisky, 1969). The latter undergoes hydrolysis to the pyridoxaminephosphate form of the enzyme, thus completing catalysis of one half of the overall reaction. The process then proceeds in reverse with a different keto acid, regenerating the pyridoxal 5′-phosphate enzyme and thereby completing catalysis of the overall reaction.

An interesting model for this enzyme has been proposed by Ivanov and Karpeisky (1969) from Braunstein's laboratory. One of the features it seeks to explain is that the nitrogen of the amino acid substrate occupies the same position as that of the lysine amino group it replaces with respect to the pyridine ring, but transamination occurs only with the former. They suppose that these two nitrogen atoms occupy different positions with respect to the catalytic groups of the

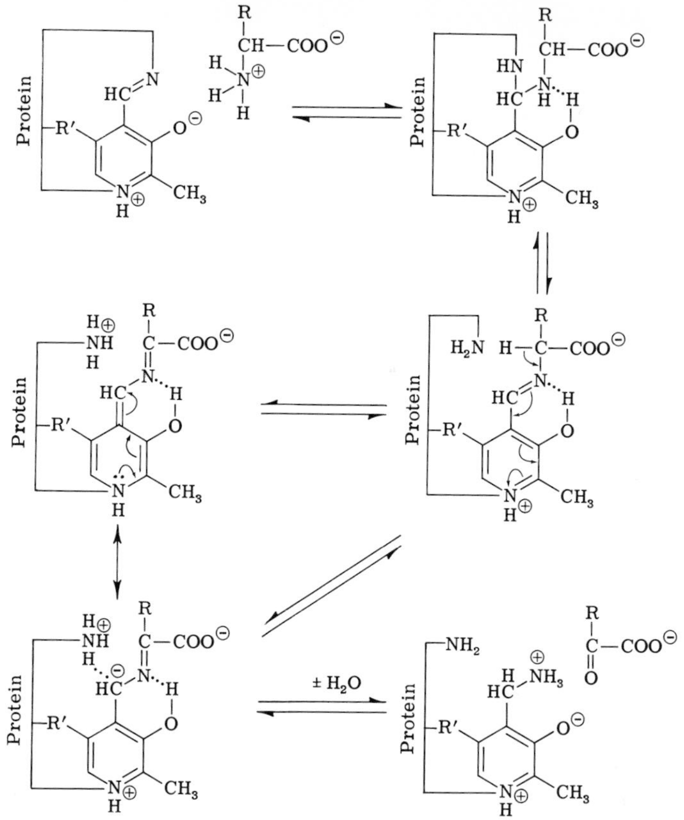

FIG. 6. Enzymic transamination. From Snell and Di Mari (1970).

protein, so that the pyridoxal moves with respect to the protein. They suppose that it rotates about an axis joining its 2-methyl and 5-methylenephosphate groups. This movement could explain a change of electron affinity of the ring (transmitted thence to the imine group) in the course of the reaction, as the distance of the pyridine nitrogen from a charged group in the enzyme varied.

A pecularity of the enzymic reactions of pyridoxal phosphate, in which they differ from the nonenzymic ones discussed in Section II,C,2, is that each enzyme catalyzes the breakage of only one of the bonds to the 2-carbon of the amino acid substrate. Dunathan (1966) has pointed out that a part of this specificity may be due to the orientation with which these three groups of the substrate are bound. It is only a group whose bond to this carbon atom lies in a plane at right angles to the plane of the pyridine ring that can easily leave, because only in this orientation can the electron pair it leaves behind conjugate effectively with the electron-withdrawing aromatic system.

A molecule

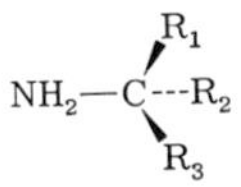

$$NH_2—C(R_1)(R_2)(R_3)$$

will have the group R_1 labilized in its imine only in the two following arrangements:

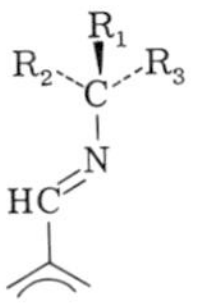

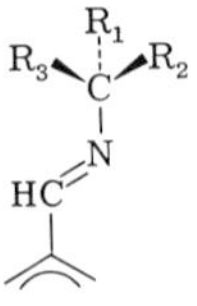

$$R_2, R_1, R_3 \text{ on } C—N{=}CH— \qquad R_3, R_1, R_2 \text{ on } C—N{=}CH—$$

(Note: In both of these arrangements the C-R_1 bond lies in a plane perpendicular to the plane of the conjugated system that includes the ring, i.e., to the plane of the paper.) This is because the electrons that the departing group R_1 leaves behind must conjugate with the planar ring system. They do so most efficiently if they enter from a direction perpendicular to the plane. R_2 and R_3 must be near the plane so that they can easily move into it as R_1 leaves to form:

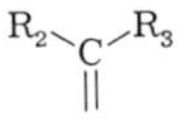

$$R_2—\underset{\|}{C}—R_3 \quad \text{or} \quad R_3—\underset{\|}{C}—R_2$$

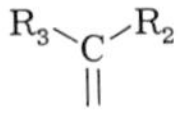

3. *Enzymes That Contain Carbonyl Groups Other Than Pyridoxal 5′-Phosphate*

In addition to pyridoxal 5′-phosphate enzymes, a number of other enzymes are known to contain a functionally essential carbonyl group (Hodgins and Abeles, 1967; Rosenthaler *et al.*, 1965; Hansen and Havir, 1969; George and Phillips, 1970; Recsei and Snell, 1970). Histidine decarboxylase from *Lactobacillus 30a* is a decameric enzyme conposed of 5 subunits of MW 28,000–29,000 and 5 subunits of MW 9000 (Riley and Snell, 1970). Pyruvate is bound in amide linkage at the amino terminus of each of the five larger subunits and appears to function as the prosthetic group of the enzyme by undergoing Schiff base formation with the substrate L-histidine. Recsei and Snell (1973) have isolated a catalytically inactive zymogen, which possesses only one type of subunit (MW 37,000) and no pyruvoyl group. On activation, each proenzyme subunit is split into two smaller units with the generation of a terminal pyruvoyl group on the larger (Eq. 20). Since the biological origin of the pyruvoyl group is

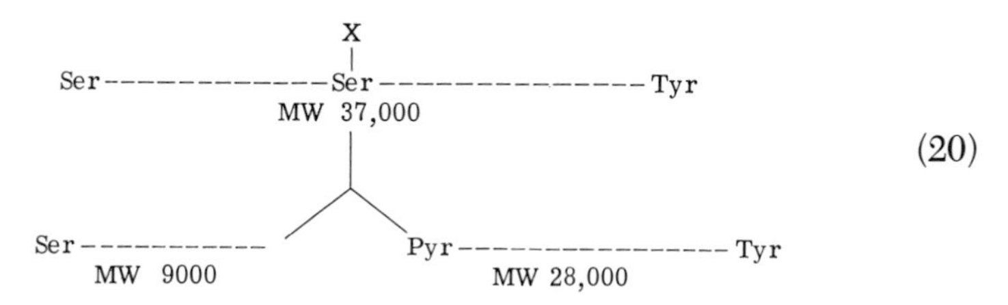

(20)

serine (Riley and Snell, 1970), serine or modified serine must be present beside the bond cleaved. This is reminiscent of the ease with which nisin splits at its 2-aminoacrylic acid residue (also probably derived from serine) to yield a fragment with a terminal pyruvoyl group (Gross and Morell, 1971; Gross *et al.*, 1973).

4. *Enzymes in Which the Schiff Base is Formed by an Amino Group of the Enzyme and a Carbonyl Group of the Substrate*

Classical reports in this area have primarily come from Westheimer and colleagues (reviewed by Westheimer, 1963). Hamilton and Westheimer (1959) suggested the formation of ketimine between acetoacetate decarboxylase and substrate. Fridovich and Westheimer (1962) showed later that the enzyme was inactivated by $NaBH_4$ in the presence of substrate and that radioactively labeled substrate was covalently linked to the protein upon reduction. The mechanism of the reaction proceeds via Schiff base formation of acetoacetate with an ϵ-aminolysine on the protein (Warren *et al.*, 1966). The release of CO_2 is activated by the protonated Schiff base since it is highly electrophilic, thereby polarizing the carbon–carbon bond in the β-position. A resonance-stabilized enamine is left behind after CO_2 release (Fig. 7). In a model reaction, aniline in acid solution catalyzes the decarboxylation of acetoacetate.

Fructose-1,6-diphosphate aldolase from animal muscle (Grazi *et al.*, 1962) as well as transaldolase (Venkataraman and Racker, 1961; Pontremoli *et al.*, 1961) have been shown to form Schiff bases between an amino group of the enzyme and a carbonyl group of the substrate. The intermediates again can be trapped by $NaBH_4$ reduction. For the mechanism of the aldolase reaction, the principles described in Section II,C,3 apply. The protonated Schiff base activates the proton at the α-carbon, thereby supplying the driving force for reaction with the aldehyde carbon (Fig. 8).

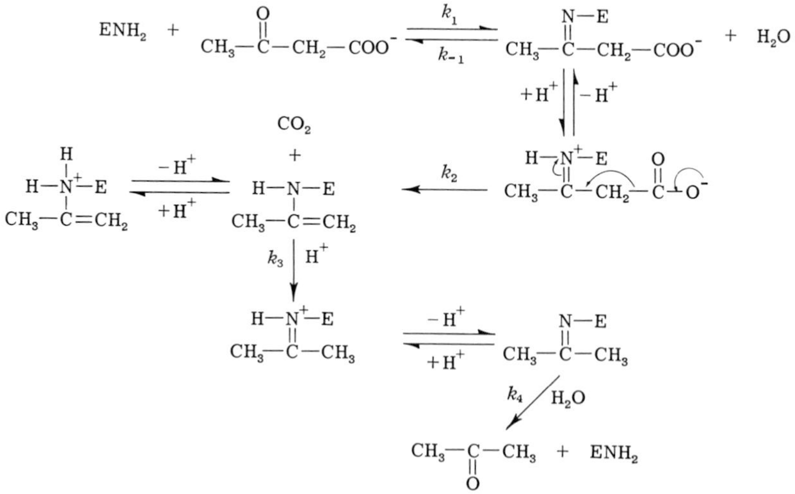

$$\mathrm{ENH_2 + CH_3{-}C({=}O){-}CH_2{-}COO^- \underset{k_{-1}}{\overset{k_1}{\rightleftharpoons}} CH_3{-}C({=}N{-}E){-}CH_2{-}COO^- + H_2O}$$

$$\mathrm{CH_3{-}C({=}N{-}E){-}CH_2{-}COO^- \underset{-H^+}{\overset{+H^+}{\rightleftharpoons}} CH_3{-}C({=}\overset{+}{N}H{-}E){-}CH_2{-}C({=}O){-}O^-}$$

$$\mathrm{CH_3{-}C({=}\overset{+}{N}H{-}E){-}CH_2{-}C({=}O){-}O^- \xrightarrow{k_2} CO_2 + CH_3{-}C({=}CH_2){-}NH{-}E}$$

$$\mathrm{CH_3{-}C({=}CH_2){-}\overset{+}{N}H_2{-}E \underset{+H^+}{\overset{-H^+}{\rightleftharpoons}} CH_3{-}C({=}CH_2){-}NH{-}E}$$

$$\mathrm{CH_3{-}C({=}CH_2){-}NH{-}E \xrightarrow[k_3]{H^+} CH_3{-}C({=}\overset{+}{N}H{-}E){-}CH_3 \underset{+H^+}{\overset{-H^+}{\rightleftharpoons}} CH_3{-}C({=}N{-}E){-}CH_3}$$

$$\mathrm{CH_3{-}C({=}N{-}E){-}CH_3 \xrightarrow[k_4]{H_2O} CH_3{-}C({=}O){-}CH_3 + ENH_2}$$

FIG. 7. Acetoacetate decarboxylation. ENH_2 = enzyme. From Warren *et al.* (1966).

Formation of a protonated imine is not the only way in which the electron attraction of a carbonyl group can be increased and thus labilize the bond from the adjacent carbon to carbon or hydrogen as in decarboxylases and aldolase. Binding of a metal ion to the carbonyl oxygen has the same effect. Thus bivalent cations catalyze the decarboxylation of oxaloacetate in a model reaction. In addition to the kind of aldolase mentioned, there is a further kind which requires a metal ion (it is inhibited by chelating agents) but does not form a Schiff base (it is insensitive to borohydride in the presence of

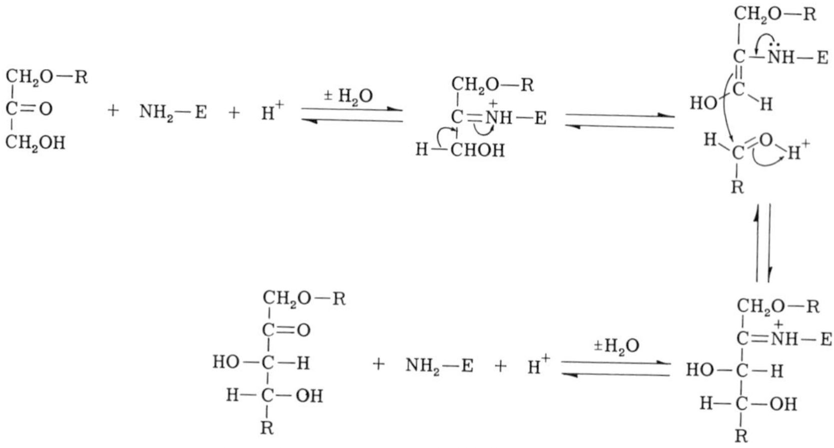

$$\mathrm{CH_2O{-}R{-}C({=}O){-}CH_2OH + NH_2{-}E + H^+ \underset{}{\overset{\pm H_2O}{\rightleftharpoons}} CH_2O{-}R{-}C({=}\overset{+}{N}H{-}E){-}CHOH{-}H \rightleftharpoons CH_2O{-}R{-}C(\ddot{N}H{-}E){=}CH(OH) + H{-}C({=}O{\cdots}H^+){-}R}$$

$$\mathrm{\rightleftharpoons CH_2O{-}R{-}C({=}\overset{+}{N}H{-}E){-}C(HO)(H){-}C(H)(OH){-}R \underset{}{\overset{\pm H_2O}{\rightleftharpoons}} CH_2O{-}R{-}C({=}O){-}C(HO)(H){-}C(H)(OH){-}R + NH_2{-}E + H^+}$$

FIG. 8. Mechanism of aldolase action.

substrate). These classes have been reviewed by Lai and Horecker (1972).

The enzyme δ-aminolevulinic acid dehydratase catalyzes the condensation of two molecules of δ-aminolevulinic acid to a pyrrole, phorphobilinogen (Nandi and Shemin, 1968). The mechanism of the reaction probably involves a Schiff base intermediate of one molecule of substrate with the enzyme, which can be trapped by borohydride. The enamine intermediate thus formed attacks a second mol-

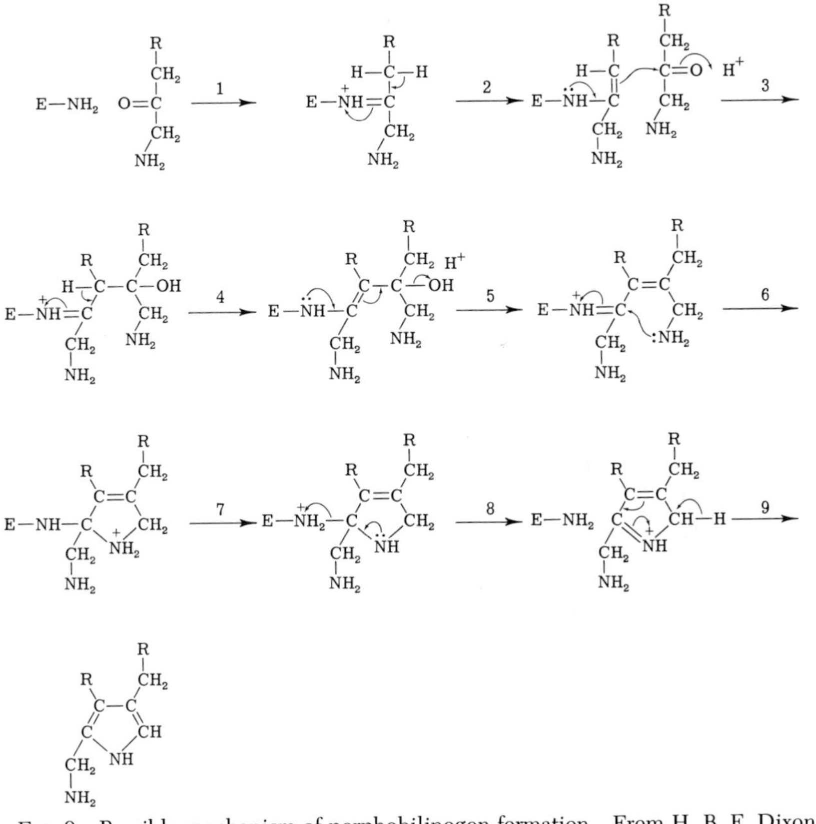

FIG. 9. Possible mechanism of porphobilinogen formation. From H. B. F. Dixon, personal communication, 1973. Remarks: (1) Borohydride inactivates the enzyme in the presence of substrate, but not in its absence. (2) Borohydride incorporates into the protein label that is added in the substrate. (3) The advantage of formation of the protonated imine (step 1) could be the catalysis of the loss of the carbon-bound proton (step 2) by avoiding the need to form a carbanion (cf. aldolase) and catalysis of the subsequent aldol condensation (step 3). (4) A similar pair of reactions (steps 4 and 5) would explain the dehydration that forms the double bond. (5) Displacement of one amine by another in an imine (steps 6–8) may be compared with displacement of the side chain of lysine by the α-amino group of an amino acid when a transaminase combines with its amino acid substrate.

ecule of substrate, and the adduct loses water. Transimination and deprotonation finally yield the product porphobilinogen (Fig. 9).

J. M. La Nauze and H. B. F. Dixon (personal communication) have shown that phosphonatase from *Bacillus cereus*, which hydrolyzes 2-phosphonoacetaldehyde, $CHO\text{-}CH_2\text{-}PO_3H_2$, to form acetaldehyde and orthophosphate (La Nauze *et al.*, 1970), may act in a similar way. Borohydride similarly inactivates the enzyme, provided that substrate or acetaldehyde is present. This may explain the earlier observations that cyanide, sulfide, and sulfite inactivate the enzyme. Presumably the increased electron-withdrawing effect of the protonated imine labilizes the carbon–phosphorus bond, just as carbon–carbon and carbon–hydrogen bonds were labilized in acetoacetate decarboxylase, various aldolases, and porphobilinogen synthetase.

B. The Visual Process

1. The Central Role of a Carotenoid in Vision

The visual process is probably the biochemical event in which formation and cleavage of a Schiff base attains one of its highest levels of molecular sophistication. A detailed compendium has recently been published by specialists in the field (Dartnall, 1972). Of the many fine earlier investigations, those of Selig Hecht and George Wald have proved to be highly significant. Hecht and Williams (1922) early observed that the degree of color fading of retina was proportional to the amount of light striking the retina. Various observations had also appeared relating vitamin A deficiency to a difficulty of seeing in the dusk or night (nutritional night blindness). But it was several years later before Wald (1935) showed that the bleaching of visual purple formed a carotenoid, retinal, and that this carotenoid could be related to vitamin A. Somewhat later (Morton and Goodwin, 1944; Ball *et al.*, 1948), it was shown that retinal was the aldehyde corresponding to vitamin A_1 and another carotenoid, 3-dehydroretinal, was the aldehyde corresponding to vitamin A_2. Since the vitamin A compounds were alcohols, an obviously vital link to the visual process was a system oxidizing the alcohol to the aldehyde, and this was soon found to be an NAD-dependent dehydrogenase (Wald, 1950).

2. The Chemical Nature of the Visual Pigment

Most visual pigments consist of retinal or 3-dehydroretinal linked to a membrane-bound protein called the opsin. For bovine rhodopsin, the most commonly investigated visual pigment, molecular weights ranging from 24,800 (Shields *et al.*, 1967) to ~40,000 (Hubbard, 1953–1954) have been reported, with phospholipids accounting for as much as 45% of the total weight. The apparent discrepancy was the result of different isolation methods. Since rhodopsin is a structural component of a membrane, it can only be solubilized with detergents. Heller (1968) has shown that phospholipids can be completely separated from rhodopsin in preparations of retina rod outer segments. The pure bovine rhodopsin is a glycoprotein with a molecular weight of about 28,000 containing two prosthetic groups: retinal and a moiety with 6 residues of carbohydrate (Heller and Lawrence, 1970).

Model studies with Schiff bases from retinal and aliphatic amines (Pitt *et al.*, 1955) led to the conclusion that 11-*cis*-retinal might form a protonated Schiff base with an amino group in the opsin. In subsequent investigations the binding of retinal to the opsin has tentatively been shown to be a Schiff base between the retinal carbonyl and a protein ε-lysylamino group (Bownds and Wald, 1965; Bownds, 1967; Akhtar *et al.*, 1967). Rhodopsin treated with $NaBH_4$ in the presence of light yields irreversibly bound retinal at an ε-aminolysyl site (Heller, 1968).

In other studies, however, it was claimed "that the prosthetic chromophore in native bovine rhodopsin exists as a chromolipid Schiff base, *N*-retinylidenphosphatidylethanolamine (N-RPE)" (Poincelot *et al.*, 1970). This would imply that in the visual cycle a transfer of retinal occurs from the phospholipid to the polypeptide part of the opsin. The apparent discrepancy over the retinal binding site at the opsin has convincingly been shown to be a matter of different isolation procedures used by the different laboratories (Anderson, 1972). It was demonstrated that *N*-RPE was an artifact formed by transimination during the extraction procedures. Therefore, all the evidence available at present points to an ε-aminolysyl group as the binding site of retinal in the opsin.

3. *The Rhodopsin Cycle*

A simplified version of the overall reaction is given in Fig. 10. When rhodopsin, the Schiff base of the 11-*cis* retinal with opsin, is exposed to light, a primary photoreaction is followed by a series of dark reactions resulting in the liberation of all-*trans*-retinal and opsin. An isomerase converts all-*trans*-retinal into the all-*cis* form, which reacts with opsin, forming rhodopsin to complete the cycle.

The molecular changes on both the protein and retinal level during the rhodopsin cycle are not too well understood. It is as-

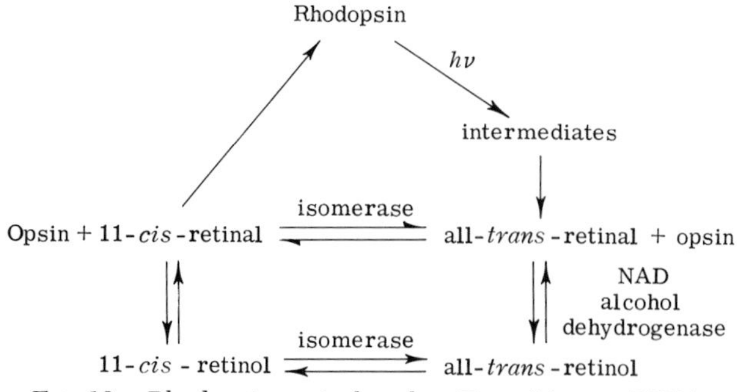

FIG. 10. Rhodopsin–retinal cycle. From Morton (1972).

sumed that 11-*cis*-retinal fits very closely into the protein structure whereas the all-*trans* form fits so loosely that it is readily detachable (Morton and Pitt, 1955; Hubbard, 1958). In the process of the all-*trans*-retinal liberation, the protein itself undergoes conformational changes. Sulfhydryl groups that are "buried" in native rhodopsin become exposed and thereby vulnerable to chemical modification (Wald and Brown, 1951–1952).

4. *Absorption Spectra*

When solutions of rhodopsin are bleached with light, the strong maximum at approximately 500 nm disappears and a new absorption

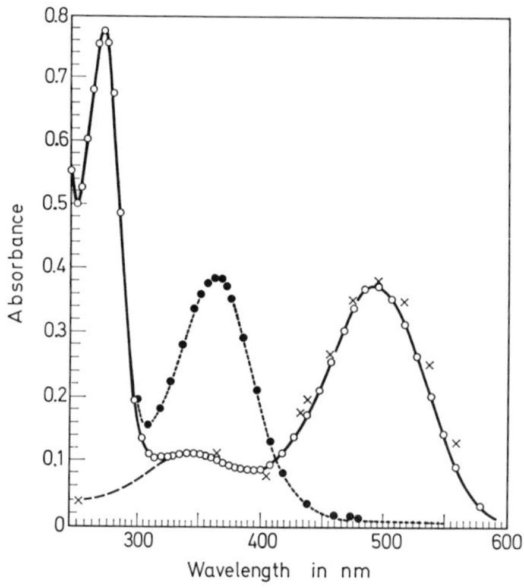

FIG. 11. The spectral absorbance of a nearly pure extract of cattle rhodopsin at pH 9.2 (○—○); and of the bleached extract (●····●) ×, Photosensitivities of frog rhodopsin adjusted to correspond to the absorbance curve at 500 nm. The dashed interpolation (---) gives the probable contribution to absorbance below 320 nm by the rhodopsin chromophore. From Morton (1972).

band at approximately 385 nm is formed (Fig. 11). This type of shift appears to be the major overall spectral change in the rhodopsin cycle, but there is a variety of intermediates many of which can be trapped only at low temperatures. The different spectra are not yet clearly understood in terms of molecular structure (Fig. 12). The absorption maxima of rhodopsins from different animal species may vary between 480 and 560 nm (Bridges, 1967; Hubbard, 1969). This means that visual proteins not only shift the optical absorption of retinal to long wavelengths, but, in addition, the shift in wavelengths can be regulated according to the requirements of different species. Based on the assumption that the common chromophore in all rhodopsins is a Schiff base of retinal with an ε-aminolysyl group of the opsin, several theoretical and experimental investigations have been carried out in an effort to improve the understanding of the bathochromic shift in rhodopsins. The numerous theories have recently been thoroughly discussed (Mantione and Pullman, 1971).

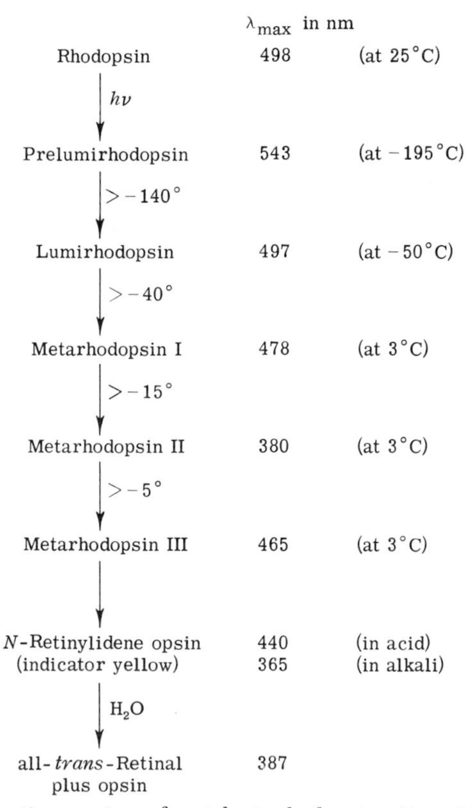

FIG. 12. Absorption maxima of vertebrate rhodopsin. From Morton (1972).

The most attractive of these theories is the "point charge perturbation theory" originally described by Morton and Pitt (1955) as well as Kropf and Hubbard (1958). They postulate that the chromophore is derived from the conjugate acid of the Schiff base (protonated Schiff base) which has been shown to absorb at around 440 nm (Ball *et al.*, 1949). The additional red shifts in the spectrum would be provided by the interaction of the protonated Schiff base with reactive groups on the opsin being able to "increase the mobility of electrons within the molecular orbitals of the chromophore" (Hubbard, 1969).

Quantum mechanical calculations by Mantione and Pullman (1971) show that differential stabilization of the ground and excited states of rhodopsin could be achieved by electrostatic interaction between an external anion with the protonated Schiff base. The magnitude of the predicted red shift is of the right order, and variations of the shift are obtained by variations in local arrangements.

These results are in accord with earlier calculations by Wiesenfeld and Abrahamson (1968) and more recent ones by Waleh and Ingraham (1973). In Fig. 13, the results of these calculations are summarized. The approximation of the negative charge to the protonated nitrogen should result in a hypsochromic shift whereas bathochromic shifts of the longest wavelength absorption are predicted

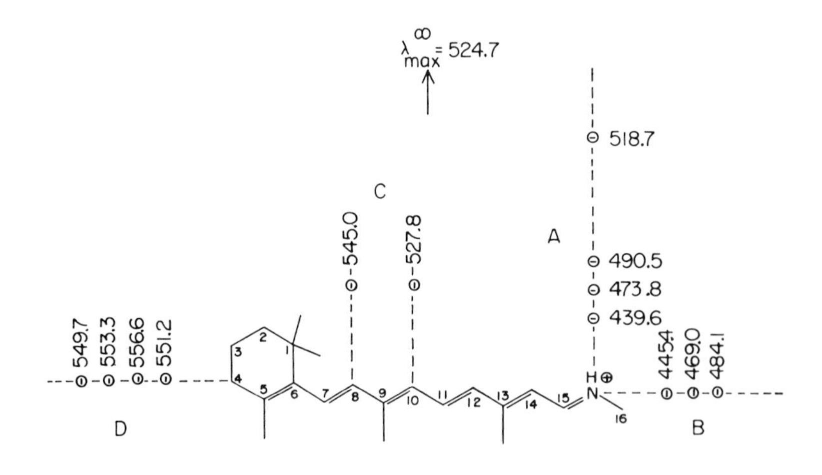

FIG. 13. Calculated absorption maxima for all-*trans*-*N*-retylidene-methylammonium ions when a single negative charge is present in one of the several positions marked on the diagram. From Waleh and Ingraham (1973).

with increasing distance of the negative charge from the nitrogen center along the conjugated polyene chain. Thus the position of the longest wavelength absorption of different rhodopsins and different intermediates in the rhodopsin cycle might be governed by different conformations of the protein(s), thereby placing negative charge(s) at different sites on the retinal moiety. It must be pointed out, however, that the results discussed above, do not exclude alternative mechanisms as suggested by Dartnall and Lythgoe (1965). Only further experimental evidence will prove which molecular interactions between opsin and its retylidene cofactor determine the absorption maxima of its electronic spectra.

C. Collagen and Elastin

Collagen and elastin, two structural proteins occurring in animal tissues, contain cross-links originating from interactions of carbonyl and amino groups, both contributed by or originating from amino acids. These proteins have been studied intensively because of their great importance in human medicine as well as in biology. Abnormalities or deteriorations in the structural elements of tissues are responsible for many debilitating diseases and are particularly critical in growth, in development, and in aging processes. Consequently, the cross-links have been extensively investigated because they have been found to be essential for the maintenance of the unique rigidity or elasticity of these proteins (Weis-Fogh and Andersen, 1970).

Many of the properties and functions of collagen and elastin have been reviewed during the past few years (Bailey, 1968; Franzblau, 1971; Traub and Piez, 1971; Gallop *et al.*, 1972; Tanzer, 1973). Collagen is distributed throughout all animal phyla. Although the cross-links are composed of various types of molecular structures, both the amino acid sequences and the overall physical and chemical properties are highly conserved. Collagen is nevertheless present in such different substances as basement membranes, skin, and the dentine of teeth. A key feature of its structure is that it exists in fibrils that are composed of constituent subunits joined by various types of cross-links. The strength and rigidity of collagen make it an important element in the gross structure and the organs of animals.

Elastin, in contrast to collagen, is apparently restricted to the vertebrates. Its mechanical properties are very different from collagen. While collagen forms rigid structures, elastin forms flexible, rubberlike structures. The cross-links in elastin must therefore provide limits of stretching rather than maintaining rigidity.

Cross-link	Origin	Comments
α,β-Unsaturated "aldol"		
NH_2—CH—COOH $(CH_2)_3$ HC = O=C—C H $(CH_2)_2$ NH_2—CH—COOH	Aldol condensation product of two α-aminoadipic acid δ-semialdehydes	Found only near NH_2-terminal of collagen chains, acting as an intramolecular cross-link; present in elastin
Lysinonorleucine		
NH_2—CH—COOH $(CH_2)_4$ NH $(CH_2)_4$ NH_2—CH—COOH	Reduced Schiff base product of lysine and α-aminoadipic acid δ-semialdehyde	Common to both collagen and elastin
Hydroxylysinonorleucine		
NH_2—CH—COOH $(CH_2)_4$ NH CH_2 CHOH $(CH_2)_2$ NH_2—CH—COOH	Reduced Schiff base product of either lysine and δ-hydroxy, α-aminoadipic acid δ-semialdehyde, or hydroxylysine and α-aminoadipic acid δ-semialdehyde	Abundant in most collagens
N^{ϵ}-Hexosylhydroxylysine		
NH_2—CH—COOH $(CH_2)_2$ CHOH CH_2 NH CH_2 $(CHOH)_4$ CH_2OH	Reduced Schiff base product of hydroxylysine and a hexose	Origin partially from collagen; present in connective tissues which contain polysaccharides
Hydroxymerodesmosine		
NH_2—CH—COOH $(CH_2)_3$ CH = CHOH—CH_2—NH—CH_2—C $(CH_2)_2$ $(CH_2)_2$ NH_2—CH—COOH NH_2—CH—COOH	Reduced Schiff base product of unsaturated "aldol" and hydroxylysine	Analogous to merodesmosine of elastin

Structure	Origin	Occurrence
Dihydroxylsinonorleucine		
NH_2—CH—COOH $(CH_2)_2$ CHOH CH_2 NH CH_2 CHOH $(CH_2)_2$ NH_2—CH—COOH	Reduced Schiff base product of hydroxylysine and δ-hydroxy, α-amino-adipic acid δ-semialdehyde	Most abundant in mineralized collagens
N^{ϵ}-Glucitol-lysine, N^{ϵ}-mannitol-lysine		
NH_2—CH—COOH $(CH_2)_4$ NH CH_2 $(CHOH)_4$ CH_2OH	Reduced Schiff base product of lysine and either glucose of mannose	Origin unknown; present in "older" connective tissues
Aldol-histidine[c]		
NH_2—CH(COOH)—CH_2— [imidazole ring, N—] HC—$(CH_2)_3$—CH(NH_2)—COOH HC—CH(CH_2OH)—$(CH_2)_2$—CH(NH_2)—COOH	Michael addition product of unsaturated "aldol" and histidine, isolated after reduction	Abundant only in cow skin collagen
Histidino-hydroxymerodesmosine		
NH_2—CH(COOH)—CH_2— [imidazole ring, N—] HC—$(CH_2)_3$—CH(NH_2)—COOH HC—CH—$(CH_2)_2$—CH(NH_2)—COOH CH—CH_2—NH—CH_2—CHOH—$(CH_2)_2$—CH(NH_2)—COOH	Reduced Schiff base product of aldol-histidine and hydroxylysine	Abundant in most collagens, isolated as two isomeric forms

[a] From Tanzer (1973). Copyright 1973 by the American Association for the Advancement of Science.

[b] The substances shown are in several cases formed by borohydride reduction of the natural, unsaturated cross-links.

[c] There is also some doubt whether histidine adds to the aldol condensation product *in vivo* (Robins and Bailey, 1973).

The Nature of the Cross-links in Collagen and Elastin

From the standpoint of comparative biochemistry, the cross-links in collagen and elastin are good examples for the development of a specialized molecular structure from the building blocks commonly present in biological materials. Each has a special type of cross-link derived from lysine residues.

In collagen, the enzyme lysyl oxidase converts lysine or hydroxylysine to the corresponding α-aminoadipic acid δ-semialdehydes (Fig. 14). The lysine residue at position 5 from the amino terminus of the polypeptide chain (three such chains exist in a helix to form a long straight molecule) is one of those sensitive to oxidation and cross-linkage, but they can cross-link intermolecularly with lysine residues in other positions. Within the collagen molecule, two of the aldehydes may form an α,β-unsaturated aldol condensation product to generate an intramolecular cross-link. This product is stable at physiological pH and relatively stable to base hydrolysis, but it is destroyed under the conditions of acid hydrolysis. The unsaturated aldol can form cross-links, most probably between adjacent molecules, in a Michael addition reaction with histidine residues (Table IV) (Tanzer *et al.*, 1973; Hunt and Morris, 1973).

The majority of the cross-links in both collagen and elastin are formed by carbonyl–amine reactions (Table IV). The amino acid

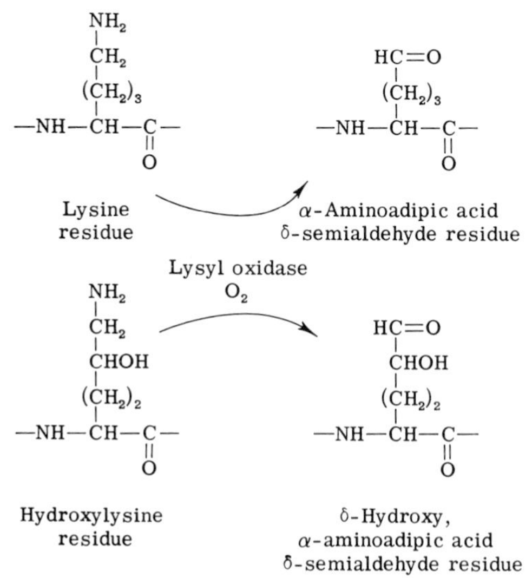

FIG. 14. Formation of collagen aldehydes by enzymic action of lysyl oxidase on collagen molecules. From Tanzer (1973).

residues taking part in the cross-links have been found to be lysine and hydroxylysine. The carbonyl compounds identified so far, include α-aminoadipic acid δ-semialdehyde, its δ-hydroxy derivative, and the sugars glucose and mannose (Tanzer, 1973). In elastin, a series of carbonyl–amine and aldol condensations can finally lead to the formation of the very stable pyridinium compounds desmosine and isodesmosine. They are not present in collagen (Fig. 15). Since the Schiff base cross-links derived from the carbonyl–amine reaction are readily reducible by exogenous reagents, it is important to know whether they may also become naturally reduced. This appears to be the case in both natural and reconstituted collagen fibrils. In the former case, isotope dilution studies, in which $NaBD_4$ was used in conjunction with mass spectral analysis, showed that 25–50% of hydroxylysinonorleucine and dihydroxylysinonorleucine became reduced *in vivo* (Tanzer, 1973). In studies of reconstituted fibrils, in which collagen had been biosynthetically labeled with [^{14}C]lysine, four nonreducible cross-links developed progressively and one of the cross-links incorporated protons from the aqueous solvent.

The understanding of the biological processes leading to the cross-linking in the collagen molecule has been greatly helped by investigation of a nutritional disease in man and animals, lathyrism. It is caused particularly by the consumption of the sweet pea, *Lathyrus odoratus*. β-Aminopropionitrile was shown to be the toxic component, causing defects in the structure of connective tissue by the formation of abnormal cross-links (Levene and Gross, 1959; Tanzer, 1965). In further studies it could be demonstrated that β-aminopropionitrile inhibits the enzyme, lysine oxidase, thereby preventing the formation of α-aminoadipic acid δ-semialdehyde, the major aldehyde component present in the normal cross-links (Pinnell and Martin, 1968). Another inhibitor in the biosynthesis of collagen was found to be penicillamine (Nimni, 1968; Deshmukh and Nimni, 1969). This compound possibly prevents the formation of the natural cross-links by competing with lysyl residues present in collagen for lysine-derived aldehydes (Schiffmann and Martin, 1970).

Investigations on the formation of elastin have proved that the ultimate synthesis of desmosine and isodesmosine is retarded by a copper deficiency (Carnes, 1971). Most likely copper deficiency lowers the activity of the enzyme lysine oxidase, thereby preventing the formation of α-aminoadipic acid δ-semialdehyde.

In order to get further insight into the structure–function relationship for proteins like collagen and elastin, it would be highly desirable to learn more about the location of the cross-links in the

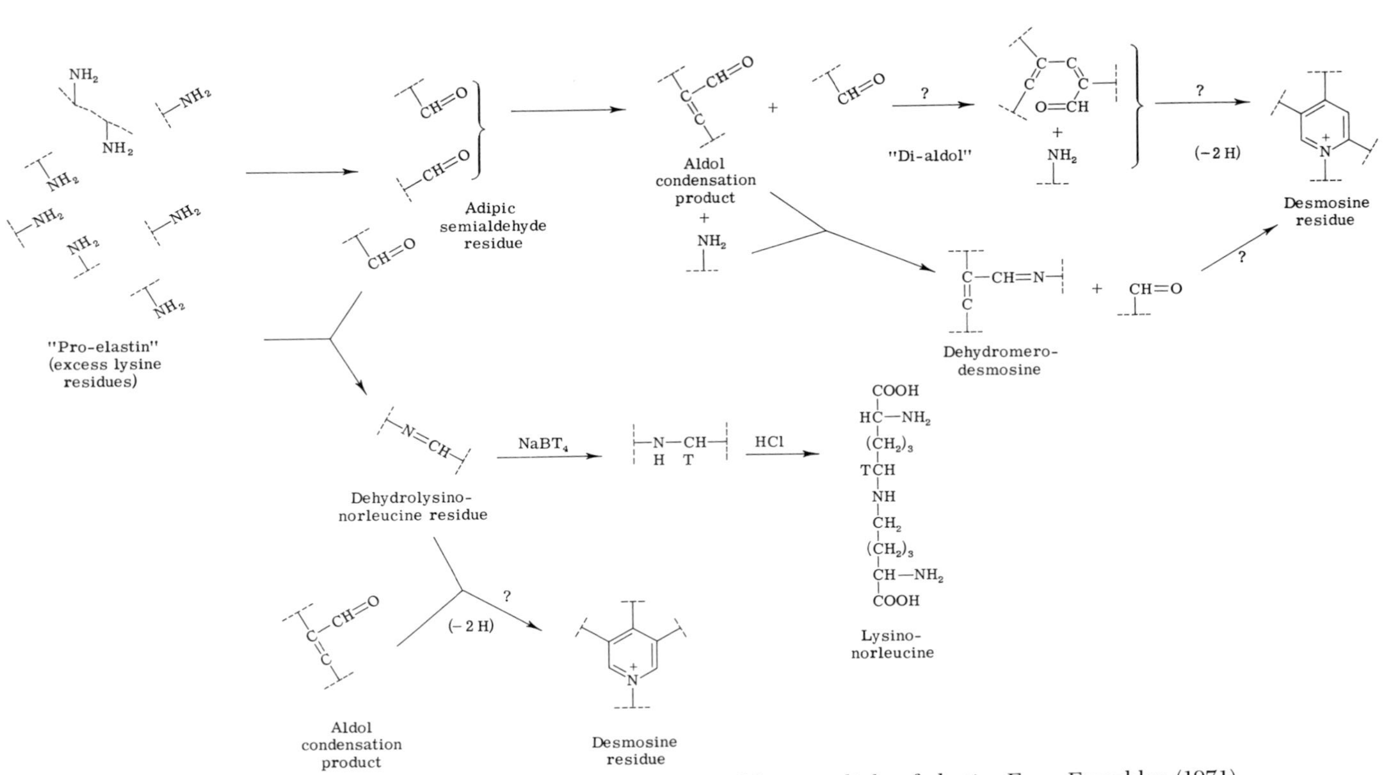

FIG. 15. Possible interrelated scheme for biosynthesis of the cross-links of elastin. From Franzblau (1971).

polypeptide structure and their abundance in the protein. Only then it will be possible to understand their different mechanical properties on the molecular level.

V. Naturally Occurring Deteriorative Reactions

There are many deteriorative reactions of biological materials which involve carbonyl-amine reactions. Most of these occur between the carbonyl groups of carbohydrates and the amino groups of amino acids in proteins, but there are also many that occur between other carbonyl compounds and various types of naturally occurring amines. There is a vast literature on this subject, most of which is not usually encountered by protein chemists unless they have been associated with areas related to food research or industrial products. But some of the interactions also occur in biological tissues of living organisms and may be of great importance to the organisms' general physiology and well-being.

A. *The Maillard Reaction*

Interactions of sugars with proteins to produce insoluble substances have been known for many years. The chemical nature of the interaction was apparently first discerned by Maillard (1912). He described the production of darkly colored compounds when solutions of sugars and amino acids were heated. He then extended these studies to different sugars and amino acids and found that it was the reducing group in the sugar that was important and that different sugars and different amino acids reacted at very different rates (Maillard, 1916, 1917). It was shown nearly two decades later that proteins reacted similarly (Ramsey *et al.*, 1933) and still later that the free amino groups of proteins were necessary (Mohammed *et al.*, 1949). The reaction has been named the Maillard reaction; other names used are nonenzymic browning and the browning reaction.

There are many articles and reviews concerning the Maillard reaction. Most of the investigations have been directed at controlling the Maillard reaction in the handling and storage of foods, in particular, dehydrated foods (Reynolds, 1969; Gottschalk, 1972). The Maillard reaction usually causes undesirable effects, such as large changes in solubility and color. However, some of the side reactions are responsible for the production of desirable as well as undesirable flavors and odors (Hodge, 1967; Streuli, 1967).

1. *Various Steps Leading to Maillard Products*

The relative importance of many different steps and side reactions encountered in the Maillard reaction has been extensively studied and many of the key intermediates have been characterized, but the later stages of the reaction are still not well understood. In describing the sequence of reactions, we prefer to use a relatively simple classification given by Reynolds (1969). She has suggested the following sequence of events: (1) The reversible formation of glycosylamine. (2) The rearrangement of the glycosylamine to the ketosamine, 1-amino-1-deoxyketose (Amadori rearrangement), or to the aldosamine, 2-amino-2-deoxyaldose (Heyns rearrangement). (3) The formation of a diketosamine or a diamino sugar. (4) Degradation of the amino sugar, usually initiated by the loss of one or more molecules of water to form amino or carbonyl intermediates. (5) Reaction of amino groups with the intermediates formed in step 4 and subsequent polymerization of these products to brown pigments and other substances.

2. *The Mechanism of the Maillard Reaction*

Reaction mechanisms for steps 1–4 have been established for the Maillard reaction, but the terminal step 5 leading to the various browned products is apparently so complicated that no satisfactory mechanism has been agreed upon. Initially, the reaction is represented by the primary step of most carbonyl–amine reactions (Section II,C,1), the formation of an α-amino alcohol, an N-substituted glucosamine in the case of glucose. The reaction principally (see Section II,C,1) has its highest rate under weakly acidic conditions. Amino acids, however, provide their own acid catalyst, and the reaction is rapid even in the absence of added acid. N-substituted glycosylamines readily hydrolyze in dilute acetic acid; many even hydrolyze relatively fast when dissolved in water at room temperature.

The next step in the Maillard sequence is uncommon to most carbonyl–amine reactions. This is either the Amadori or Heyns rearrangement. These closely related rearrangements are key reactions in the Maillard sequence and are shown in Eqs. (21) and (22) (Reynolds, 1965). The aldosamines are converted to a 1-1 deoxyketose, while the ketosamines are transformed to 2-amino-2-deoxyaldoses. The rearrangement reactions are acid catalyzed, the carboxyl groups of amino acids providing the internal acid catalyst. In the Amadori rearrangement (Fig. 16) formation of protonated Schiff base with subsequent prototropic shifts is the critical step for ketosamine formation. A ketosamine derived from a primary amine can react with another molecule of aldose. The product can undergo an Amadori rearrangement to give a diketosamine. Apparently the only

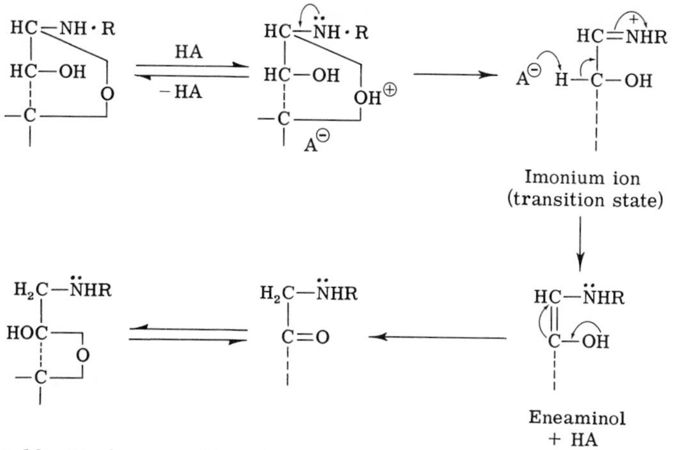

Imonium ion (transition state)

Eneaminol + HA

FIG. 16. Mechanism of Amadori rearrangement. From Gottschalk (1972).

crystalline compound of this type isolated at present is difructose glycine. The reactions of diketosamines are very similar to those of ketosamines with the notable exception that diketosamines readily decompose in water (Fig. 17).

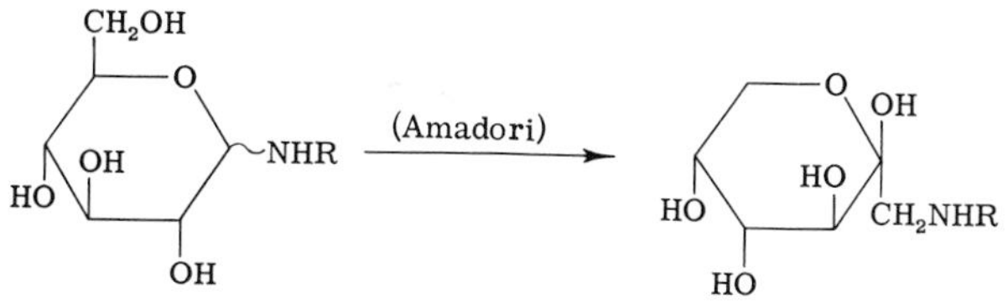

(Amadori) (21)

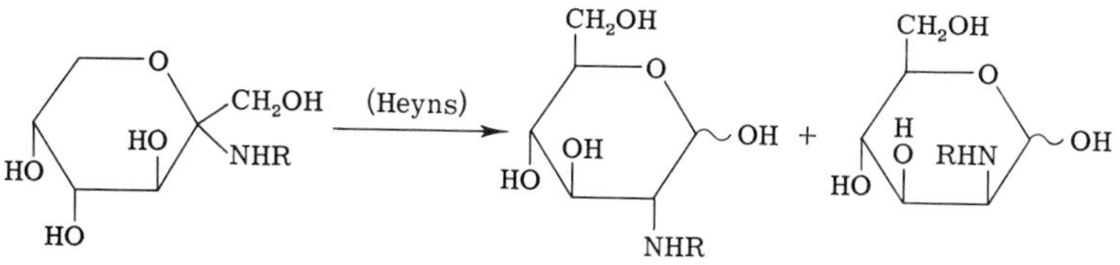

(Heyns) (22)

Dialdehydes and α,β-unsaturated aldehydes have been identified as degradation products of amino sugars. Their condensation with unreacted amino groups in the protein, mainly ϵ-aminolysine residues, ultimately leads to high molecular weight brown polymers and pigments.

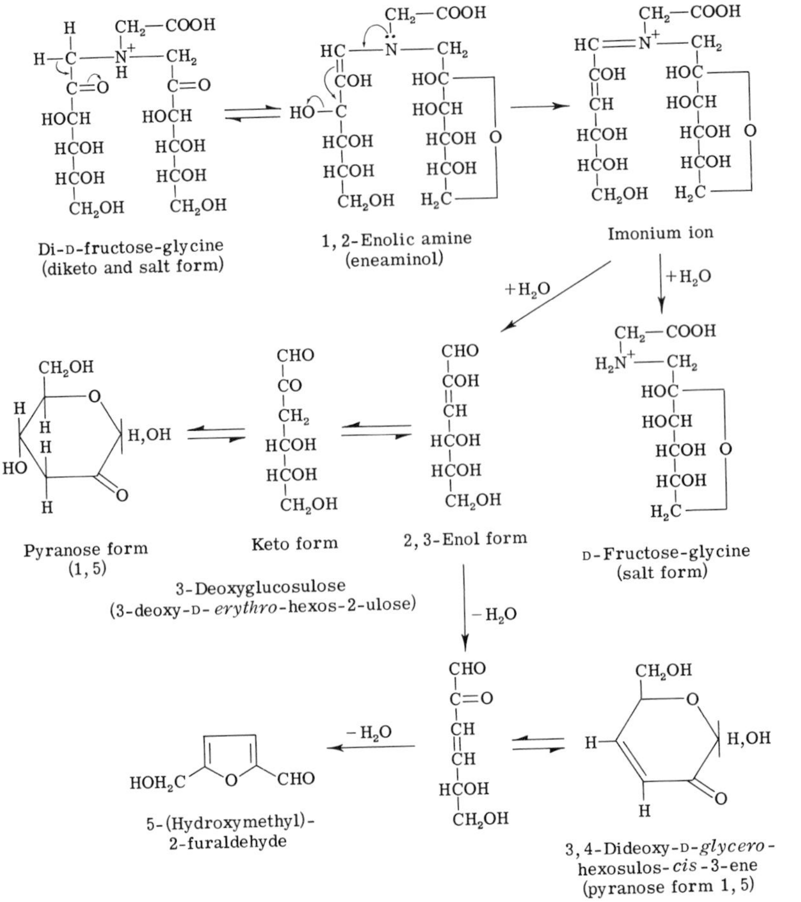

Di-D-fructose-glycine (diketo and salt form)

1,2-Enolic amine (eneaminol)

Imonium ion

$+H_2O$

$+H_2O$

Pyranose form (1,5)

Keto form

2,3-Enol form

3-Deoxyglucosulose (3-deoxy-D-*erythro*-hexos-2-ulose)

D-Fructose-glycine (salt form)

$-H_2O$

$-H_2O$

5-(Hydroxymethyl)-2-furaldehyde

3,4-Dideoxy-D-*glycero*-hexosulos-*cis*-3-ene (pyranose form 1,5)

FIG. 17. Reaction mechanism of the degradation of di-D-fructoseglycine to 5-(hydroxymethyl)-2-furaldehyde and D-fructoseglycine. From Gottschalk (1972).

B. Reactions of Sugars in Foods

The Maillard reactions have potentially devastating effects on the properties of proteins (Reynolds, 1965). In food products the Maillard reaction usually causes large changes in solubility as well as the formation of fluorescent and darkly colored products.

1. Food Proteins

The reaction of sugars, including glucose and fructose, and aldehydes with some of the more common food proteins have been studied. When casein is treated with glucose, most of the lysyl residues are modified whereas only a small percentage of other amino acids react (Table V). They are arginine, histidine, tyrosine, and methionine (Lea and Hannan, 1950a,b). In the earlier stages of the modification, when the lysine is probably present as the glycosylamine (Fig. 16), it is not nutritionally available, but is liberated on acid hydrolysis. Hence amino acid analysis gives a falsely high estimate of the nutritionally available lysine. Since lysine deficiency is one of the main causes of protein malnutrition, this error can be important. Carpenter (1960) has shown that a chemical test based on the ability of the lysine residues to react with fluorodinitrobenzene correlates much better with the amount of lysine that is nutritionally available than does the total content after acid hydrolysis (for review, see Carpenter, 1973).

Changes in solubility and browning of casein have also been observed after treating it with glucose. Bjarnason and Carpenter (1969, 1970) have studied ways of blocking losses of lysine occurring on

TABLE V
Destruction or Combination of Amino Acid Residues in the Reaction between Glucose and Casein[a,b]

Amino acid	Initial content moles/10^5 g of dry casein[b]	Moles combined or destroyed/10^5 g of dry casein			
		Estimated in protein		Estimated in acid hydrolyzate	
		Sample 5D[c]	Sample 30D[d]	Sample 5D[c]	Sample 30D[d]
Lysine	56.4	37.8	51.3	10.7	17.5
Arginine	22.1	1.8	15.7	1.5	15.3
Histidine	20.6	0.6	5.8	0.8	5.8
Tyrosine[e]	34.9	4.2	12.9	2.1[f]	4.2[f]
Methionine	17.6	4.2	10.0	0.4	0.9
Tryptophan	7.3	0.0	0.0	0.0[f]	0.0[f]
Glutamic acid	151.0	—	—	0.0	0.0
Phenylalanine	40	0.0	0.0	—	—
		48.6	95.7	15.5	43.7

[a] Adapted from Lea and Hannan (1950b). From Reynolds (1965).
[b] Dry, ash-, and sugar-free basis.
[c] Sample 5D prepared from sodium caseinate and glucose (4 moles/free amino nitrogen) held 5 days at 37°C and 70% RH under nitrogen; excess glucose removed by dialysis; freeze-dried.
[d] Sample 30D, prepared as 5D, stored 30 days.
[e] Determined by α-nitroso-β-naphthol; Millon's reagent indicated no loss.
[f] Determined on alkali hydrolyzates.

TABLE VI

A Comparison of the Results from the Rat Growth and Metabolism Experiments, Using Acylated Materials[a]

Material	Percent of lysine ε-NH_2 groups in sample[b] unreacted (a)	Relative activity of substrates for rat kidney ε-lysine acylase (Leclerc and Benoiton, 1968) (b)	Estimated partition of the lysine from the test materials in the rat experiments: Percent available for growth (c)	Percent excreted: Fecal (d)	Percent excreted: Urinary (e)	Sum of c + d + e (f)
Formyl						
Lysine	–	(94)	–	–	–	–
Protein (lactalbumin)	28	–	77	–	–	–
Acetyl						
Lysine	0	(100)	50	18	18	86
Protein (BPA)[c]	15	–	67	–	–	–
Propionyl						
Lysine	0	(0)	0	19	41	60
Protein (lactalbumin II)	2	–	43	(–16)	41	84[d]
Control protein (BPA)	91	–	85	14	1	100
Heat-damaged protein (BPA)	24	–	13	66	4	83

[a] From Bjarnason and Carpenter (1969).

[b] For the amino acids the zero values were based on the failure to find free lysine by thin-layer chromatography; for the proteins the direct FDNB-available lysine values were taken as the measure of unreacted lysine.

[c] BPA, bovine plasma albumin.

[d] This value is the sum of c and e alone, without the negative value for d subtracted.

heating, drying, and storage of food proteins. Acylation considerably lowered the degree of the Maillard reaction. Propionylated lysines, however, were not nutritionally utilized in rat diets, but acetylated and formylated proteins were at least partly utilized (Table VI).

One of the more common methods of retarding the nonenzymic browning in food products has been sulfur dioxide treatment or the addition of sulfite, either dissolved or suspended in water. The initial product is the sugar–sulfite adduct that effectively blocks the reaction of sugars with amines. Sulfite has widely been used by the food industry. However, it is currently under extensive attack because of its possible toxic and even mutagenic effects. Sulfite can add to the 5-6 double bond of cytosine, destroying its aromatic character, and favoring imine structure in the 4-amino group, which is labilized and hydrolyzes. Departure of the sulfite completes the conversion of cytosine into uracil.

2. *Reactions in Eggs during Incubation*

Food products containing dissolved proteins may undergo Maillard reactions. Within four days of incubation of birds' eggs at 37° (Feeney *et al.*, 1964) the gel electrophoretic patterns of the egg-white proteins change significantly, indicating the loss of basic groups. This effect was most pronounced on ovotransferrin, a protein that contains a large amount of lysyl ε-amino groups (~50 amino groups per mole of protein), many of which can be readily chemically modified. The reaction was shown to be caused by the egg-white glucose. Therefore great care should be taken to avoid these reactions in biochemical and genetic investigations of egg proteins.

3. *Reaction of Sugars with Amino Groups of Lipids and Lipoproteins*

Amino groups in lipids and lipoproteins have also been found to undergo Maillard-type reactions. One of the more classical examples occurring in a food is the reaction of glucose with the amino group of ethanolamine in the phospholipids in dried eggs (Kline and Sonada, 1951; Kline *et al.*, 1951a,b). This is the reaction which is responsible for the vile flavor and odors of dehydrated eggs produced before the middle 1950s and so ingrained in the memories of many members of overseas contingents of the American armed forces in World War II.

4. *Flavors and Odors Produced by the Maillard Reaction in Food Products*

Hodge *et al.* (1972) have recently reviewed the structure of flavor compounds formed in browning reactions from cooked grains and meals, roasted malt, nuts, coffee, chicory, cocoa, fruits, and vegetables. Two classes of compounds could be characterized as having distinctive aromas. Nitrogen-heterocyclic compounds derived from sugar–amine interactions are responsible for nutty-, corny-, and bread-aroma flavors (Fig. 18). Oxygen-heterocyclic compounds are mainly responsible for caramel aroma (Fig. 19). Most of these characteristic aromas are formed through the Maillard reaction by the thermal degradation of Amadori compounds of amines and amino

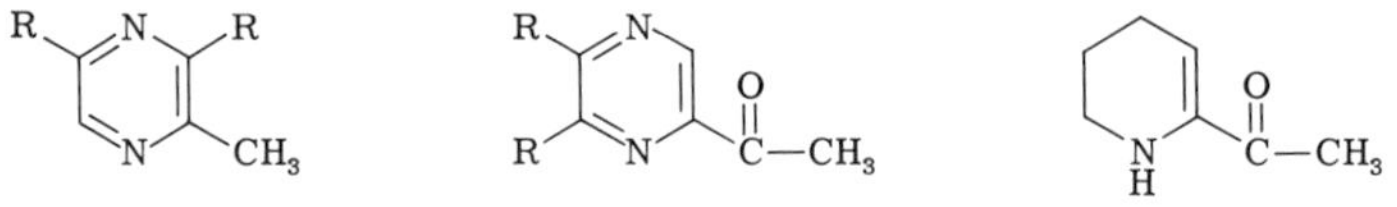

R = H, Me, or Et
Alkyl pyrazines
[roasted, nutty]

R = H or Me
2-Acetopyrazines
[popcorny]

CH-2-aceto-Δ^2-piperideine
(fresh bread, cracker)

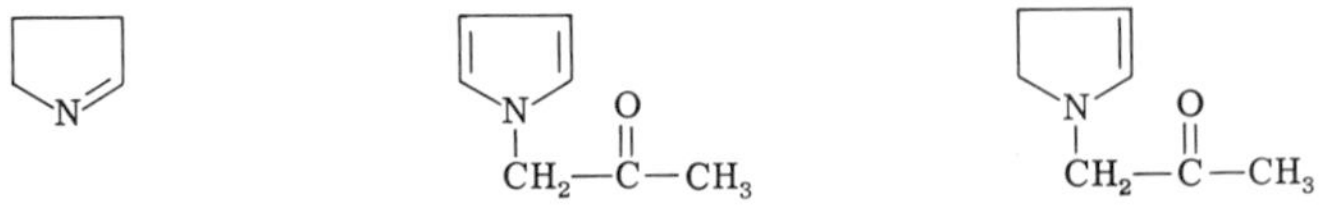

1-Pyrroline
[corny]

N-Acetonylpyrrole

N-Acetonylpyrroline
[bready]

FIG. 18. *N*-Heterocyclic nutty-, corny-, bready-aroma compounds. From Hodge *et al.* (1972).

R = H, R = OH

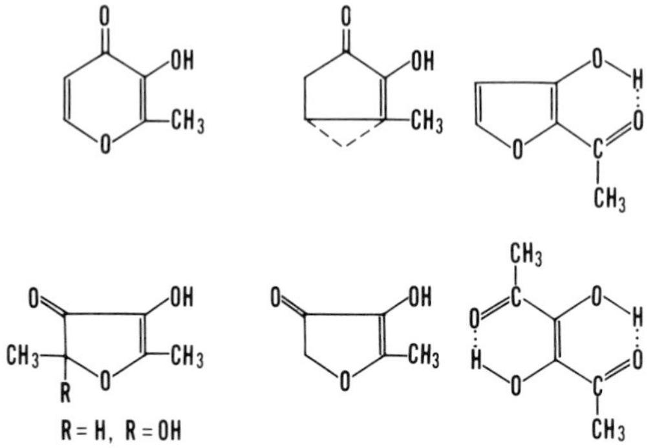

FIG. 19. *O*-Heterocyclic and alicyclic caramel-aroma compounds. From Hodge *et al.* (1972).

acids. Among the various amino acids, proline provides important compounds of bready aroma.

C. Glucosyl Peptides in Hemoglobin A_{Ic}

Holmquist and Schroeder (1966) showed that the N-terminus of the β-chains of hemoglobin A_{Ic}, a minor component of human hemoglobin, was blocked by an alkali labile group which lowered its p*K*. Since reduction by $NaBH_4$ made this group alkali stable, they suggested that these properties could be explained by the presence

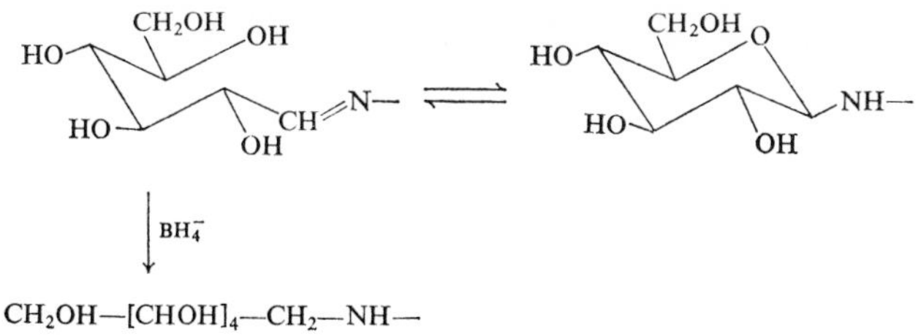

HO CH$_2$OH OH HO OH CH=N— ⇌ HO CH$_2$OH O HO OH NH—

↓ BH_4^-

$CH_2OH—[CHOH]_4—CH_2—NH—$

FIG. 20. Interconversions of imines of sugars. From Dixon (1972).

of a Schiff base. Bookchin and Gallop (1968) characterized the terminal peptide of $NaBH_4$-reduced hemoglobin A_{Ic} as possessing the $CH_2OH—[CHOH]_4—CH_2$-residue on the α-amino group of the terminal valine.

Dixon (1972) studied the reaction of glucose with the peptide valylhistidine and isoleucyltyrosine and was able to show that the reaction product of valylhistidine was identical to that described by Holmquist and Schroeder (1966). Its properties are consistent with its formulation as a glycosylamine in equilibrium with a Schiff base (Fig. 20). This formulation would explain both the sensitivity to borohydride, already ascribed to the imine structure, and, if equilibrium greatly favored the glycosylamine form, the observed stability in aqueous solution. From the estimated equilibrium constant at neutral pH for the dissociation of glycosylvalylhistidine to valylhistidine and glucose, Dixon (1972) concluded that no more powerful glycosylating agent than blood glucose has to be postulated for the *in vivo* conversion of 6% of hemoglobin A into hemoglobin A_{Ic}, which is the extent to which the latter is found in human blood.

D. *Deteriorative Processes in Tissues Related to Aging*

During the last decade several different routes of investigation have led to the conclusion that carbonyl–amine reactions are highly significant in the deterioration of biological systems. Based on studies of egg-white proteins during the incubation of eggs, our laboratory (Feeney *et al.*, 1964) suggested that the sugar–amine reaction may occur not only in proteins, but also in lipids and nucleic acids. The principal advancement, however, has been the demonstration that carbonyl compounds are formed in the aging process of lipids leading to carbonyl–amine reactions in tissues (Tappel, 1973).

1. *Lipid Autoxidation and Interactions with Proteins*

The principal products in the aging process of lipids giving rise to carbonyl–amine reactions have been shown to be dialdehydes, mainly malonaldehyde. Their protein reactions have been discussed in Section III,C.

2. *Carbonyl–Amine Reactions in Membranes Related to Autoxidation*

The complex structure of membranes and their contents of both highly unsaturated lipids, phospholipid amino groups, and protein amino groups might be expected to provide optimal conditions for deteriorative reactions involving oxidatively produced carbonyl groups and amino groups. There is now much information on this subject, particularly on the formation of lipid soluble, fluorescent products from different tissues (Tappel, 1972, 1973; Minssen and Munkres, 1973; Harmon, 1969; Passwater, 1970; Kohn, 1970; Hochschild, 1970; Bender *et al.*, 1970). Tappel (1973) has demonstrated the importance of antioxidants in the diet of rats to prevent some of the deteriorative effects in their tissue.

In an *in vitro* study, Minssen and Munkres (1973) have shown that in the handling and preparation of mitochondrial membrane proteins from *Neurospora crassa*, carbonyl–amine products can be found unless precautions are taken against lipid autoxidation. The products were identified as 1-amino-3-iminopropene groups formed by the reaction of protein amino groups with malonaldehyde, which apparently was formed by autoxidation. This reaction could be prevented by the addition of antioxidants during the isolation procedure

TABLE VII
Concentration of Malonaldehyde in Mitochondrial Membranes of Neurospora crassa *after Preparation of the Membranes with and without Antioxidant*[a,b]

Origin of sample	Replicate No:	Malonaldehyde/protein (ng/mg) 1	2	3
Preparation + antioxidant		+2	−2	−1
Preparation − antioxidant		26	25	22
Preparation − antioxidant (antioxidant was added to the preparation just before malonaldehyde analysis)		–	25	25

[a] From Minssen and Munkres (1973).
[b] Membranes were isolated in parallel simultaneous procedure with and without the antioxidant, 3,5-di-*tert*-butyl-4-hydroxybenzylalcohol, in the isolation media. Malonaldehyde was obtained from the proteins by acid hydrolysis of iminopropene bonds and determined fluorometrically after reaction with sulfonyldianiline.

(Table VII). Several of the unusual properties found in aged membrane proteins, such as low solubility, resistance to enzyme hydrolysis, and changes in isoelectric point and electrophoretic mobility, were directly attributed to the consequences of lipid autoxidation processes. As a result of this study, it was strongly suggested that precaution against oxidation must be taken in isolating mitochondrial proteins to prevent the formation of artifacts.

E. Possible Roles of Diamine, Polyamines, and Their Carbonyl Derivatives in Biological Systems

The properties and metabolisms of diamine and polyamines have received extensive study during the last decade (Tabor and Tabor, 1972), especially in their interaction with nucleic acids. Putrescine [$NH_2(CH_2)_4NH_2$], spermidine [$NH_2(CH_2)_4NH(CH_2)_3NH_2$], and spermine [$NH_2(CH_2)_3NH(CH_2)_4NH(CH_2)_3NH_2$] have been found in microorganisms, plants, and animals.

1. Oxidations of Spermine and Spermidine

Spermine and spermidine are oxidized by a beef serum amine oxidase to carbonyl derivatives by an oxidative deamination [Eq. (23)]. These carbonyl derivatives are toxic to a variety of cells.

$$\underset{\text{Spermidine}}{NH_2(CH_2)_4NH(CH_2)_3NH_2} \xrightarrow{O_2} NH_2(CH_2)_4NH(CH_2)_2CHO + NH_3 + H_2O_2$$

$$\underset{\text{Spermine}}{NH_2(CH_2)_3NH(CH_2)_4NH(CH_2)_3NH_2} \xrightarrow{2\,O_2} OHC(CH_2)_2NH(CH_2)_4NH(CH_2)_2CHO + 2\,NH_3 + 2\,H_2O_2 \qquad (23)$$

2. Toxicity of Oxidized Spermine

Oxidized spermine has recently been reported to have a phagocidal action on T_5 bacteriophage. This action was not given by decomposition products such as acrolein (Nishimura *et al.*, 1972).

F. Reactions of Gossypol from Cottonseed Meal

Gossypol is a yellow phenol found in the pigment glands of the seed of germs of cotton, *Gossypium* (Singleton and Kratzer, 1973). Chemically it is 1,1′,6,6′,7,7′-hexahydroxy-5,5′-diisopropyl-3,3′-

dimethyl-(2,2′-binaphthalene)-8,8′-dicarboxyaldehyde (XII, Singleton and Kratzer, 1973). Gossypol has become particularly important

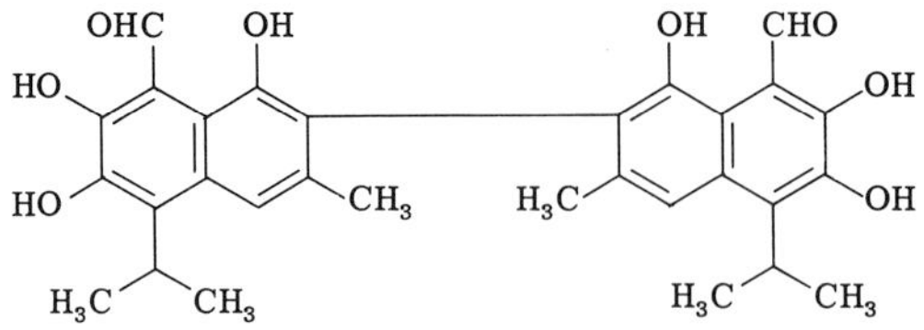

OHC OH OH CHO
HO OH
HO CH_3 H_3C OH
H_3C CH_3 H_3C CH_3
(XII)

because cottonseed meal is used as a protein supplement for animal and human diets. The Instituto de Nutricion de Centro America y Panama (INCAP) has developed, and made highly successful feeding tests with, Incaparina, a food containing 38% cottonseed flour. Unfortunately, gossypol is toxic when fed in the diets of some animals. When pigs are fed toxic levels of gossypol, they may appear normal for long periods and then suddenly develop breathing problems and die in a few days with severe anemia and other complications. Hence gossypol must either be removed from the cottonseed meal or converted into a nontoxic form when foods such as Incaparina are produced.

The main toxic effects of gossypol are thought to be caused by reaction between its aldehyde groups and the amino groups of amino acids and proteins (Singleton and Kratzer, 1973). Schiff bases of gossypol and lysine ε-amino groups in a protein have recently been characterized with pepsinogen and pepsin (Finlay *et al.*, 1973). The reaction with pepsinogen prevented its activation to pepsin and was irreversible; it was slowed by the presence of pyridoxal phosphate. The rate of inactivation was independent of gossypol concentration, so there must be a rate-determining change in pepsinogen before gossypol can combine. Reaction with gossypol also inhibited pepsin, apparently reversibly, since the inhibition was competitive with substrate (K_i about 50 μM).

VI. Commercial Applications

A. *Use of Aldehydes in Tanning*

The tanning action of certain aldehydes, particularly formaldehyde, has been known for many years. Many different recipes have been used, and significant progress has been achieved by using new aldehydes for the tanning of leather. Nevertheless, aldehyde tanning might be aptly described as an art at this time because the molecular interactions responsible for tanning are at best only partly understood. Most of the investigations performed in former years dealt with simple aldehydes, such as formaldehyde. Most other aldehydes and ketones are less reactive and have therefore not been used in industrial tanning. Straight-chain dialdehydes, however, in contrast with saturated monoaldehydes, appear to maintain tanning power as their molecular weight is increased. Oxidation with periodate converts starch into a substance of high molecular weight with repeating dialdehyde units, which has been shown to be a very useful tanning agent [Eq. (24)] (Filachione, 1960). General descriptions of the tanning action of aldehydes are available in several references (French and Edsall, 1945; Bowes, 1948; Gustavson, 1956; Fein and Filachione, 1960; Weakley *et al.*, 1961).

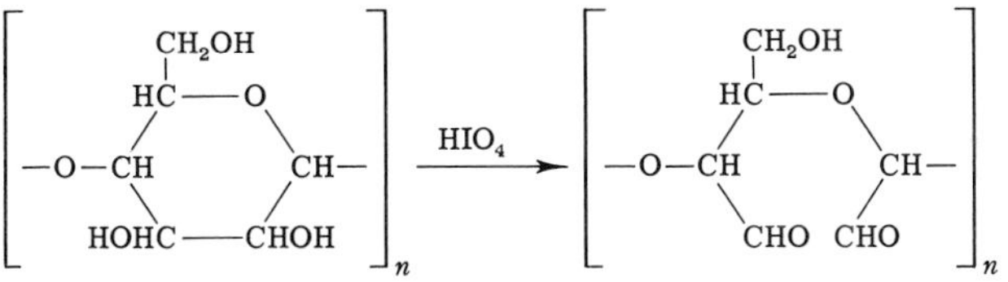

$$\left[\begin{array}{c} CH_2OH \\ | \\ HC\text{—}O \\ / \quad \backslash \\ \text{—O—}CH \qquad CH\text{—} \\ \backslash \quad / \\ HOHC\text{—}CHOH \end{array}\right]_n \xrightarrow{HIO_4} \left[\begin{array}{c} CH_2OH \\ | \\ HC\text{—}O \\ / \quad \backslash \\ \text{—O—}CH \qquad CH\text{—} \\ \backslash \quad / \\ CHO \quad CHO \end{array}\right]_n \tag{24}$$

The purpose of tanning is to improve the properties of leather in a way to make it suitable for manufacturing and use. In most cases, the product must remain pliable rather than become brittle over a comparatively wide range of temperatures. One of the main tanning agents used in industry is chromium oxide (Cr_2O_3) in alkaline solution, sometimes in conjunction with aldehydes. One of the functions of aldehydes in tanning appears to be the formation of cross-links between and within protein molecules. This would explain why formaldehyde is much more effective than its higher homologs. Dialdehydes, however, readily form cross-links, and this may be why they are all active in the tanning process. Malonaldehyde and succinaldehyde tan best between pH 3.5 and 4.5. Glutaraldehyde is the most reactive among the dialdehydes. The rates of tanning at approximately pH 8 of some aldehydes are given in Fig. 21.

B. Formation of Toxoids

One of the earlier applications of chemically modifying proteins was the use of formaldehyde by the pharmaceutical industry to modify bacterial toxins and viruses. The purpose of this treatment was to inactivate or change the virus or toxin to render it incapable of being reproduced or of causing pathological responses, while still retaining its ability to produce an immunogenic response when injected into an animal. When bacterial toxins are modified this way, they are called toxoids. Procedures similar to those used commercially nearly 40 years ago are apparently still applied. The selected reac-

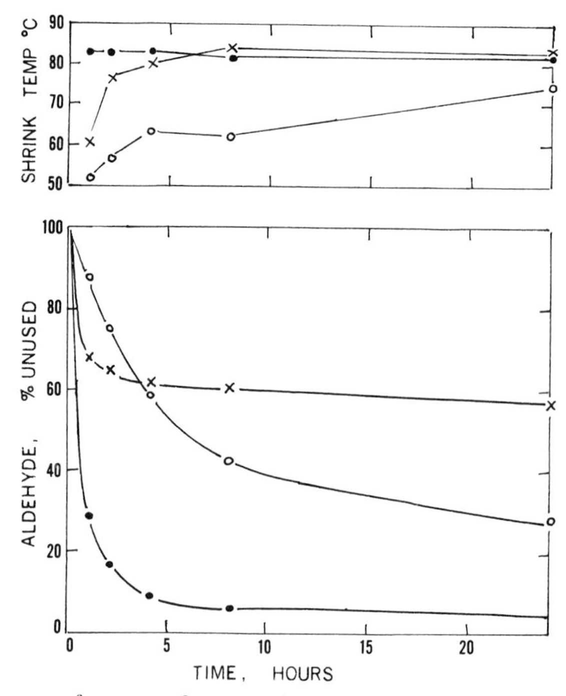

FIG. 21. The rate of tanning of various aldehydes at pH about 8. Buffer: $NaHCO_3$, pH 7.3–8.4. ×—×, Formaldehyde; ○—○, glyoxal; ●—●, glutaraldehyde. Adapted from Filachione (1960).

tion conditions were primarily a result of trial and error. In a typical process, filtrates containing toxins were treated with 0.10–0.20% formaldehyde at 37–39° for several weeks. The chemistry of the formaldehyde reaction is probably very closely related to that in the tanning process. The reaction conditions for any particular toxin must be rigorously followed because too little treatment could result in retention of toxicity, whereas too extensive treatment might result in the loss of antigenicity. Well controlled modifications are necessary for inactivating the toxins and affecting changes in physical properties desirable for fractionation and use. The widespread success in the use of millions, or perhaps hundreds of millions, of injectable doses of formaldehyde-derived toxoids is perhaps one of the highest tributes to the application of the carbonyl–amine reaction in protein chemistry.

ACKNOWLEDGMENTS

The authors gratefully acknowledge advice and assistance from many sources. The idea of this review originated while the senior author was on sabbatical leave in the

Department of Biochemistry, University of Cambridge, during 1972. Researches supplying background information were supported by grants HD00122-09 and AM1 3686-05 from the National Institutes of Health. Particular appreciation is due the many individuals who supplied materials for the manuscript and offered comments on its content, to Ms Clara Robison and Ms Billie Tordesillas for typing the manuscript, and to Ms Jody Burke and Ms Chris Howland for editorial assistance.

References

Akhtar, M., Blosse, P. T., and Dewhurst, P. B. (1967). *Chem. Commun.* **13**, 631.

Anderson, B. M., Anderson, C. D., and Churchich, J. E. (1966). *Biochemistry* **5**, 2893.

Anderson, R. E. (1972). *Biochemistry* **11**, 1224.

Bailey, A. J. (1968). *Compr. Biochem.* **26**, Part B, 297.

Ball, S., Goodwin, T. W., and Morton, R. A. (1948). *Biochem. J.* **42**, 516.

Ball, S., Collins, F. D., Dalvi, P. D., and Morton, R. A. (1949). *Biochem. J.* **45**, 304.

Barnett, R., and Jencks, W. P. (1967). *J. Amer. Chem. Soc.* **89**, 5963.

Bell, R. P., and Clunie, J. C. (1952). *Trans. Faraday Soc.* **48**, 439.

Bender, A. D., Kormendy, C. G., and Powell, R. (1970). *Exp. Gerontol.* **5**, 97.

Bjarnason, J., and Carpenter, K. J. (1969). *Brit. J. Nutr.* **23**, 859.

Bjarnason, J., and Carpenter, K. J. (1970). *Brit. J. Nutr.* **24**, 313.

Boeker, E. A., and Snell, E. E. (1972). *In* "The Enzymes" (P. D. Boyer, ed.), 3rd ed., Vol. 6, pp. 217–253. Academic Press, New York.

Bookchin, R. M., and Gallop, P. M. (1968). *Biochem. Biophys. Res. Commun.* **32**, 86.

Borch, R. F., Bernstein, M. D., and Durst, H. D. (1971). *J. Amer. Chem. Soc.* **93**, 2897.

Bowes, J. H. (1948). *In* "Progress in Leather Science: 1920–1945," pp. 501–518. Brit. Leather Mfr.'s Res. Ass., London.

Bownds, D. (1967). *Nature (London)* **216**, 1178.

Bownds, D., and Wald, G. (1965). *Nature (London)* **205**, 254.

Braunstein, A. E., and Shemyakin, M. M. (1953). *Biokhimia* **18**, 393.

Bresler, S., and Firsov, L. (1968). *J. Mol. Biol.* **35**, 131.

Bridges, C. D. B. (1967). *Compr. Biochem.* **27**, 31.

Bruice, T. C., and Benkovic, S. J. (1966). "Bioorganic Mechanisms." Benjamin, New York.

Buttkus, H. (1967). *J. Food Sci.* **32**, 432.

Carnes, W. H. (1971). *Fed. Proc., Fed. Amer. Soc. Exp. Biol.* **30**, 995.

Carpenter, K. J. (1960). *Biochem. J.* **77**, 604.

Carpenter, K. J. (1973). *Nutr. Abstr. Rev.* **43**, 423.

Chio, K. S., and Tappel, A. L. (1969a). *Biochemistry* **8**, 2821.

Chio, K. S., and Tappel, A. L. (1969b). *Biochemistry* **8**, 2827.

Cordes, E. H., and Jencks, W. P. (1962). *Biochemistry* **1**, 773.

Cortijo, M., and Shaltiel, S. (1972). *Eur. J. Biochem.* **29**, 134.

Crawford, D. L., Yu, T. C., and Sinnhuber, R. O. (1967). *J. Food Sci.* **32**, 332.

Dartnall, H. J. A., and Lythgoe, J. N. (1965). *Vision Res.* **5**, 81.

Dartnall, H. J. A., ed. (1972). "Handbook of Sensory Physiology," Vol. VII, Part I. Springer-Verlag, Berlin and New York.

Deppert, W., Hucho, F., and Sund, H. (1973). *Eur. J. Biochem.* **32**, 76.

Deshmukh, K., and Nimni, M. E. (1969). *J. Biol. Chem.* **244**, 1787.

Dixon, H. B. F. (1972). *Biochem. J.* **129**, 203.

Dixon, H. B. F., and Fields, R. (1972). *In* "Methods in Enzymology" (C. H. W. Hirs and S. N. Timasheff, eds.), Vol. 25, Part B, p. 409. Academic Press, New York.

Dixon, H. B. F., and Weitkamp, L. R. (1962). *Biochem. J.* **84**, 462.
Dunathan, H. C. (1966). *Proc. Nat. Acad. Sci. U. S.* **55**, 712.
Feeney, R. E., Clary, J. J., and Clark, J. R. (1964). *Nature (London)* **201**, 192.
Fein, M. L., and Filachione, E. M. (1960). U. S. Patent 2,941,859.
Filachione, E. M. (1960). *Leather Mfr.* **77**, 14.
Finlay, T. H., Dharmgrongartama, E. D., and Perlmann, G. E. (1973). *J. Biol. Chem.* **248**, 4827.
Fischer, E. H. (1964). *In* "Structure and Activity of Enzymes" (T. W. Goodwin, J. I. Harris, and B. S. Hartley, eds.), pp. 111–120. Academic Press, New York.
Florkin, M., and Stotz, E. H., eds. (1971). *Compr. Biochem.* **26**, Part C, 595–898.
Forcina, B. G., Ferri, G., Zapponi, M. C., and Ronchi, S. (1971). *Eur. J. Biochem.* **20**, 535.
Franzblau, C. (1971). *Compr. Biochem.* **26**, Part C, 659.
French, D., and Edsall, J. T. (1945). *Advan. Protein Chem.* **2**, 277.
Frey, P. A., Kokesh, F. C., and Westheimer, F. H. (1971). *J. Amer. Chem. Soc.* **93**, 7266.
Fridovich, I., and Westheimer, F. H. (1962). *J. Amer. Chem. Soc.* **84**, 3208.
Gallop, P. M., Blumenfeld, O. O., and Seifter, S. (1972). *Annu. Rev. Biochem.* **41**, 617.
George, D. J., and Phillips, A. T. (1970). *J. Biol. Chem.* **245**, 528.
Gottschalk, A. (1972). *Glycoproteins* **5B**, 141–157.
Grazi, E., Cheng, T., and Horecker, B. L. (1962). *Biochem. Biophys. Res. Commun.* **7**, 250.
Gross, E., and Morell, J. L. (1971). *J. Amer. Chem. Soc.* 93, 4634.
Gross, E., Kiltz, H. H., and Nebelin, E. (1973). *Hoppe-Seyler's Z. Physiol. Chem.* **354**, 810.
Gustavson, K. H. (1956). "The Chemistry of Tanning Processes." Academic Press, New York.
Habeeb, A. F. S. A., and Hiramoto, R. (1968). *Arch. Biochem. Biophys.* **126**, 16.
Hamilton, G., and Westheimer, F. H. (1959). *J. Amer. Chem. Soc.* **81**, 6332.
Hansen, K. R., and Havir, E. A. (1969). *Fed. Proc., Fed. Amer. Soc. Exp. Biol.* **28**, 602.
Harmon, D. (1969). *J. Amer. Geriat. Soc.* **17**, 721.
Hecht, S., and Williams, R. E. (1922). *J. Gen. Physiol.* **5**, 1.
Hedrick, J. L., Shaltiel, S., and Fischer, E. H. (1966). *Biochemistry* **5**, 2117.
Hedrick, J. L., Shaltiel, S., and Fischer, E. H. (1969). *Biochemistry* **8**, 2422.
Heinert, D., and Martell, A. E. (1963). *J. Amer. Chem. Soc.* **85**, 183.
Heller, J. (1968). *Biochemistry* **7**, 2906.
Heller, J., and Lawrence, M. A. (1970). *Biochemistry* **9**, 864.
Heyl, D., Luz, E., Harris, S. A., and Folkers, K. (1951). *J. Amer. Chem. Soc.* **73**, 3430.
Hochschild, R. (1970). *Gerontology* **10**, Part II, 29.
Hodge, J. E. (1967). *In* "Symposium on Foods: The Chemistry and Physiology of Flavors" (H. W. Schultz, E. A. Day, and L. M. Libbey, eds.), p. 465. Avi, Westport, Connecticut.
Hodge, J. E., Mills, F. D., and Fisher, B. E. (1972). *Cereal Sci. Today* **17**, 34.
Hodgins, D. S., and Abeles, R. H. (1967). *J. Biol. Chem.* **242**, 5158.
Holmquist, W. R., and Schroeder, W. A. (1966). *Biochemistry* **5**, 2503.
Hubbard, R. (1953–1954). *J. Gen. Physiol.* **37**, 381.
Hubbard, R. (1958). *In* "Visual Problems of Colour," Proc. Nat. Phys. Lab. Symp., No. 8 (National Physical Laboratory, Teddington, England, corp. ed.), pp. 153–169. HM Stationery Office, London.
Hubbard, R. (1969). *Nature (London)* **221**, 432.

Hucho, F., Markau, U., and Sund, H. (1973). *Eur. J. Biochem.* **32**, 69.
Hunt, E., and Morris, H. R. (1973). *Biochem. J.* **135**, 833.
Ivanov, V. I., and Karpeisky, M. Ya. (1969). *Advan. Enzymol.* **32**, 21.
Jencks, W. P., ed. (1969). "Catalysis in Chemistry and Enzymology." McGraw-Hill, New York.
Kahan, L., and Kaltschmidt, E. (1972). *Biochemistry* **11**, 2691.
Kallen, R. G., and Jencks, W. P. (1966). *J. Biol. Chem.* **241**, 5864.
Kline, L., and Sonoda, T. T. (1951). *Food Technol.* **5**, 90.
Kline, L., Gegg, J. E., and Sonoda, T. T. (1951a). *Food Technol.* **5**, 181.
Kline, L., Hanson, H. L., Sonoda, T. T., Gegg, J. E., Feeney, R. E., and Lineweaver, H. (1951b). *Food Technol.* **5**, 323.
Koehler, K., Sandstrom, W., and Cordes, E. H. (1964). *J. Amer. Chem. Soc.* **86**, 2413.
Kohn, R. R. (1970). *Gerontology* **10**, Part II, 28.
Kokesh, F. C., and Westheimer, F. H. (1971). *J. Amer. Chem. Soc.* **93**, 7270.
Kon, H., and Szent-Györgyi, A. (1973). *Proc. Nat. Acad. Sci. U. S.* **70**, 3139.
Korn, A. H., Feairheller, S. H., and Filachione, E. M. (1972). *J. Mol. Biol.* **65**, 525.
Kropf, A., and Hubbard, R. (1958). *Ann. N. Y. Acad. Sci.* **74**, 226.
Kwon, T. W., and Brown, W. D. (1965). *Fed. Proc., Fed. Amer. Soc. Exp. Biol.* **24**, 592.
Lai, C. Y., and Horecker, B. L. (1972). *Essays Biochem.* **8**, 149.
La Nauze, J. M., Rosenberg, H., and Shaw, D. C. (1970). *Biochim. Biophys. Acta* **212**, 332.
Lea, C. H., and Hannan, R. S. (1950a). *Biochim. Biophys. Acta* **4**, 518.
Lea, C. H., and Hannan, R. S. (1950b). *Biochim. Biophys. Acta* **5**, 433.
Leclerc, J., and Benoiton, L. (1968). *Can. J. Chem.* **46**, 471.
Levene, C. I., and Gross, J. (1959). *J. Exp. Med.* **110**, 771.
Lewin, S. (1956). *Biochem. J.* **64**, 30P.
Lin, Y., Means, G. E., and Feeney, R. E. (1969). *J. Biol. Chem.* **244**, 789.
Maillard, L. C. (1912). *C. R. Acad. Sci.* **154**, 66.
Maillard, L. C. (1916). *Ann. Chim. (Paris)* [9] **5**, 258.
Maillard, L. C. (1917). *Ann. Chim. (Paris)* [9] **7**, 113.
Mantione, M.-J., and Pullman, B. (1971). *Int. J. Quantum Chem., Symp.* **5**, 349.
Marfey, P. S., Nowak, H., Uziel, N., and Yphantis, D. A. (1965). *J. Biol. Chem.* **240**, 3264.
Means, G. E., and Feeney, R. E. (1968). *Biochemistry* **7**, 2192.
Means, G. E., and Feeney, R. E. (1971). "Chemical Modification of Proteins." Holden-Day, San Francisco, California.
Metzler, D. E., and Snell, E. E. (1955). *J. Amer. Chem. Soc.* **77**, 2431.
Metzler, D. E., Ikawa, M., and Snell, E. E. (1954). *J. Amer. Chem. Soc.* **76**, 648.
Minssen, M., and Munkres, K. D. (1973). *Biochim. Biophys. Acta* **291**, 398.
Mohammed, A., Olcott, H. S., and Fraenkel-Conrat, H. (1949). *Arch. Biochem.* **24**, 270.
Morton, R. A. (1972). *In* "Handbook of Sensory Physiology" (H. J. A. Dartnall, ed.), Vol. VII, Part 1, pp. 33–68. Springer-Verlag, Berlin and New York.
Morton, R. A., and Goodwin, T. W. (1944). *Nature (London)* **153**, 405.
Morton, R. A., and Pitt, G. A. J. (1955). *Biochem. J.* **59**, 128.
Mühlrad, A., Ajtai, K., and Fábián, F. (1970). *Biochim. Biophys. Acta* **205**, 355.
Nandi, D. L., and Shemin, D. (1968). *J. Biol. Chem.* **243**, 1236.
Nimni, M. E. (1968). *J. Biol. Chem.* **243**, 1457.
Nishimura, K., Komano, T., and Yamada, H. (1972). *Biochim. Biophys. Acta* **262**, 24.
Ottesen, M., and Svensson, B. (1971a). *C. R. Trav. Lab. Carlsberg* **38**, 171.

Ottesen, M., and Svensson, B. (1971b). *C. R. Trav. Lab. Carlsberg* **38**, 445.
Paik, W. K., and Kim, S. (1972). *Biochemistry* **11**, 2589.
Passwater, R. A. (1970). *Gerontology* **10**, Part II, 28.
Patai, S., ed. (1966). "The Chemistry of the Carbonyl Group," Vol. I. Wiley (Interscience), New York.
Patai, S., ed. (1968). "The Chemistry of the Amino Group." Wiley (Interscience), New York.
Patai, S., ed. (1970). "The Chemistry of the Carbonyl Group," Vol. II. Wiley (Interscience), New York.
Pinnell, S. R., and Martin, G. R. (1968). *Proc. Nat. Acad. Sci. U. S.* **61**, 708.
Piszkiewicz, D., Landon, M., and Smith, E. L. (1970). *J. Biol. Chem.* **245**, 2622.
Pitt, G. A., Collins, E. D., Morton, R. A., and Stok, P. (1955). *Biochem. J.* **59**, 122.
Poincelot, R. P., Millar, P. G., Kimbel, R. L., Jr., and Abrahamson, E. W. (1970). *Biochemistry* **9**, 1809.
Pontremoli, S., Prandini, B. D., Bonsignore, A., and Horecker, B. L. (1961). *Proc. Nat. Acad. Sci. U. S.* **47**, 1942.
Quiocho, F. A., and Richards, F. M. (1966). *Biochemistry* **5**, 4062.
Raetz, C. R. H., and Auld, D. S. (1972). *Biochemistry* **11**, 2229.
Ramsey, R. J., Tracy, P. H., and Ruehe, H. A. (1933). *J. Dairy Sci.* **16**, 17.
Rasched, I., Joernvall, H., and Sund, H. (1974). *Eur. J. Biochem.* **41**, 603.
Recsei, P. A., and Snell, E. E. (1970). *Biochemistry* **9**, 1492.
Recsei, P. A., and Snell, E. E. (1973). *Biochemistry* **12**, 365.
Reiss, U., Tappel, A. L., and Chio, K. S. (1972). *Biochem. Biophys. Res. Commun.* **48**, 921.
Reynolds, T. M. (1965). *Advan. Food Res.* **14**, 168–283.
Reynolds, T. M. (1969). *In* "Symposium on Foods: Carbohydrates and their Roles" (H. W. Schultz, R. F. Cain, and R. W. Wrolstad, eds.), pp. 219–252. Avi, Westport, Connecticut.
Rice, R. H., and Means, G. E. (1971). *J. Biol. Chem.* **246**, 831.
Richards, F. M., and Knowles, J. R. (1968). *J. Mol. Biol.* **37**, 231.
Riley, W., and Snell, E. (1970). *Biochemistry* **9**, 1485.
Rippa, M., and Pontremoli, S. (1969). *Arch. Biochem. Biophys.* **133**, 112.
Roberts, J. D., and Caserio, M. C. (1964). "Basic Principles of Organic Chemistry." Benjamin, New York.
Robins, S. P., and Bailey, A. J. (1973). *Biochem. J.* **135**, 657.
Ronchi, S., Zapponi, M. C., and Ferri, G. (1969). *Eur. J. Biochem.* **8**, 325.
Rosenthaler, J., Guirard, B. M., Chang, G. W., and Snell, E. E. (1965). *Proc. Nat. Acad. Sci. U. S.* **54**, 152.
Schiff, H. (1900). *Justus Liebig's Ann. Chem.* **310**, 25.
Schiff, H. (1901). *Justus Liebig's Ann. Chem.* **319**, 59.
Schiffmann, E., and Martin, G. R. (1970). *Arch. Biochem. Biophys.* **138**, 226.
Shaltiel, S., Hedrick, J. L., and Fischer, E. H. (1966). *Biochemistry* **5**, 2108.
Shaltiel, S., Hedrick, J. L., and Fischer, E. H. (1969). *Biochemistry* **8**, 2429.
Shapiro, S., Enser, M., Pugh, E., and Horecker, B. L. (1968). *Arch. Biochem. Biophys.* **128**, 554.
Shields, J. E., Dinovo, E. C., Henriksen, R. A., Kimbel, R. L., Jr., and Millar, P. G. (1967). *Biochim. Biophys. Acta* **147**, 238.
Singleton, V. L., and Kratzer, F. H. (1973). *In* "Toxicants Occurring Naturally in Foods" (F. M. Strong, ed.), pp. 309–345. Nat. Acad. Sci., Washington, D. C.
Smith, E. L., Landon, M., Piszkiewicz, D., Brattin, W. J., Langley, T. J., and Melamed, M. D. (1970). *Proc. Nat. Acad. Sci. U. S.* **67**, 724.

Snell, E. E. (1962). *Brookhaven Symp. Biol.* **15,** 32.
Snell, E. E., and Di Mari, S. J. (1970). *In* "The Enzymes" (P. D. Boyer, ed.), 3rd ed., Vol. 2, pp. 335–370. Academic Press, New York.
Sörensen, S. P. L. (1908). *Biochem. Z.* **7,** 45.
Streuli, H. (1967). *In* "Aroma-und Geschmacksstoffe in Lebensmitteln" (J. Solms, and H. Neukom, eds.), p. 119. Forster-Verlag A. G., Zurich.
Subramanian, A. R. (1972). *Biochemistry* **11,** 2710.
Tabor, H., and Tabor, C. W. (1972). *Advan. Enzymol.* **36,** 203–268.
Tanzer, M. L. (1965). *Int. Rev. Connect. Tissue Res.* **3,** 91.
Tanzer, M. L. (1973). *Science* **180,** 561.
Tanzer, M. L., Housley, T., Berube, L., Fairweather, R., Franzblau, C., and Gallop, P. M. (1973). *J. Biol. Chem.* **248,** 393.
Tappel, A. L. (1972). *Ann. N. Y. Acad. Sci.* **203,** 12.
Tappel, A. L. (1973). *In* "Pathological Aspects of Cell Membranes" (B. F. Trump and A. Arstils, eds.), Vol. 1. Academic Press, New York (in press).
Traub, W., and Piez, K. R. (1971). *Advan. Protein Chem.* **25,** 243.
Udenfriend, S., Stein, S., Böhlen, P., Dairman, W., Leimgruber, W., and Weigle, M. (1972). *Science* **178,** 871.
Venkataraman, R., and Racker, E. (1961). *J. Biol. Chem.* **236,** 1876.
Wada, H., and Snell, E. E. (1962). *J. Biol. Chem.* **237,** 127.
Wald, G. (1935). *J. Gen. Physiol.* **18,** 905.
Wald, G. (1950). *Biochim. Biophys. Acta* **4,** 215.
Wald, G., and Brown, P. K. (1951–1952). *J. Gen. Physiol.* **35,** 797.
Waleh, A., and Ingraham, L. L. (1973). *Arch. Biochem. Biophys.* **156,** 261.
Walker, J. F. (1964). "Formaldehyde." Van Nostrand-Reinhold, Princeton, New Jersey.
Warren, S., Zerner, B., and Westheimer, F. H. (1966). *Biochemistry* **5,** 817.
Weakley, F. B., Mehltretter, C. L., and Rist, C. E. (1961). *Tappi,* **44,** 456.
Weis-Fogh, T., and Andersen, S. O. (1970). *Nature* (*London*) **227,** 718.
Westheimer, F. H. (1963). *Proc. Chem. Soc., London,* pp. 253–261.
Wiesenfeld, J. R., and Abrahamson, E. W. (1968). *Photochem. Photobiol.* **8,** 487.
Williams, J. N., Jr., and Jacobs, R. M. (1966). *Biochem. Biophys. Res. Commun.* **22,** 695.
Williams, J. N., Jr., and Jacobs, R. M. (1968). *Biochim. Biophys. Acta* **154,** 323.
Wold, F. (1967). *In* "Methods in Enzymology" (C. H. W. Hirs, ed.), Vol. 11, p. 617. Academic Press, New York.

EXPERIMENTAL AND THEORETICAL ASPECTS OF PROTEIN FOLDING

By C. B. ANFINSEN and H. A. SCHERAGA

National Institute of Arthritis, Metabolism and Digestive Diseases,
Bethesda, Maryland, and Department of Chemistry,
Cornell University, Ithaca, New York

I. Introduction

Proteins are synthesized in cells by a stepwise process in which amino acids are added, one by one, from the NH_2-termini of the chains (Bishop *et al.*, 1960; Dintzis, 1961; Canfield and Anfinsen, 1963; Naughton and Dintzis, 1962). The folding of such nascent chains can occur spontaneously without the need for any additional information beyond that contained in the amino acid sequence and its surroundings; the same is true of "structureless" unaggregated polypeptides produced by denaturation of native molecules. The experimental evidence that supports this conclusion has been reviewed in some detail elsewhere (Epstein *et al.*, 1963; Anfinsen, 1966, 1973; Wetlaufer and Ristow, 1973). The term "protein folding" refers to the processes involved in the conversion of an ensemble of newly synthesized (or denatured) polypeptide chain conformations to the unique three-dimensional conformation of the native protein. When present, disulfide bonds serve to stabilize the native conformation by reducing the structural fluctuations of alternative (denatured) forms (Schellman, 1955; Flory, 1956; Scheraga, 1963). In some cases, it has been found that the attainment of the final native conformation depends on the presence of specific ligands, frequently metal ions or prosthetic groups (although with many enzymes these may be the substrate molecules themselves). The presence of ligands does not necessarily lead to a significantly different *conformation* of the folded protein, but helps to stabilize the native form (Schechter *et al.*, 1969; Taniuchi *et al.*, 1969). It is currently believed that the three-dimensional structure of the native protein in a given environment (solvent, pH, ionic strength, presence of other components, temperature, etc.) is the one in which the Gibbs free energy of the *whole system* is a minimum with respect to all degrees of freedom, i.e., that the native conformation is determined by the various interatomic interactions and hence by the amino acid sequence, in a *given environment*. Whether or not this "local" minimum in the free energy hyperspace is the global one is at present an unsettled question.

Since it seems quite reasonable, on the basis of the available experimental data, to accept the hypothesis of spontaneous folding to a unique structure, it is of interest to consider the way in which such a conversion from "one" to "three" dimensions occurs, and the nature and magnitude of the interactions necessary to determine a specific structure. It has been suggested (Levinthal, 1968) that the structure of a native protein might occasionally represent a unique metastable state, not necessarily the one of lowest free energy. Such a view would require that the folding of the chain occur by means of a specific pathway which could lead to the metastable state only by following an ordered sequence of events. One suggestion (Phillips, 1967; Chantrenne, 1961; Dunnill, 1965, 1967) proposes that the chain folds as it is synthesized from the NH_2- to the COOH-terminus. Both stereochemical considerations and experiments on protein folding in solution make these ideas unlikely. Despite the possible presence of high energetic barriers between conformational states, the conformational space seems to be sufficiently large for pathways around these barriers to exist, enabling the flexible polypeptide chain to attain its most stable conformation (Hantgan *et al.*, 1974).

To be more precise, one should distinguish between proteins that consist of a single chain and those containing more than one, noncovalently bonded chain (subunits). In the former case, the amino acid sequence is presumably such that no decrease in free energy from *inter*molecular interactions is possible at low temperature, and the protein is stable in a single-chain (folded) form. In other cases, the amino acid sequence has evolved (to achieve a particular biological advantage) so that favorable intermolecular interactions (in addition to the intramolecular ones) can decrease the free energy of the system by forming multisubunit (folded) aggregates, and even larger organized structures.[1]

Evidence is now accumulating to suggest that nearest-neighbor, short-range interactions play the dominant role in determining the conformational preferences of the *backbones* of the various amino acids, but that next-nearest neighbor (medium-range) interactions and, to a lesser extent, long-range interactions involving the rest of the protein chain are required to provide the incremental free energy to stabilize the backbone of the native structure (Scheraga, 1973b, 1974a,b). The basis for this view comes from helix-probability profiles of denatured proteins and their correlation with helical regions in native proteins, from the ability to predict with some success the location of helix, extended structures, and β-turns in native proteins, from conformational energy calculations on short oligopeptides, and from experiments on the folding of protein fragments and sets of complementary protein fragments. In contrast, the conformations of the *side chains* of a native protein *may* be influenced more by long-, rather than by short-range, interactions. Thus, it is thought that short-range interactions lead to one or more *backbone* nucleation sites[2] along the chain, where there is a tendency to form specific structures resembling the native conformations; these various sites acquire additional stability from medium-range interactions and are

[1] It may be noted that even a single-chain protein can undergo a decrease in free energy, for example, by heating to form an aggregated complex. Several possibilities then exist. If the native protein is the more stable one at the lower temperature, the denatured aggregate may either revert spontaneously to the native protein, or be prevented from doing so by kinetic barriers, when the temperature is lowered. Alternatively, if the denatured aggregate is the more stable form at the lower temperature, kinetic barriers could prevent the denaturation of the native protein at a low temperature. In any case, it is postulated that a reasonable amount of conformation space is accessible in going from the newly synthesized chain to the folded, native structure, without interference from intervening barriers, but that such barriers may exist to prevent the conversion of the protein from the native to an *aggregated* denatured form (conversion to an unaggregated denatured form upon heating is postulated to be reversible when the temperature is lowered).

[2] The term "nucleation site" is used here to designate a portion of the amino acid sequence in which the "native format" or conformation can form, i.e., a region in which a specific *local* backbone conformation has a *tendency* to form because of short- and medium-range interactions. However, it becomes stabilized in the specific con-

directed toward each other (possibly by formation of β-turns at intervening sites) to bring on the long-range interactions required for stabilization of the whole (backbone and side-chain) structure. One can thus envisage several possible pathways for folding, involving many intermediate states.

The location of the nucleation sites and their stabilization by medium-range interactions (i.e., the propensity for certain portions of the chain to adopt the "native format"), and the specific long-range interactions brought about, say by β-turns, depend on the amino acid sequence and on the environment of the chain. The amino acid sequence, in turn, is selected during evolution to lead to a particular biological function; the resulting sequence enables proper folding to take place according to the dictates of thermodynamics—that is, the free energy of the system must be a minimum. In prebiotic synthesis, a very large number of amino acid sequences must have been generated, and the probability that one of them would fold into a compact globular structure with biological activity was essentially zero. However, a limited number of amino acid sequences have been selected to lead to a particular biological function. Selection may take place not only for biological function but also for stability of the molecule chosen for such function, as, for example, in the case of the enzymes of thermophilic bacteria.

Since the folding of any sequence in a given environment follows thermodynamic requirements, it has been advantageous to use synthetic polypeptides (which are simple, well-characterized models of natural, more complex polypeptides) to determine the magnitudes of interactions that are operative not only in the synthetic system but, by inference, also in the natural one. The helix-coil transition in synthetic homopolymers and copolymers, spectroscopic studies of the conformations of oligopeptides, the study of reversible protein denaturation, and the folding of recombined protein fragments (natural, and synthetically modified) are the basic experimental approaches for determining the nature and magnitude of the interactions which determine how the natural polypeptide chain folds into the native conformation of a globular protein.

---

formation only when long-range interactions come into play. In using this term, nothing is implied about the temporal sequence of folding; that is, nucleation sites may become stabilized in their native form at various stages in the folding sequence. Such nucleation sites may be regular—for example, α-helices, extended structures, and β-turns—or nonregular, but specifically folded forms. One of the first nucleation sites in the folding of ribonuclease A has been postulated to be nonregular (Burgess and Scheraga, 1975b); see Fig. 20.

This chapter correlates the growing body of experimental and theoretical evidence with a model for the process of protein folding which involves, for any particular molecule in solution, several portions of the amino acid sequence of the completed polypeptide chain, each participating in an equilibrium between a random set of states and the "native format" which it possesses when it is part of the ordered three-dimensional structure of the protein. The successful transition from a random polypeptide chain to a native protein then involves further interactions between two or more such regions which have a tendency to adopt the "native format." Nucleation of the folded conformation would be highly favored in portions of the amino acid sequence that have a high probability of forming the "native format" in aqueous solution.

II. Spontaneous Folding of Proteins

It has been known for many years that proteins can be made to undergo reversible denaturation. In his classical review in *Advances in Protein Chemistry*, Anson (1945) pointed out: "Hemoglobin which has been denatured in a variety of ways can be converted back into native protein which has the same solubility as the original protein, is crystallizable, has the characteristic spectrum of native hemoglobin, can combine loosely with oxygen, has the same relative affinities for carbon monoxide and oxygen, and is not readily digested by trypsin." These experiments (Anson and Mirsky, 1934b) and many others, such as those of Northrop (1932) and Anson and Mirsky (1934a) on trypsin and chymotrypsin and their zymogens, clearly showed that the biological properties of the protein molecule in solution are directly related to its chemical properties, and that refolding of a denatured protein to the native form can occur spontaneously. An abundance of later studies have supported this idea.

The early work was hampered by the almost total lack of knowledge of the chemical structure of proteins, and conclusions about renaturation were necessarily phenomenological. Two developments helped to convert the study of protein folding into a field in which a quantitative description could be attempted: first, the basic experiments of Sanger (1956), Moore and Stein and their colleagues (Hirs, 1960; Smyth *et al.*, 1962, 1963), and others on amino acid sequences and covalent cross-linkages in protein molecules; and second, the examination of the refolding of proteins containing disulfide bonds after these had been broken by reductive cleavage in denaturing solvents (Sela *et al.*, 1957; White, 1961; Anfinsen *et al.*, 1961). The latter studies helped to dramatize the precision of the natural

TABLE I

The Number of Ways in Which 2n Sulfhydryl Groups Can Combine to Form j Disulfide Bonds[a,b]

Number of bonds	Number of combinations
1	1
2	3
3	15
4	105
5	945
6	10395
7	135135
8	2027025
9	34459425
10	654729075
11	13749310575
12	316234143225
13	7905853580625
14	213458046676875
15	6190283353629375
16	191898783962510625
17	6332859870762850625
18	221643095476699771875
19	8220079453263789155937 5
20	319830986772877770815625
21	13111307045768798860344 0625
22	563862029680583509947946875
23	25373791335626257947657609375
24	1192568192774434123539907640625
25	58435841445947272053455474390625

$$N_{2n}^{\ j} = \frac{(2n)!}{2^j(2n-2j)!j!}$$

[a] From Anfinsen (1966).

[b] The data listed in the table are for the special care where $j = n$.

phenomenon by showing that only a single set of disulfide bonds was formed when a protein chain bearing numerous SH groups was allowed to undergo air oxidation.[3] The statistics of the process of random formation of n SS bonds from $2n$ SH groups (Kauzmann, 1959a; Sela and Lifson, 1959) are shown in Table I. With pancreatic ribonuclease, $n = 4$ and one of 105 possibilities represents the native

[3] Most protein chains do *not* contain disulfide bonds or other covalent cross-linkages. Although, in those cases studied to date (Epstein *et al.*, 1971a) proteins without cross-links appear to fold more rapidly than those containing disulfide bonds, it is not known whether this is a general phenomenon *under physiological conditions* where (correct) disulfide bond pairing could conceivably increase the rate. In either case, the rate would be affected by the extent to which the chain explores the states of

state. With an immunoglobulin containing 23 SS bonds, on the order of 2×10^{28} possibilities exist. In spite of the small chance for random formation of the correct set of disulfide bonds, close to quantitative yields can be obtained from a number of fully reduced proteins, e.g., native ribonuclease (Anfinsen and Haber, 1961) and a specific antibody (Freedman and Sela, 1966), when care is taken to provide the proper environmental conditions during the oxidation process.

The precision of the factors that select for one set of SS bonds is also shown by experiments on the interchange of incorrectly formed internal disulfide bonds (Haber and Anfinsen, 1962). If reduced ribonuclease is oxidized in 8 *M* urea, a product is obtained which contains a large number (within the constraints of avoiding steric hindrance) of the 105 possible isomeric forms. When the urea is removed, exposure of such "scrambled" material to a small amount of a sulfhydryl group-containing reagent such as mercaptoethanol at neutral or slightly alkaline pH, or exposure to a disulfide-interchange enzyme, leads to interchange of half-cystine residues with essentially quantitative production of the native enzyme (Fig. 1).

Actually, the formation of the disulfide bonds of ribonuclease does not lead immediately to the native conformation, as can be seen from the data of Fig. 2, which pertain to disulfide-bond formation in the presence of oxidized and reduced glutathione (Hantgan *et al.*, 1974). The disappearance of free SH groups is much more rapid than the regain of properties associated with the native conformation. This indicates that there is an initial formation of (probably many) wrong SS bonds, which slowly anneal to the correct SS-bond arrangement. This conclusion is supported by the results of peptide analysis (showing which sulfhydryl groups are paired) during the course of SS-bond formation (Hantgan *et al.*, 1974). In the case of lysozyme, it appears that only a very limited number of structures are formed in the early stages of the reoxidation (Ristow and Wetlaufer, 1973) and, in the case of bovine pancreatic trypsin inhibitor, it has been possible to detect several transient intermediates with an incomplete set of disulfide bonds (Creighton, 1974). These experiments, together

internal rotation about the single bonds of the polypeptide (subject to such directing influences as, for example, the tendency to form helical sections, extended structures, β-turns, etc.). Earlier work (Anfinsen *et al.*, 1961) focused attention on the precision with which the total interactions between amino acid side chains lead to a unique thermodynamically stable structure, and current studies (e.g., Sachs *et al.*, 1972a,b,c) have been concerned with determining those aspects of nucleation and folding that make possible the extreme rapidity with which native protein structures are formed.

FIG. 1. The spontaneous conversion of a randomly cross-linked protein derivative to the native form under conditions favoring disulfide interchange. Regions of the amino acid sequence that are involved in the active center are indicated by cross-hatching. From Anfinsen (1966).

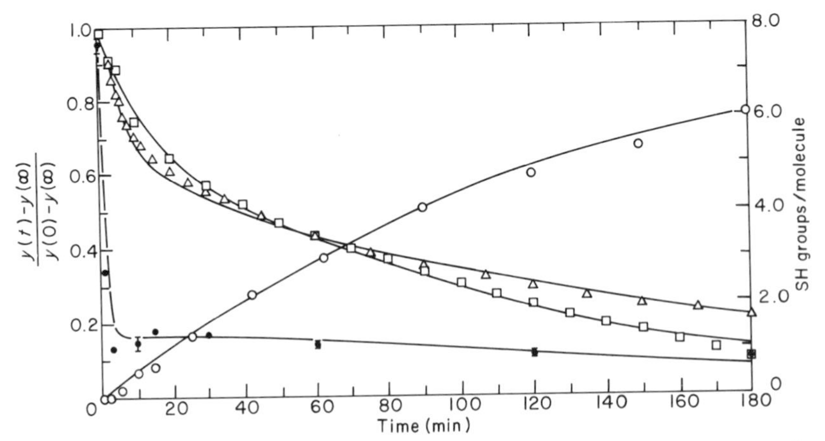

FIG. 2. Time dependence of the reoxidation of ribonuclease in the presence of oxidized and reduced glutathione. The left-hand ordinate is the fraction of activity (○), absorbance (□), or fluorescence (△). The right-hand ordinate (●) is the number of sulfhydryl groups per protein molecule. From Hantgan *et al.* (1974).

with those of Haber and Anfinsen (1962) cited above, demonstrate the strong thermodynamic force which drives the conformation from a "scrambled" set of isomers (of high free energy) to the native structure (of lower free energy) and also indicate that the native conformation is attainable from many different conformations; the latter observation supports the argument that the native conformation may not be a metastable one.

The study of the folding process in single-chain disulfide bond-containing proteins has been reviewed extensively (Epstein *et al.*, 1963; Wetlaufer and Ristow, 1973), and the so-called "thermodynamic hypothesis," stated in the Introduction, appears to be generally applicable. This hypothesis is also thought to apply to globular proteins that contain more than one chain (subunits), where folding of the subunits and the formation of the multichain complex is determined by *both* intrachain and interchain interactions. However, in this review we shall limit our discussion primarily to a consideration of the factors that underlie the folding of a newly synthesized, single-chain protein, and not treat proteins with subunits.

It is not necessary that the amino acid sequence be a unique one to obtain a particular three-dimensional structure or a closely homologous one. The genetic information encoded in the primary sequence of a protein may be modified naturally by mutations or experimentally in a number of ways without destroying the capacity of the resulting chain (or chain fragments) to assume the native or similar conformation. For example, the structures of some of the hemoglobins, myoglobins, cytochrome *c*'s, etc., of different species are apparently unaltered by many such substitutions (Perutz and Lehmann, 1968). On the other hand, some nonlethal substitutions occasionally lead to an altered conformation of the native protein, as occurs in several mutant hemoglobins and in immunologically different variants of tryptophan synthetase (Helinski and Yanofsky, 1963). (The alteration in structure is usually inferred from an alteration in biological activity, although this need not be so; that is, a conformational change could occur without a change in biological activity.) The ability of the chain to tolerate such substitutions in some cases, but not in others, is now understood (Lewis and Scheraga, 1971a) in terms of the preservation (or nonpreservation) of similar interatomic interactions (preserving specific local conformations), when one amino acid is substituted for another (see Section VIII,B,5). In fact, different proteins, which have a high degree of similarity in their amino acid sequences [such as chymotrypsin, trypsin, elastase, and thrombin (Hartley, 1970), or lysozyme and α-

lactalbumin (Brew and Campbell, 1967; Brew *et al.*, 1970)], have, or are thought to have, similar three-dimensional structures. It has also been shown that functionally related proteins (e.g., the NAD binding dehydrogenases) have similar structures (Bränden *et al.*, 1973; Hill *et al.*, 1972; Jörnvall, 1973; Rossmann *et al.*, 1971). Such amino acid substitutions (produced either naturally or synthetically) can be more easily accommodated in the native structure if they occur on the "outside" of the protein at the protein–solvent interface (Perutz and Lehmann, 1968), presumably because more redundancy is permissible in the conformation-preserving interactions involving the outside residues (which may serve primarily to keep the protein soluble in water). On the "inside," the long-range interactions that determine the packing of the side chains could be more easily disrupted by substitutions. Of course, if a residue at an active site is modified, then the biological specificity will be altered.

An illustrative experiment carried out some years ago (Cooke *et al.*, 1963), on the ease with which modifications of the residues at the protein–solvent interface can be accommodated, is summarized in the schematic drawing shown in Fig. 3. Ribonuclease can be converted from its native, globular state with four disulfide bonds to an open, random coil by reduction of these SS bonds with mercaptoethanol in urea solution. As mentioned earlier, reoxidation with molecular oxygen yields the original native structure containing the properly paired SH groups in disulfide linkage. When, as indicated in Fig. 3, eight of the eleven free amino groups are modified by the addition of (water-soluble) poly-DL-alanine chains containing as many as eight alanine monomers per chain, the enzyme retains its full activity as well as many of its native, physical properties. Such derivatives also retain the capacity to undergo completely reversible reduction and reoxidation of the four SS bonds. Thus, the long chains of poly-DL-alanine neither interfere with the "correct" packing of side chains in the interior of the molecule nor introduce stereochemical constraints that alter the interactions that lead to the correct structure of the unmodified sequence. Polyalanylated ribonuclease bears the same number of positive charges as the original native protein on which the chains were grown; other modifying groups that confer a negative charge, or that substitute a neutral for a positive charge, were also attached to the ϵ-amino groups; the resulting products were, again, capable of reversible refolding (Epstein *et al.*, 1963).

The X-ray structures of homologous proteins also show that many of the amino acid substitutions occur on the surface (Perutz *et al.*,

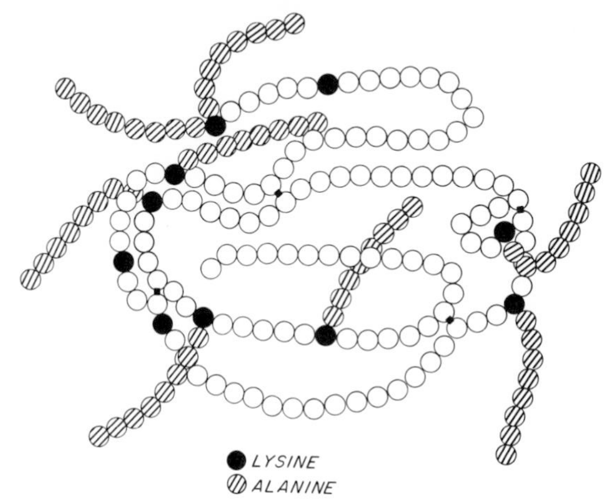

FIG. 3. Schematic representation of a fully active poly-DL-alanylribonuclease molecule. The crosshatched circles indicate alanyl residues, attached in chains to ε-amino groups. From Cooke *et al.* (1963).

1965). In fact, hairpin turns, which generally occur on the surface, are often sites of insertions or deletions of amino acids; such modifications are very easily accommodated on the surface, but the new sequence must preserve the tendency toward formation of the required turn (Light, 1974; Warme *et al.*, 1974).

The early studies of Perutz and his colleagues, and the flood of new sequence and crystallographic data are generally consistent with the idea that mutations leading to replacement of nonpolar by polar residues are not permissible when they occur in the internal, more nonpolar portion of the folded molecule.[4]

Occasionally, double mutations occur in which biological function is first lost and then regained. Examples of this arise in the work of Yanofsky and his colleagues (Helinski and Yanofsky, 1963), who isolated a number of mutants of *Escherichia coli* that produce inactive tryptophan synthetase. In these species, point mutations have led to the synthesis of tryptophan synthetase molecules in which a single amino acid replacement has dictated an altered three-dimen-

[4] Lee and Richards (1971) have shown that the side chains in lysozyme, ribonuclease, and myoglobin that are accessible to solvent molecules are contributed by nonpolar and polar amino acids to the extent of approximately 45% and 55%, respectively. Thus, the generalization that the "inside" of a globular protein molecule in solution is mainly nonpolar, and the "outside" polar, may be an oversimplification. They suggest that assignment of amino acid side chains to "buried" and "exposed" categories is of questionable value in attempts at empirical structure prediction.

sional structure (indicated by the fact that the mutant enzyme does not react with antibodies against the native enzyme) lacking biological activity. Second-site mutants, which produce active enzyme, were then isolated. This regain of function is achieved by introducing a second amino acid replacement such that the side chains of the two new amino acids presumably interact in a way similar to that of the original residues to preserve the three-dimensional structure, and hence biological activity, of the native enzyme. The lesion introduced by the initial mutation has thus been "cured" by the second one.

The kinds of alterations that can be made in the amino acid sequence without loss of folding capacity varies greatly from protein to protein, and it is not possible at present to suggest general rules. For example, removal of the NH_2-terminal 20 residues of bovine pancreatic ribonuclease (Fig. 4) produces RNase-S-protein (Richards, 1958) which is an intrinsically unstable structure lacking, in its reduced form, the interactions necessary to determine the three-dimensional arrangement of its chain or the pairing of its 8 half-cys-

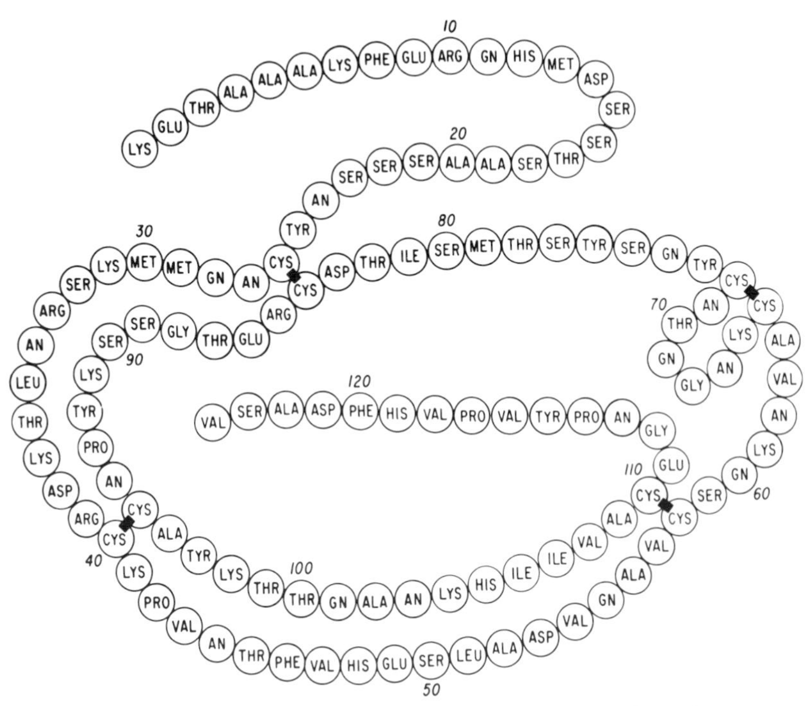

FIG. 4. The amino acid sequence of bovine pancreatic ribonuclease.

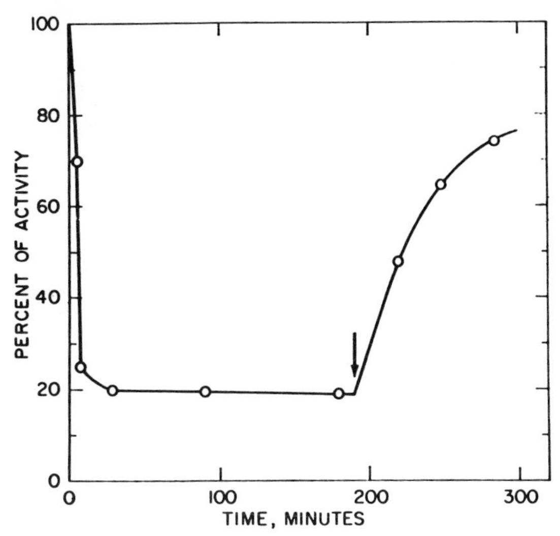

FIG. 5. Inactivation and disulfide interchange of native RNase-S-protein catalyzed by prereduced interchange enzyme (Kato and Anfinsen, 1969b; Fuchs *et al.*, 1967). The arrow indicates the time of addition of RNase-S-peptide (1.3 equivalents relative to S-protein) to the reaction mixture. Since RNase-S-protein is itself catalytically inactive, RNase-S-peptide (1.3 equivalents) was added to aliquots taken prior to the time marked by the arrow, and the mixtures were assayed for RNase activity.

tine residues in disulfide linkage. Conditions favorable for disulfide interchange lead to "scrambling" of the disulfide pairing, and restoration of the native pairing of SH groups and the normal three-dimensional structure takes place only when the S-peptide is added to the solution (Kato and Anfinsen, 1969b). In the experiments summarized in Fig. 5, S-protein was exposed, at zero time, to an enzyme isolated from microsomal membranes that catalyzes disulfide interchange (Goldberger *et al.*, 1963; Venetianer and Straub, 1963; De Lorenzo *et al.*, 1966; Fuchs *et al.*, 1967). In such a system, the free energy can be lowered by the interchange among disulfide bonds to form a scrambled set; the enzyme catalyzes interchange between alternative disulfide pairings but, since there is presumably no uniquely stable set of SS bonds on the basis of the interactions among the amino acid sequence in the truncated ribonuclease chain, only a chaotic mixture of isomers results. When the interactions for correct SS pairing and for the determination of the entire enzyme structure are restored by the addition of the S-peptide portion, the catalyzed disulfide interchange leads to the thermodynamically most stable structure with its proper pairing of sulfhydryl groups.

In contrast to the case of RNase-S, five amino acid residues may be removed from the NH_2-terminus of staphylococcal nuclease (Bohnert and Taniuchi, 1972; Cone *et al.*, 1971) (Fig. 6) without loss of either activity or, on the basis of a variety of physical measurements, of its three-dimensional conformation. This derivative is capable of undergoing completely reversible denaturation (Taniuchi *et al.*, 1967; Taniuchi and Anfinsen, 1968). Three additional residues, numbers 6, 7, and 8, are probably also dispensable, as suggested by the studies on semisynthetic nuclease (Ontjes and Anfinsen, 1969) discussed in Section V.

Removal of a tetrapeptide sequence from the COOH-terminus of pancreatic ribonuclease by pepsin digestion (Anfinsen, 1956) produces a derivative which is enzymically inactive (pepsin-inactivated ribonuclease, PIR), and addition of the removed peptide does not restore activity (as it does when RNase-S-peptide is added to RNase-S-protein). Inspection of the molecular structure (Kartha *et al.*, 1967; Wyckoff *et al.*, 1970) of ribonuclease shows clearly that RNase-S-peptide, whose partial helical character exists only at low temperature when alone in solution (Klee, 1968; Brown and Klee, 1969, 1971; Silverman *et al.*, 1972), can assume its partially helical character and bind tightly to the RNase-S-protein moiety through a number

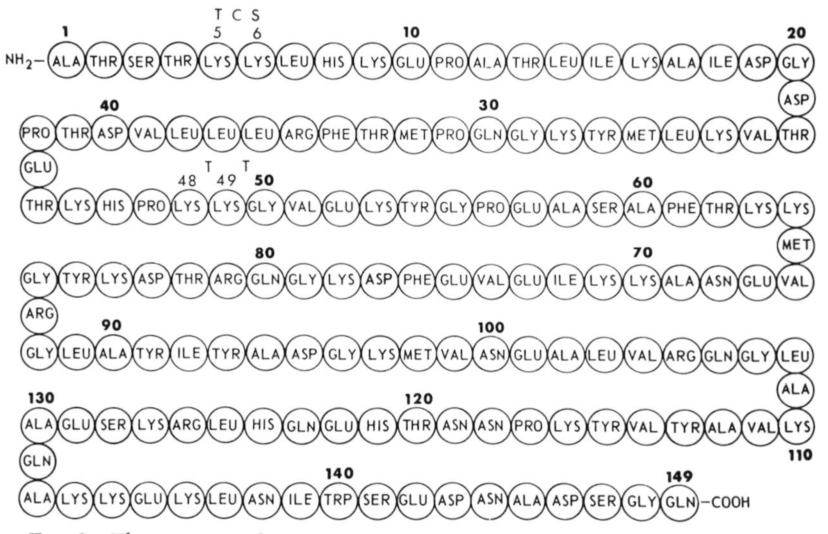

FIG. 6. The amino acid sequence of an extracellular nuclease of *Staphylococcus aureus* (Cone *et al.*, 1971; Bohnert and Taniuchi, 1972). Specific points of cleavage, during digestion in the presence of deoxythymidine 3′,5′-diphosphate and calcium ions by trypsin (T), chymotrypsin (C), and subtilisin (S) are indicated.

of side-chain interactions, even though the 20–21 peptide bond is not re-formed. The COOH-terminal tetrapeptide fragment, however, is apparently too small to interact with a large enough binding energy to provide the additional interactions necessary for determination of the stable structure of RNase (although, as we shall discuss in Section III, complementation with a longer COOH-terminal fragment of 14 residues *will* regenerate an active structure). Reduced PIR cannot be reoxidized to yield the correct half-cystine pairs, and conditions of disulfide interchange cause randomization of the native pairing (Taniuchi, 1970).

An example of a protein that does not arise spontaneously from its reduced form—that is, that appears to be the metastable product of a stable precursor—is the naturally occurring hormone, insulin. Experience with the disulfide interchange enzyme had suggested that it possessed the capacity to "probe" protein structures for stability of particular arrangements of SS bonds, and for whether the pairing of half-cystine residues was specifically determined by the covalent structure of the polypeptide chains (Givol *et al.*, 1965). Addition of this enzyme in its reduced (SH) form to insulin catalyzed a rapid disulfide interchange, and a mixture of randomly cross-linked A and B chains precipitated.

After the isolation of proinsulin (Fig. 7) by Steiner (1967) and his colleagues and the determination of the sequences of proinsulins from several species (e.g., Chance *et al.*, 1968), it became clear that the correct three SS bonds in insulin are determined by the *intact* single-chain structure of the precursor molecule, and that cleavage at two points to yield the commonly observed insulin structure occurs after initial formation and storage of this substance; that is, proinsulin is a thermodynamically stable molecule (with respect to the reduced one), but insulin is not. In this respect, the proinsulin–insulin system resembles other zymogen systems found in the pancreas—for example, chymotrypsinogen, which can refold and form correct SS bonds after prior reduction, and chymotrypsin, which cannot (Givol *et al.*, 1965).

While the experiments on ribonuclease S discussed above and on staphylococcal nuclease discussed in Section III indicate that several fragments can associate in the refolding process, in general the folding process is more easily accomplished if it is an *intra*molecular one. Intramolecular folding must work against conformational entropy. However, folding which involves *inter*molecular interaction between two or more fragments must, in addition, work against a loss of translational and rotational entropy, and will thus be difficult

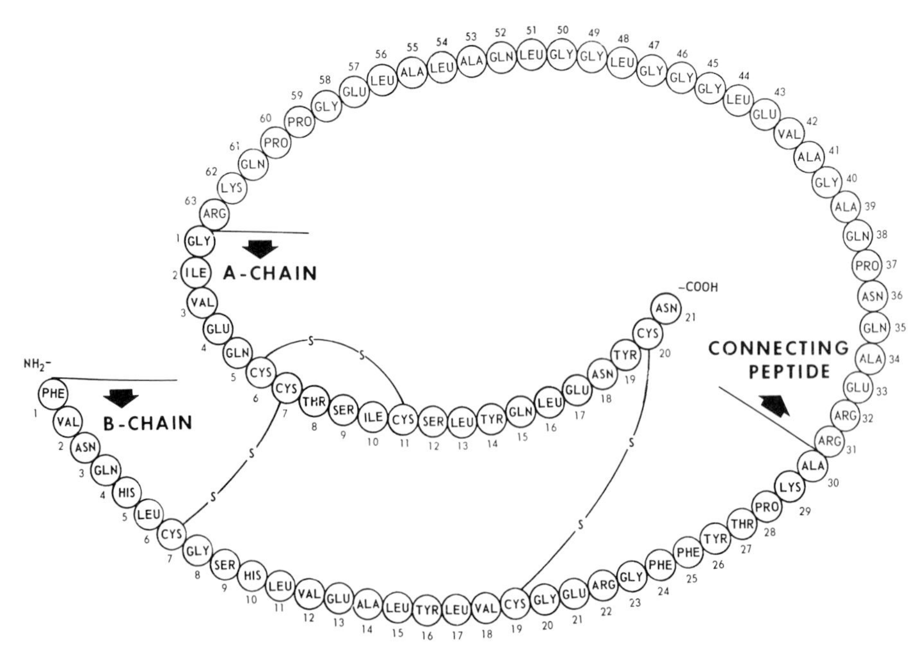

FIG. 7. The structure of porcine proinsulin, including the amino acid sequence of the connecting peptide that joins the B chain to the A chain. From Chance *et al.* (1968). Copyright 1968 by the American Association for the Advancement of Science.

unless accompanied by strong specific association between the fragments as occurs in the cases of ribonuclease S and staphylococcal nuclease. This is presumably the reason why the three cyanogen-bromide peptide fragments of myoglobin do not regain the native conformation easily when mixed together (Epand and Scheraga, 1968); also, perhaps the charged end groups, which are removed further from the site of binding when longer, overlapping peptides are used, interfere with the folding of myoglobin.

In some cases, either or both of the fragments in a complementing pair derived by proteolytic digestion of a protein chain may be degraded without loss of the capacity to regenerate a biologically active complex. For example, RNase-S-peptide may be degraded with carboxypeptidase A, with removal of 5 COOH-terminal residues, without losing its ability to attach to, and activate, RNase-S-protein (Potts *et al.*, 1963). A number of synthetic derivatives of RNase-S-peptide, shortened at either or both the NH_2- or COOH-termini, have been prepared; these are also able to produce active variants of RNase-S (Hofmann *et al.*, 1966; Scoffone *et al.*, 1967).

On the other hand, the low yield of insulin regenerated from the two reduced, separated chains upon combination in solution under oxidizing conditions is a good example of the effect of extensive removal of amino acid residues from the structure of a protein molecule. The reduced chain of the insulin precursor, proinsulin, regenerates the native, disulfide-cross-linked precursor in high yield (Steiner, 1967).

Additional insight into the spontaneous folding of proteins is provided by a consideration of the energetics of the polypeptide chain. Conformational energy calculations, discussed in Section VIII, suggest that the interactions among the various portions of the chain (primarily local interactions, but also medium- and long-range ones) contribute in dictating the three-dimensional structure that the polypeptide will adopt in a given solvent. Two sources of entropy also contribute; these are the solvation entropy (contributing to a preference for nonpolar groups to be in the interior and for polar groups to be on the surface of a protein in water) and the conformational entropy, the latter arising from fluctuations around the native conformation (primarily from vibrations around single bonds). The large conformational entropy characteristic of the denatured form can be reduced by the introduction of disulfide cross-links, thereby stabilizing the native structure relative to the denatured form (see Section IV). Ligands can also stabilize native proteins, despite a loss of conformational entropy, by providing binding energy (see Section IV).

III. *In Vitro* Complementation of Protein Fragments

Experiments on ribonuclease S, involving association of two complementary peptide fragments to form a biologically active enzyme, have been discussed in Section II. In this section we shall describe similar results with staphylococcal nuclease and with other derivatives of bovine pancreatic ribonuclease. These experiments help to delineate those parts of polypeptide chains that provide the interactions required for folding to yield a functional three-dimensional structure.

From crystallographic studies, it appears that certain residues may not be crucial for maintaining a native structure. Thus, looseness of the chain in the neighborhood of residues on either side of residues 19–21 of ribonuclease-S is indicated by the poor resolution of these residues in electron density maps (Wyckoff *et al.*, 1970). Similarly, the NH_2- and COOH-terminal portions of the sequence of staphylococcal nuclease, which are not visible on the electron density map (Arnone *et al.*, 1971), are probably not important for structural stabili-

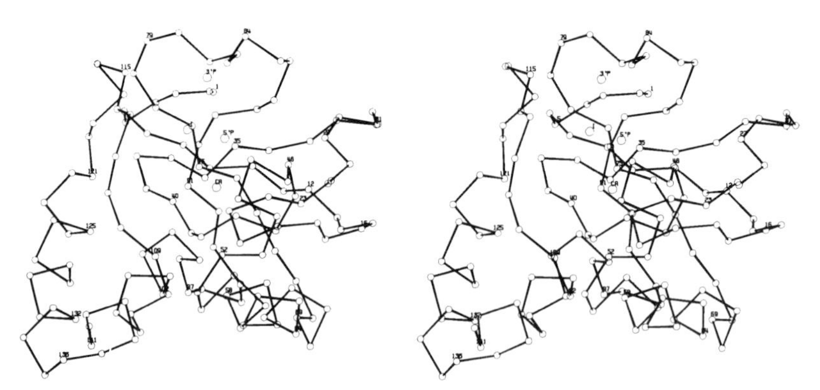

FIG. 8. A stereo drawing of the backbone of staphylococcal nuclease, showing the locations of the essential calcium ion and the 5′iodo-, 3′phosphate, and 5′phosphate groups on the heavy atom ligand, 3′,5′,5-iodothymidine diphosphate, that was used in crystallographic studies. From Arnone *et al.* (1971).

zation. This conclusion is also reached on the basis of synthetic studies in which residues 1–8 (Ontjes and Anfinsen, 1969) and 141–149 (Parikh *et al.*, 1971) have been omitted from semisynthetic nuclease (see Section V), with retention of significant activity, and the physical properties of the native enzyme.

The overall geometric features of staphylococcal nuclease are summarized in Fig. 8. This stereoscopic representation of the polypeptide backbone is based on the coordinates of the α-carbon atoms of each amino acid in the chain (Arnone *et al.*, 1971). Nuclease may be cleaved (Taniuchi *et al.*, 1967; Taniuchi and Anfinsen, 1968) with trypsin in the presence of calcium ions and a tightly bound competitive inhibitor, 3′,5′-thymidine diphosphate (pdTp), to yield complementing fragments (Figs. 8 and 9). Residues 1 through 5 are first rapidly cleaved off to yield nuclease-(6–149), which has the same specific activity and general physicochemical behavior as native nuclease. The bonds between residues 48 and 49 or 49 and 50 in the sequence -Pro-Lys-Lys-Gly- (residues 47–50) are then cleaved somewhat more slowly to yield two fragments which interact noncovalently to form a stable complex called nuclease-T. When cleavage occurs between 49 and 50, residue 49 is subsequently removed from the (6–49) fragment by the further action of trypsin. The final complex consists of two species: nuclease-T-(6–48) combined with either nuclease-T-(49–149) or nuclease-T-(50–149). In the presence of calcium ions, both of these complexes exhibit enzyme activity—about 8–10% that of the native enzyme. Nuclease-T possesses a helix content, as indicated by its circular dichroism or optical rotatory dispersion properties, which is approximately 80% that

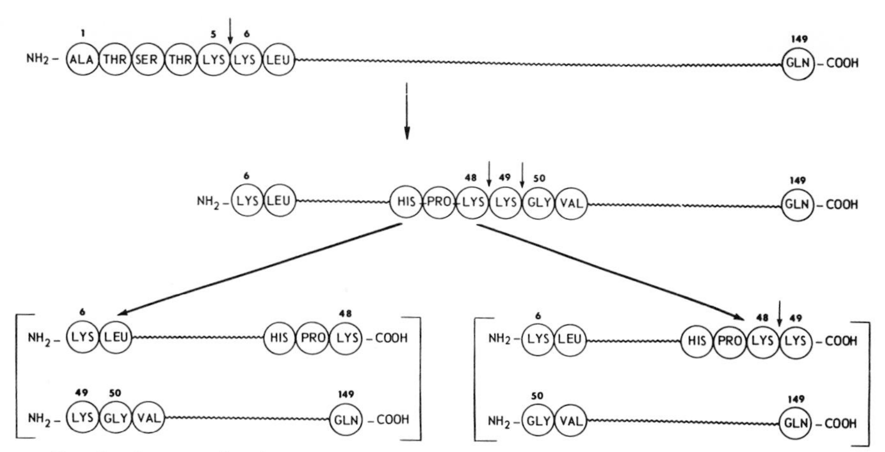

FIG. 9. Steps in the cleavage of staphylococcal nuclease by trypsin in the presence of the stabilizing ligands, Ca^{2+} and 3′,5′-thymidine diphosphate. From Taniuchi and Anfinsen (1968).

of native nuclease (ca. 18% of the chain) and exhibits the same ultraviolet spectral anomalies in the tyrosine region. Its tryptophan residue is similarly buried in a nonpolar environment as judged by measurements of the fluorescence emission spectrum upon activation with light at 295 nm. The nuclease-T complex consisting of residues (6–48) and (49–149) has been crystallized in the presence of calcium ions and pdTp and is isomorphous with similarly liganded native nuclease (Taniuchi *et al.*, 1972).

Several other sets of complementing fragments of staphylococcal nuclease have been studied (Taniuchi and Anfinsen, 1971). Figure 10 gives a schematic summary of these fragments in terms of their origins within the sequence of the native chain of 149 residues. The three helical regions in nuclease are located in the (49–149) fragment, discussed above as part of nuclease-T, and antiparallel pleated sheet structures are located between residues 12–35 and (not indicated in Fig. 10) approximately 73–95.

If nuclease is trifluoroacetylated (TFA), brief trypsin digestion yields a derivative, from which the TFA groups may be removed under mild conditions, consisting of residues 1–126 (Taniuchi and Anfinsen, 1969). This material is essentially inactive[5] and struc-

[5] Even after exhaustive purification by ion exchange and affinity chromatography, fragment (1–126) shows about 0.12% of the activity of native nuclease. This activity is intrinsic to the molecule (Sachs *et al.*, 1974) and not due to persistent contamination with native enzyme. This is demonstrated by the fact that an immunoglobulin fraction, isolated from antinuclease antiserum, which is specific for the antigenic determinant supplied by the fragment (127–149), inactivates native nuclease but has no effect on the slight activity associated with (1–126).

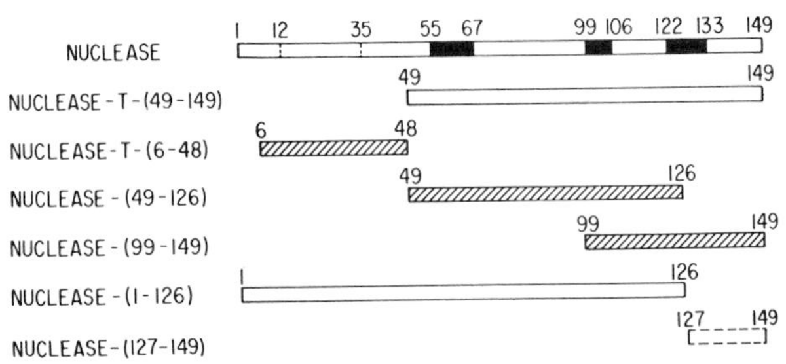

FIG. 10. Diagram of linear relationship of the amino acid sequences of nuclease and its fragments which yield productive complementation. The X-ray crystallographic study (Arnone *et al.*, 1971) has shown β structures between residues 12–35 and 73–95 and three α-helical parts (black bars) between residues 55–67, 99–106, and 122–133, which are indicated on the nuclease line. Nuclease-(127–149) does not bind to nuclease-(1–126). From Taniuchi and Anfinsen (1969).

tureless as judged by hydrodynamic and spectral measurements, and is not activated, or folded, by the addition of the (127–149) fragment which has been removed. In terms of the interactions required for folding, this latter behavior is analogous to that of the mixture of the tetrapeptide, containing residues 121–124 of pancreatic ribonuclease, with des-(121–124)-ribonuclease (PIR) which is also inactive and structurally unstable (Anfinsen, 1956). However, (1–126) may be activated by the addition of the peptide, (49–149), or of a fragment produced by CNBr digestion of native nuclease, containing residues (99–149). In Fig. 11, A and B, fragment (1–126) is shown in two roles, first donating the (6–48) portion of the nuclease sequence to complement (49–149) to produce a nuclease-T-like structure and, second, furnishing the interactions required for complementation with the COOH-terminal portion of (49–149) to yield a stable nucleaselike structure. The activities of the complexes shown in Fig. 11A and 11B, which are sufficiently stable to permit their isolation on phosphocellulose columns, are approximately 8% and 15%, respectively, that of the native enzyme.

As mentioned above, the addition of calcium ions and pdTp stabilizes the nuclease structure against trypsin cleavage except at certain specific peptide bonds that are present in accessible portions of the three-dimensional structure. If these ligands are added together with trypsin to a *freshly prepared* equimolar mixture of (1–126) and (49–149), presumably containing both of the complementation complexes shown in Fig. 11, the protease "clips" off the redundant portions of the sequence (Taniuchi and Anfinsen, 1969). After digestion, the fragment (6–48) is left from (1–126), and the fragment

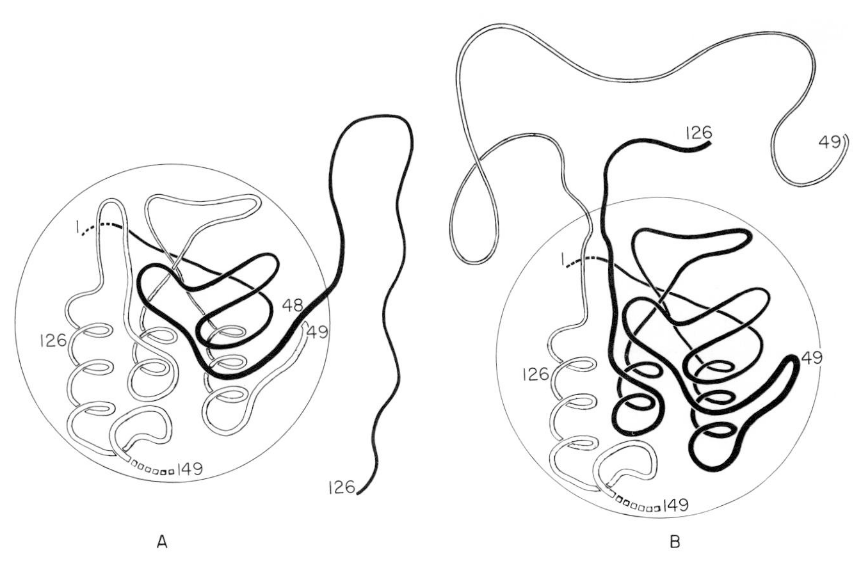

FIG. 11. Schematic representation of the two types of complementation formed by noncovalent interactions of nuclease-(1–126) (filled structures) and nuclease-T-(49–149) (open structures). The ordered structures are surrounded by the circles. Residue numbers are also shown. Type A: (1–48) + (49–149); type B: (1–126) + (111–149). From Taniuchi and Anfinsen (1971).

(111–149) from (49–149). These two new fragments are produced in approximately equal amounts, indicating that the initial formation of the two kinds of complementing structures in Fig. 11A and B occurs to about the same extent. With longer preincubation of the mixture of (1–126) and (49–149) before trypsin addition, the ratio of the complex [(6–48) + (49–149)] to [(1–126) + (111–149)] becomes about 2:1, indicating the greater thermodynamic stability of the former pair of fragments (Taniuchi and Bohnert, 1973).

The complementing system involving (1–126) and (99–149) is of interest in connection with the ability of fragments of protein chains to form stable, cooperative structures, such as helices. As in the experiments just described, careful trypsin digestion of this complex in the presence of calcium ions and pdTp removed a number of the redundant residues to yield a "trimmed" derivative containing (1–126) and (111–149) (Taniuchi and Anfinsen, 1971). Residues 99–106 form one of the three helical regions in native nuclease. Since residues 99–110 were removed from (99–149), they were presumably not involved in long-range interactions that might have hindered proteolytic digestion of the complex. Furthermore, the helix-forming residues, 99–106, are at the NH_2-terminus of the (99–149) component and, as suggested a number of years ago by Schellman (1955), would have the incomplete hydrogen bonding arrangement characteristic of the end of a chain. The helical section in the complex must therefore be contributed by (1–126) and must be stabilized by long-range interactions.

One may ask how extensively a polypeptide chain may be cleaved and still retain, in its fragments, the capacity for complementation. Recent experiments by Andria *et al.* (1971) on staphylococcal nuclease and by Lin *et al.* (1970) and Gutte *et al.* (1972) on bovine pancreatic ribonuclease have shown that even *three* fragments can complement effectively to yield biologically active complexes, provided that the fragments are large enough to provide enough interaction energy to overcome the loss of translational and rotational entropy, and to stabilize nucleation sites; *long* fragments may also be required to avoid end effects of charged groups. The former group prepared the fragments (6–48), (49–126), and (99–149) and demonstrated that these bind to one another, in an equimolar amount, to form a complex with low but significant nuclease activity. The complex bound to an agarose-pdTp affinity column in the presence of calcium ions, but not in the absence of this metal ion. When any two of the three components of the complex were passed through such a column, no binding took place.

Lin and his colleagues showed that the ternary complex of RNase-S-peptide (residues 1–20 of ribonuclease), des-(121–124)-RNase-S-protein and a synthetic peptide containing residues 111–124 of the protein exhibited approximately 30% of the activity of the native enzyme. The intact disulfide bonds in the largest component of this complex may be responsible for its surprisingly high stability and activity. In the various derivatives of the disulfide bond-free staphylococcal nuclease, activity and stability to heat decrease markedly as one goes from the single-chain native nuclease to the double-chain nuclease-T or the three-component structure described above.

To summarize, a polypeptide sequence can frequently be shortened, cleaved into fragments, and otherwise manipulated without destroying the interactions required for proper folding. Furthermore, experiments such as those just described (see Fig. 11) indicate that the three-dimensional native structure of a protein, inferred from the presence of biological activity, can be attained in more than one way. The eliminated or modified portions in these experiments may be essential in the living cell (one might expect that natural selection would otherwise have gotten rid of them by now), but may be redundant or superfluous *in vitro*. The experimental determination of the minimum portion of the chain required to form a nativelike structure may help to simplify the ultimate calculation of the three-dimensional structure from amino acid sequence data alone. However, from the fact that less than 100% activity is sometimes attained in the complementation experiments, it appears that the portions that have been removed probably provide significant long-range interactions, required to stabilize a *fully active* enzyme.

IV. Flexibility of Proteins in Solution: Effect of Cross-Links and Ligands

A native globular protein molecule in solution is not a static entity, but rather one which undergoes conformational fluctuations about its most stable conformation. These fluctuations arise primarily from the ease of rotation about the single bonds of the backbone and side chains of the protein molecule. It has, of course, been known for many years that globular proteins are denatured when exposed to extremes of pH or to urea or guanidine (see Anson and Mirsky, 1934a,b). The significance of this phenomenon in terms of conformational fluctuations was not appreciated since proteins were treated as essentially rigid bodies. However, the demonstration that proteins could swell in appropriate solvents implied that some degree of hindered rotation about single bonds, manifesting itself in conformational flexibility, was possible. Early evidence for such flexibility was obtained from light scattering (Doty and Katz, 1950) and hydrodynamic data (Scheraga and Mandelkern, 1953), which showed that a globular protein molecule could swell in the presence of certain additives, such as urea.

The internal rotations about single bonds should lead to conformational fluctuations, which constitute a conformational entropy contribution to the free energy, and hence should increase the stability of the average native conformation of a protein (Go and Scheraga, 1969) compared to what it would be if the native protein were rigid.

One form of direct evidence for conformational fluctuations comes from hydrogen-exchange experiments. The exchangeable hydrogens of a protein have been observed to fall into several classes on the basis of their rates of exchange, viz., those with rapid or instantaneous rates, intermediate rates, and very slow rates (Hvidt and Linderstrøm-Lang, 1954; Hvidt and Nielsen, 1966). The rate of exchange depends on many factors including interactions involving hydrogens, and those which shield hydrogens from the solvent.

Since the extent of interaction or shielding may fluctuate as the conformation fluctuates, the rate of exchange will thus be influenced also by the flexibility of the protein molecule; however, it will usually be difficult to discriminate between flexibility of the molecule and a range of accessibilities of its protons.

Flexibility can be considerably influenced by covalent cross-links and by noncovalently bound ligands. Let us consider first the case of noncovalent ligand binding. If the apoprotein is in equilibrium with partially unfolded forms, the totality of which makes the protein appear to be flexible, the binding of a ligand will shift the equilibrium in favor of a single form. Thus, flexibility is reduced but, despite this loss of conformational entropy, the protein is stabilized in essentially one conformation because of the binding energy between the ligand and the protein. For example, the conformational flexibility of apomyoglobin is reduced by complexing with a heme group, that of staphylococcal nuclease by complexing with calcium ion and pdTp, and that of pancreatic ribonuclease by complexing with an anion, such as a phosphate ion. The effects of such ligands are shown in Figs. 12 and 13 for staphylococcal nuclease and myoglobin, respectively. In the absence of ligands, both proteins (in which all exchangeable hydrogens had previously been replaced by tritium) lost most of the tritium atoms at pH 7–8 at an extremely rapid rate (Schechter *et al.*, 1969), indicating that the structures were quite flexible. In the presence of the ligands, hematin or Ca^{2+} and pdTp, the flexibility was reduced to such an extent that, in myoglobin and staphylococcal nuclease, respectively, 25 and 34 tritium atoms became essentially nonexchangeable even after many hours in aqueous solution. Presumably, the rapidly exchanging tritium atoms are near the "outside" of the protein, while the slowly exchanging ones are on the "inside."

In contrast to ligand binding, covalent cross-links, such as disulfide bonds, which appear in *specific* sites in a globular protein, stabilize the protein against denaturation because such cross-links lower the entropy of the *denatured* or random-coil form of the protein. A randomly coiled chain has considerable entropy, and the effect of cross-links on its entropy depends on how the cross-links are introduced. The cross-links can be introduced either at random in the random coil or, alternatively, by ordering the chains, cross-linking the ordered system, and then melting it to obtain a cross-linked random coil. These two types of random coil differ in entropy, the latter having a lower entropy than the former. For the former, the entropy is essentially unaffected since the entropy loss

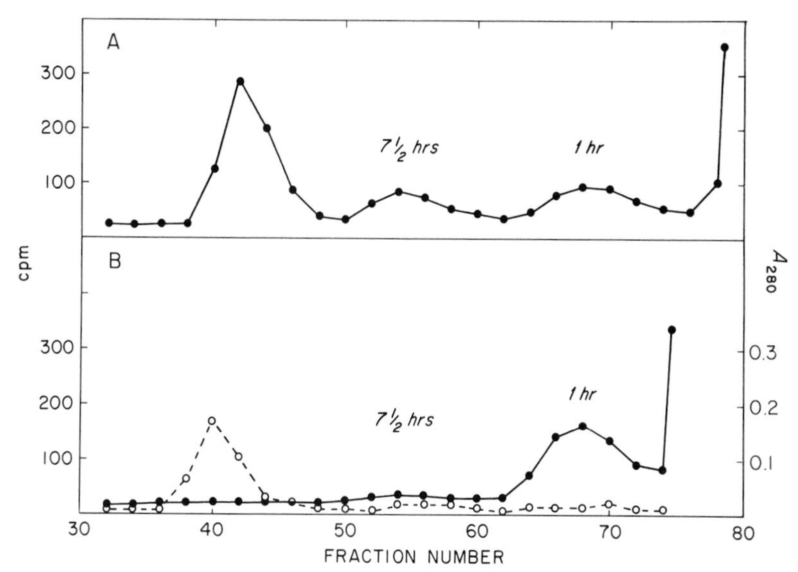

FIG. 12. Interrupted gel filtration of nuclease. Aliquots containing 3.4 mg of tritium-labeled nuclease in 0.5 ml of 0.05 *M* Tris·HCl, pH 8.1, were filtered on a Sephadex G-25 column (2 × 100 cm) in the same buffer at room temperature. (A) 3′,5′-Thymidine diphosphate and calcium chloride were added to the protein shortly before application, and to the column buffer, to make final concentrations of 6×10^{-4} *M* and 1×10^{-2} *M*, respectively; (B) no ligands added. ○, Optical density at 280 nm; ●, radioactivity. The rapid elution of the protein was interrupted after the effluent volume was 40 and then 70 ml, for 1 and 7.5 hours, respectively. The radioactivity in tubes 35–50 represents tritium atoms still bound to the protein after 8.5 hours of out-exchange on the column; the ligands have "trapped" about 34 tritium atoms per molecule of nuclease. The radioactivity under peaks labeled "1 hr" and "7½ hr" represents out-exchange during these periods of flow interruption. From Schechter *et al.* (1969).

from the restriction of chain mobility by the cross-links is compensated by an entropy gain from the random placement of cross-links. However, if the cross-linking is carried out in the ordered state, the resulting random coil exhibits only the entropy loss from the restriction of chain mobility; for this case, Flory (1956) has shown how the entropy loss of the random coil depends on the number of cross-links (introduced in the ordered state), and thus has provided a quantitative basis for the well-known observation (Kauzmann, 1959b) that protein stability increases as the number of disulfide bonds increases. For example, if one cross-link occurs at every 28 residues, Flory's (1956) theory leads to a decrease of 8.5 EU for the denatured form (Scheraga, 1963). A model polymeric system which illustrates the effect of placing cross-links in specific sites when the chains are

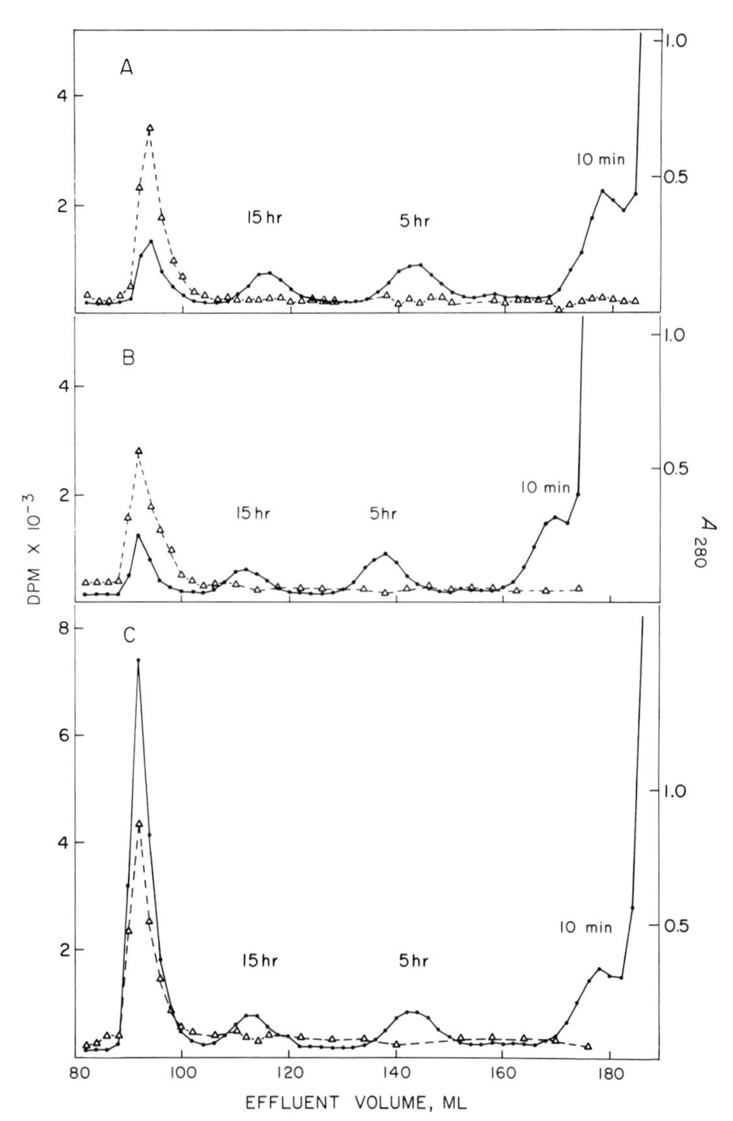

FIG. 13. Gel filtration of myoglobin, apomyoglobin, and reconstituted metmyoglobin at room temperature. Aliquots containing 2 mg of protein labeled at pH 8.8 were filtered on a Sephadex G-25, fine grade, column (2 × 100 cm) in 0.05 *M* phosphate, pH 7.0. As described for Fig. 12, the elution of the protein at 60 ml per hour was interrupted after the volume of the effluent was 20 ml, 50 ml, and 75 ml, for 10 minutes, 5 hours, and 15 hours, respectively. Fractions of 2.0 ml were collected, and optical density at 280 nm and radioactivity were assayed. (A) The protein was labeled and filtered as metmyoglobin; (B) the protein was labeled and filtered as apomyoglobin; (C) the protein was labeled as apomyoglobin, reconstituted with hematin, and filtered as metmyoglobin. ●, Radioactivity (disintegrations per min); △, optical density at 280 nm. From Schechter *et al.* (1969).

ordered is racked rubber (Roberts and Mandelkern, 1958); this is rubber which is rendered fibrous by stretching and cross-linking *while in the stretched state.* Roberts and Mandelkern (1958) showed that such cross-linked racked rubber undergoes a phase transition, the melting properties depending on the degree of cross-linking, according to Flory's (1956) predictions. Disulfide bonds of native proteins, being in specific sites, are the analogs of the cross-links that are introduced into rubber when it is in the stretched (ordered) state. Thus, disulfide bonds serve a primary role in reducing conformational fluctuations in the *denatured* form, thereby stabilizing native proteins relative to denatured ones. Disulfide bonds also reduce conformational fluctuations in the native protein, but the larger effect is their influence on the entropy of the denatured form.

In summary, native apoproteins have some degree of flexibility which can be reduced by ligand binding. In some proteins, the conformational fluctuations are reduced by using disulfide bonds in specific locations to destabilize the denatured form relative to the native form.

V. Synthetic Analogs of Proteins

In Section II, we discussed the effect of variations in amino acid sequence on the three-dimensional structure when these variations occur as a result of mutations (e.g., the various hemoglobins), or as compensatory double mutations (e.g., tryptophan synthetase), or are observable in proteins that have homologous sequences (e.g., chymotrypsin and trypsin). Such variations in sequence provide information as to which portions of the polypeptide chain are essential for biological activity and, therefore, presumably for achieving the particular three-dimensional structure.

Unfortunately, natural mutations can provide only a limited number of altered proteins for study. This is because a large fraction of the mutations that can occur in a particular gene are lethal if the protein concerned is an essential one for the physiology of the cell. Furthermore, mutations might cause such severe changes in the interactions involved in the amino acid sequence that an unrecognizable protein might result that could not be isolated for study. For these reasons, recent developments in the field of peptide chain synthesis become of great importance in the production of chemical "mutants" in which the organic chemist can deliberately introduce an alteration in the sequence.

Development of new side-chain protecting groups and coupling procedures have made it possible to synthesize quite long peptide chains, even with the "classical" techniques of peptide synthesis in solution. Such methods have been used for the synthesis of numerous analogs of RNase-S-peptide (Hofmann *et al.*, 1966; Scoffone *et al.*, 1967). Complete syntheses have also been carried out for insulin (Yu-Cang *et al.*, 1961; Jiang *et al.*, 1963; Katsoyannis *et al.*, 1963, 1964; Meienhofer *et al.*, 1963; Zahn *et al.*, 1964; Ching-I. *et al.*, 1964), adrenocorticotropic hormone (Schwyzer and Sieber, 1963), mellitin (Wünsch *et al.*, 1968a,b), vasopressin (Katsoyannis *et al.*, 1957), and oxytocin (du Vigneaud *et al.*, 1954). The entire glucagon molecule (Wünsch *et al.*, 1968a,b) as well as thyrocalcitonin (Guttmann *et al.*, 1971) and secretin (Bodanszky and Williams, 1967; Bodanszky *et al.*, 1967) have also recently been synthesized by these careful stepwise procedures. These molecules are of such a length that the synthesis of analogs might be realistically undertaken using the traditional methods of coupling and purification. Except for insulin, it is not yet known whether the polypeptides listed, all hormones with the exception of RNase-S-peptide, form unique ordered three-dimensional structures in aqueous solution in the absence of their receptors; thus, correlations of amino acid sequence with three-dimensional structure cannot be made.

Since synthetic protein chains should presumably be capable of folding to the native structure, analogs of such chains could, in principle, be produced that would give useful information on the alterations that are permissible and those that are not. However, only in the case of RNase-S-protein, having 104 amino acid residues, has an essentially classical, stepwise synthesis been achieved (Hirschmann, 1971), and the difficulty of the task makes synthesis of further analogs a major undertaking. For this reason, the development of the solid phase method by Merrifield and his colleagues (Merrifield, 1965) has introduced new optimism into this field of research. The method is still far from perfect, suffering from an unpredictable incompleteness of coupling at many steps and from side reactions that may be introduced during deprotection and removal from the resin support. Nevertheless, in cases where products could be purified by "functional" purification, a number of relatively homogeneous polypeptides have been made which exhibited activity and physical properties quite similar to those of the native molecule. Gutte and Merrifield (1969), for example, prepared totally synthetic ribonuclease which, after being allowed to fold into its three-dimensional form with the attendant purification resulting from the restrictions imposed by the pairing of half-cystine residues, was further purified by digestion with trypsin; that is, the molecules with incorrect amino acid sequences were digested with trypsin, whereas native ribonuclease is resistant to tryptic digestion. The final material showed a specific activity of 80% that of native enzyme. Even with this success, however, the effort involved in the synthesis of analogs would still be very great, and no such synthetic "mutants" have yet been made.

Greater synthetic flexibility is afforded by peptides of somewhat more modest size than RNase. As mentioned above, RNase-S-peptide has been extensively studied by the amino acid replacement approach, using traditional peptide synthesis, and many of the side-chain interactions that are responsible for productive interaction with RNase-S-protein have thus been established. Stepwise coupling in solution begins to be less applicable when polypeptides reach 30 residues in length, or thereabouts, owing in large part to the great insolubility that develops with peptides of this size and to the slowness of coupling reactions. The synthesis of a large variety of fragments of staphylococcal nuclease, ranging from about 30 up to 43 residues in length, has been attacked using the solid phase method since the products could be "functionally" purified by complementation with a natural fragment (Ontjes and Anfinsen, 1969; Kato and Anfinsen, 1969a). For example, the complex formed between synthetic fragment (6–47) and natural (49–149) can be purified essentially to complete identity with natural nuclease-T by digestion with trypsin in the presence of the stabilizing ligands, calcium ions, and pdTp (Chaiken, 1971). The capacity of such synthetic fragments to produce activity when added to the complementing natural fragment could also be greatly enhanced by adsorption on, and subsequent elution from, an affinity chromatography column of agarose to which the natural fragment had been attached. In this way, a number of analogs of (6–47) were purified by passage through agarose-(49–149) columns (Ontjes and Anfinsen, 1969). Such experiments lead successfully to purification only when the synthetic fragment complements the immobilized, natural portion of the protein.

Synthetic analogs of nuclease-T have been prepared for two reasons, first as a means of gaining some understanding of the side-chain functional groups that are involved in substrate or inhibitor binding and in the actual catalytic process of phosphodiester cleavage, and second to study the interactions between different parts of the protein that help stabilize the native structure. As illustrated in Fig. 14, the various critical dicarboxylic amino acids (one of the four, aspartic acid residue 19, is not shown in this figure), suggested by crystallographic analysis to be involved in calcium binding, were

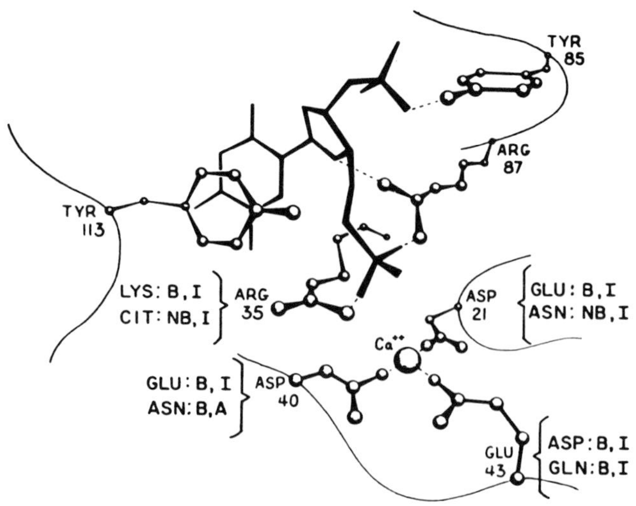

FIG. 14. Correlation of results obtained with synthetic active site analogs with a schematic representation of the deoxythymidine 3′,5′-diphosphate and Ca^{2+} binding site region of nuclease as obtained by X-ray crystallographic studies (Arnone *et al.*, 1971; see also stereo view in Fig. 8). The symbols B and NB indicate whether the synthetic 6–47 analog was or was not bound to nuclease-T-(49–149). The symbols I and A indicate whether the resulting complexes were enzymically inactive or active (Chaiken and Anfinsen, 1971).

replaced by the corresponding amide or by a residue with a longer or shorter side chain (Chaiken and Anfinsen, 1971). In all cases but one (aspartic acid residue 40, where substitution with asparagine still left some residual activity), such changes in charge or chain length resulted in total loss of activity in the complex formed by complementation with (49–149) isolated from the native enzyme. In experiments on the importance of the guanido group at position 35, substitution of arginine with lysine led to inactivation. All the analogs of (6–47) shown in Fig. 14 were capable of *binding* to the larger fragment except for those containing citrulline at position 35 and asparagine at position 21. Thus, in spite of loss of function, the modified fragments could furnish the interactions required to attain the three-dimensional structure of nuclease, or at least most of its characteristic solution properties. Only a direct crystallographic comparison (now in progress by H. Taniuchi and his colleagues) of native nuclease-T and the active or inactive semisynthetic complexes will permit the required quantitative evaluation as to how close these structures are to the native one. At the moment, the extent of similarity to authentic nuclease-T rests on indirect tests such as measurement of the degree of fluorescence emission shown by tryptophan residue 140, which is greatly enhanced when this residue is buried in the hydrophobic pocket formed upon the folding of the nuclease chain (Figs. 15 and 16). The stability of complexes may also be estimated by optical rotatory measurements and by the resistance of the complexes to trypsin digestion in the presence of stabilizing ligands.

An example of the usefulness of synthesis in the investigation of the principles of folding and function is afforded by a comparison of the complex containing synthetic (6–44) and natural (49–149) with that in which the first fragment has been replaced with (6–43) or with an analog of (6–44) having an alanine residue in the COOH-terminal

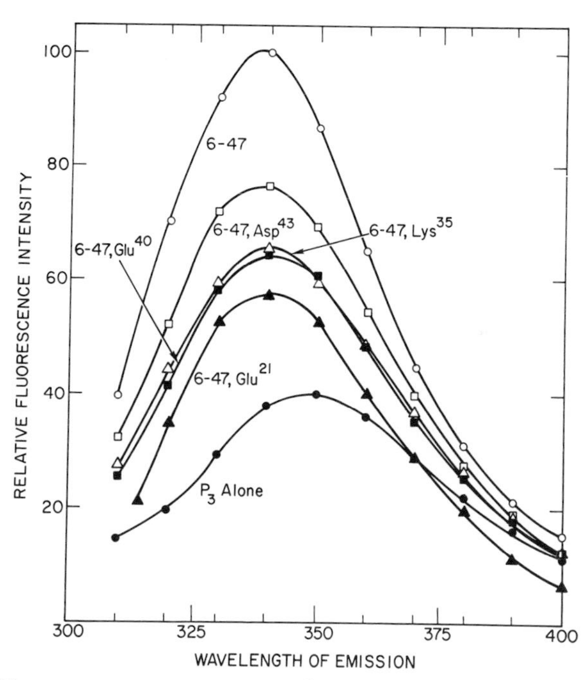

FIG. 15. Fluorescence emission spectra for mixtures of charge-preserved synthetic analog peptides with nuclease fragment (49–149). For each trial, synthetic peptide (0.35 μmole) was mixed with 0.018 μmole of (49–149) in 0.5 ml of 0.05 *M* Tris, pH 8, containing 0.01 *M* $CaCl_2$ and 0.0001 *M* deoxythymidine 3′,5′-diphosphate. After incubation for 1 hour at room temperature, spectra were measured with an Aminco-Bowman recording spectrophotofluorometer. Spectra for synthetic peptide + (49–149) mixtures were corrected for the emission measured for synthetic peptide alone. The spectrum for (49–149) alone was corrected for the small emission given with buffer. From Chaiken and Anfinsen (1971).

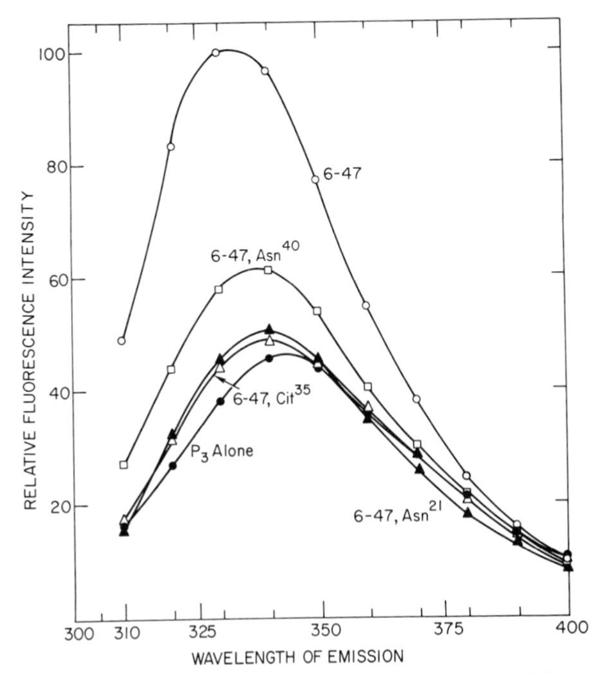

FIG. 16. Fluorescence emission spectra for mixtures of "amide" synthetic analog peptides with nuclease fragment (49–149). Samples were prepared and spectra obtained as described in Fig. 15. From Chaiken and Anfinsen (1971).

position rather than the naturally occurring threonine (Sanchez *et al.*, 1973). The complex with (6–44), whose activity is the same as that of the complex containing (6–47), is stable to trypsin in the presence of ligands and shows native-type optical and fluorescence properties. Replacement of threonine with alanine causes no significant change in any of these properties. However, when glutamic acid 43 (see Figs. 6 and 14) is made the COOH-terminal residue, activity is essentially completely destroyed. The complex formed shows somewhat less stability than that of the (6–44) complex, but it is clear that a native-type complex *does* form since trypsin digestion is much more rapid when the stabilizing ligands are omitted. Thus, even this considerably modified and inactive analog of nuclease can supply the proper interactions required for the formation of a macromolecule bearing a unique binding site for calcium and pdTp and possessing a remarkably high degree of stability in solution. Experiments of this type serve to identify the portions of the chain that are essential for folding to the native conformation.

VI. Folding of Fibrous Proteins

As indicated in the Introduction, some globular proteins exist as single-chain units, whereas others form multisubunit aggregates. Fibrous proteins, such as myosin, fibrin, keratin, collagen, generally are most stable in *aggregated* forms; that is, their native conformations are determined by *both* intermolecular and intramolecular interactions. The possibility could exist that a thermodynamically stable state of a fibrous protein, such as collagen (in the form in which it is generally isolated for study), might be a mixture of many imperfectly matched aggregated chains, as is observed with double-stranded *synthetic* polynucleotides [see Poland and Scheraga (1970) for discussion of stability of such structures]. If perfectly matched structures are to be attained, a large set of specific interactions are required. For example, the newly synthesized chains could be properly aligned by connecting covalent bonds, and the connecting segment be then removed by proteolysis. If such a mechanism were operative, the resulting molecule might not be thermodynamically stable with respect to unaggregated chains. Unfortunately, except for collagen, very little is known about these additional interactions or the mechanisms that lead to perfect matching in the formation of fibrous structures.

When the myosin "molecule" [now known to consist of two long polypeptide chains (Slayter and Lowey, 1967; Godfrey and Harrington, 1970) together with smaller chains (Dreizen *et al.*, 1967; Weeds, 1969) which must be present for ATPase activity in the myosin complex] is completely dissociated in 5 *M* guanidine hydrochloride solution, removal of the denaturing agent by dialysis at neutral pH leads to the regeneration of an entity having the hydrodynamic properties of the original myosin (Godfrey and Harrington, 1970). Enzyme activity is not recovered, possibly because of lack of proper incorporation of the smaller "light" chains. No data are available on the kinetics of the regeneration process, either *in vivo* or *in vitro*, and hence a discussion of myosin folding is not possible at the present time.

The details of the formation of fibrin clots from thrombin-activated fibrinogen (i.e., fibrin monomer) are also obscure. Considerable information is available about the geometry of aggregation of fibrin monomer, during the early stages of the reaction, in which short rodlike oligomers are formed by staggered-overlapping (Krakow *et al.*, 1972) with a first-order rate of release of heat (Sturtevant *et al.*, 1955); similarly, there is considerable information about the structure and properties of the end product, viz., the fibrin clot (Ferry and Morrison, 1947). However, there is relatively little information as to how the intermediate rodlike polymers proceed to the organized, periodic structures observed in electron micrographs of fibrin clots (Hall, 1949). For a recent review of fibrinogen and fibrin, see Doolittle (1973).

In the case of collagen, there is more information available about the formation of the native structure. The rate of refolding of denatured collagen *in vitro* is extremely slow compared to that of the globular proteins, particularly those free of disulfide bonds. Harrington and his colleagues have carried out an extensive and careful series of studies of the kinetics of the renaturation of single and cross-linked collagen chains which entail processes with half-times of many hours (Hauschka and Harrington, 1970a,b,c). These experiments provide useful examples of the crystallization of long polypeptide chains during the folding process, but the times involved are much too long to permit any interpretation of collagen folding as it occurs *in vivo*. Experiments involving pulse-labeling of collagen-producing tissue cultures, much like the early Dintzis (1961) experiments on hemoglobin synthesis in reticulocytes, show that the overall time for the assembly of triple-stranded procollagen, including time required for numerous hydroxylation and glycosylation steps, is only about 6 minutes (Speakman, 1971; Vuust and Piez, 1972; Piez, 1972) (Fig. 17).

The slow *in vitro* rate is thought to be due to the mismatching of the amino acid triplets that constitute the backbones of the α_1 and α_2 chains (a typical sequence would be -Gly-X-Y-Gly-Pro-Y-Gly-X-Hyp-Gly-Pro-Hyp-, where X and Y can be essentially any amino acid but where the occurrence of Gly at every third position is obligatory for the formation of the collagen fold). Using the long chains obtained by the denaturation of collagen, mismatching can be very severe. In shorter fragments of collagen such as the 36-residue peptide, α1-CB2, isolated from digests of the α_1 chain, the third-order dependence on concentration inherent in triple helix formation can easily be shown (Piez and Sherman, 1970). Furthermore, any mismatching of the three *short* chains, leading to the formation of "out-of-phase" triple helices with unmatched ends, could be rearranged much more easily to yield the maximally cross-linked form with lowest conformational free energy.

The relatively short time required for collagen formation *in vivo* has been rationalized on the basis of the hypothesis summarized in Fig. 17 (e.g., Speakman, 1971) involving precursor chains (Layman *et*

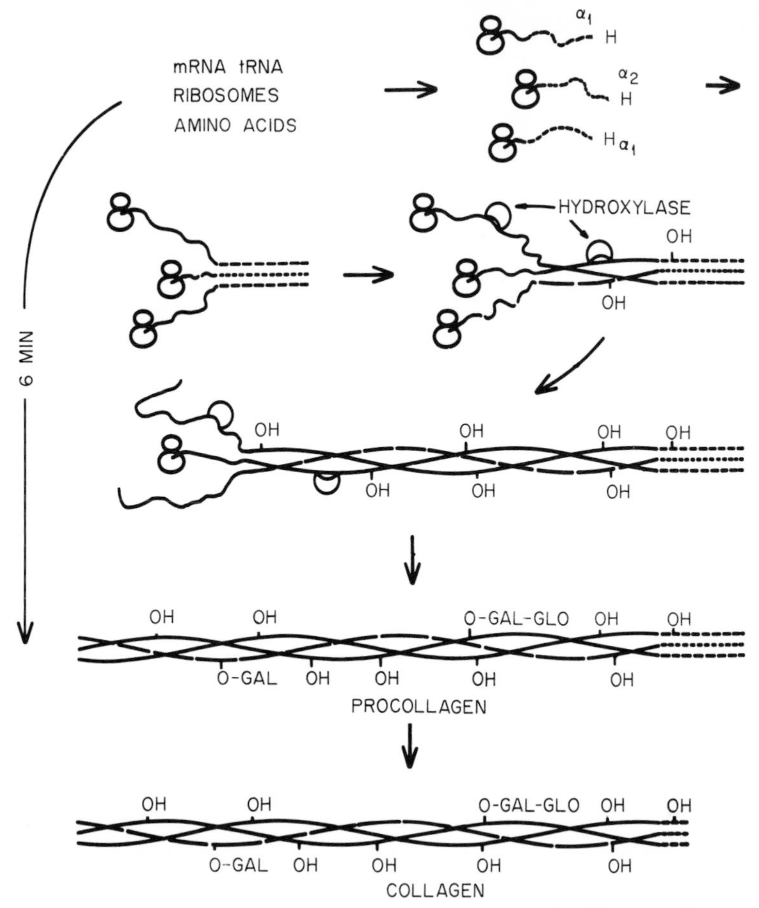

FIG. 17. A schematic representation of the biosynthesis of collagen. The NH_2-terminal dash lines represent the "procollagen" portions of the α chains, which are assumed to direct aggregation and facilitate transport, and which are lost prior to fibril formation. The hydroxylase is indicated by a three-quarter moon. See Speakman (1971).

al., 1971; Bellamy and Bornstein, 1971), especially designed to ensure that mismatching will not occur (much as proinsulin ensures the formation of the correct disulfide bonds in insulin prior to conversion, by proteolytic cleavage, from a single- to a double-stranded molecule). As illustrated schematically in Fig. 17, two pro-α_1 chains and one pro-α_2 chain become organized into an "in-phase" set of three chains which then can rapidly take on the triple helical collagen characteristics without mismatching of the repeating pattern of three amino acid residues in the backbones of the chains. Subsequent removal of the amino-terminal portions of the chains is then postulated to produce triple-stranded collagen monomers, which go on to form the more aggregated strands seen in connective tissue.

The collagen story is not yet complete. Two laboratories have discovered the presence, in fibroblast cultures, of a major component that consists of hydroxylated procollagen chains held together by disulfide bonds. The formation of these covalent cross linkages in the domain that is subsequently removed by proteolysis may be related to the mechanism for chain alignment which permits the relatively rapid synthesis and assembly of this complex macromolecular material (Smith *et al.*, 1972; Kerwar *et al.*, 1972).

In summary, the assembly of fibrous protein structures, involving several peptide chains, is dictated thermodynamically by intermolecular, in addition to intramolecular interactions. The structure of each chain in the aggregate is different from what it would be in the absence of *inter*molecular interactions (Miller and Scheraga, 1975). In addition, if mismatching of chains is to be avoided, some additional chain portions (which may be removed later by proteolysis) are required for proper alignment by specific interactions. While we are not considering them in this review, proteins consisting of several subunits (such as hemoglobins) also acquire stable native structures as a result of intermolecular as well as intramolecular interaction. Another example, in which intermolecular interactions may help stabilize different structures than would be obtained solely from intramolecular interactions, comes from crystals of polyamino acids. Thus, isolated chains of poly-β-benzyl-L-aspartate are stable as α-helices in solution, but transform to ω-helices when they acquire additional (intermolecular) interactions in a crystal (McGuire *et al.*, 1971).

VII. Experimental Approaches to the Study of Conformation

Various techniques have been used to determine the pathway of folding of a protein and also to determine the extent to which fragments of a polypeptide chain (in water and in other solvents) adopt the conformation which they possess when they are parts of a native protein. These involve the use of NMR, CD, proteolytic digestion, and immunological methods. Even if there is a tendency for a particular fragment to adopt the conformation that it has in the native protein, one would still expect the fraction of the total population, in which the *whole* fragment is in the native or nearly native conformation *in water*, to be very small; thus, physicochemical methods that measure average properties over the whole population would be hard-pressed to detect small amounts of the native conformation. On the other hand, techniques that detect specific conformations provide this information more easily.

The problem may be stated in another way. The native protein consists of many segments (in their so-called "native formats") which are stabilized by long-range interactions. The native protein may be envisaged as being in equilibrium with various unfolded forms (which are at extremely low concentrations under conditions where the protein is native). In those forms in which the long-range interactions do not exist, the various segments take on an ensemble of conformations. In these partially unfolded forms, there is still some tendency (on the basis of the short- and medium-range interactions, which are still present in each segment) for various segments to adopt the native format, even though the fraction of such folded molecules may be small. While the equilibrium position between such (separated) native formats and other (random) forms lies toward the random form, it does not take very much free energy (acquired from long-range interactions when the protein folds up; see part C of this

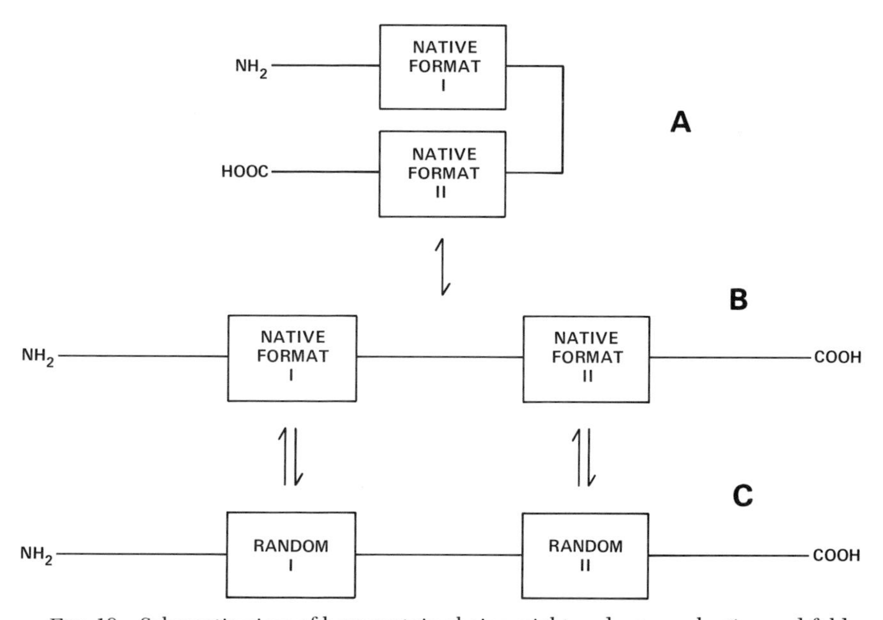

FIG. 18. Schematic view of how protein chains might undergo nucleation and folding. From Anfinsen (1973). Copyright 1973 by the American Association for the Advancement of Science.

section) to shift these equilibria and thus stabilize the entire protein molecule in its native form. These ideas are represented schematically in Fig. 18, where (only for simplicity in illustration) the native protein (A) is shown to consist of two segments, each in its native format. The native protein is represented in equilibrium with a partially unfolded form (B) in which the two segments are in the native conformation; in the equilibrium between A and B (which presumably lies far in the direction of A), the long-range interactions between native formats I and II are involved. In the additional reactions shown, the individual native formats (B) are in equilibrium with random forms (C). The equilibrium between B and C is thought to lie far in the direction of C. Whatever (small, but nontrivial) amount of B there is in equilibrium with C exists in the native format because of short- and medium-range interactions *within* the individual segments. However, the incremental free energy, acquired from long-range interactions in the equilibrium between A and B, provides the driving force to convert C to A. These equilibria can be studied with excised segments of the protein, with complementary protein fragments, and with the whole protein.

With the scheme of Fig. 18 in mind, it is of interest to determine the relative amounts of the various native formats which are present in form B, and to detect the actual equivalent of the schematic pathway C → B → A in the folding of a native protein.

A. Nuclear Magnetic Resonance Approach to Folding

Nuclear magnetic resonance (NMR) spectroscopy appears to have great potential for describing the detailed mechanism of protein folding. Resonances from many chemical groups in a protein may be resolved and identified, and thus information about the behavior of different parts of a protein may be obtained. So far, nuclear magnetic resonance experiments on protein folding have been limited primarily to equilibrium studies.

A number of years ago, McDonald and Phillips (1967), Cohen and Jardetzky (1968), and others demonstrated that high resolution NMR spectra of native and denatured proteins were sufficiently different so that structural information related to the folding process could be deduced. However, it was only after the imidazole resonances of ribonuclease (Bradbury and Scheraga, 1966; Meadows *et al.*, 1968) and several aromatic and imidazole resonances of staphylococcal nuclease (Markley *et al.*, 1968, 1970) were identified, and nuclease was selectively deuterated [by growing the staphylococci in cultures with deuterated amino acids, so as to simplify the downfield region of the NMR spectra (Putter *et al.*, 1969)], that detailed information about folding was obtained.

Jardetzky *et al.* (1971) studied the alkaline-induced unfolding of nuclease by measuring the values of the chemical shifts of histidine, tyrosine, and tryptophan residues as a function of pH. It was found that the curves for the pH-dependence of these chemical shifts were not superimposable, and this was interpreted to mean that different parts of the protein were unfolding at different pH values. An attempt was made to correlate these differences with specific parts of the nuclease molecule based on the identification that had been made of several resonances and on the high resolution X-ray structure (Arnone *et al.*, 1971). Indeed a series of diagrams of the unfolding (or folding) process was produced on this basis. These diagrams tend to imply a set of kinetically distinguishable, relatively stable intermediates. This interpretation is open to some question on the following grounds: (a) the transition studied is not reversible, (b) when the conversion of native to denatured form is slow on the NMR time scale, chemical shift, without considering the areas of the individual resonance peaks, does not provide a measure of the extent of denaturation, (c) in addition, the chemical shift depends not only on the fraction of denatured form but also on differences in pK of the several histidine and tyrosine residues, and (d) the identification of many of the resonances was somewhat equivocal. Further, studies of the various fragments alone in solution (H. Taniuchi, unpublished data) suggest that certain of the postulated stable intermediates are unlikely to exist in significant concentrations. Despite these objections, the idea that NMR might be useful for detecting equilibrium intermediates in a folding transition was established by this study.

Epstein *et al.* (1971b) measured the areas of the four histidine residues during the reversible, cooperative, *acid*-induced transition of nuclease. They found that two histidine residues underwent a conformational transition which was exactly superimposable on spectral and hydrodynamic measurements of the low-pH transition. However, a third histidine residue exhibited a complex slow exchange process with a midpoint about 0.5 pH unit higher than the overall transition. The behavior of the fourth residue was not ascertainable. In an analogous study of the acid-heat denaturation of ribonuclease, Westmoreland and Matthews (1973) and Roberts and Benz (1973) were also able to demonstrate equilibrium conformational intermediates.

Full application of the potential of NMR for describing the kinetics and mechanism of protein folding may be realized by recent advances in pulsed Fourier transform and flow technology for NMR (Sykes and Scott, 1972).

B. Proteolytic Digestion and Protein Unfolding

Many native proteins, particularly those in which conformational fluctuations are reduced by disulfide bonds or by ligands, are not attacked by such proteolytic enzymes as trypsin, α-chymotrypsin, and carboxypeptidase A. However, as the protein is unfolded reversibly by heating, e.g., ribonuclease (Hermans and Scheraga, 1961), successive peptide bonds become accessible to proteolytic attack (Mihalyi and Harrington, 1959; Ooi *et al.*, 1963; Rupley and Scheraga, 1963; Scott and Scheraga, 1963; Ooi and Scheraga, 1964; Klee, 1967). The transition curves for the reversible thermal denaturation of ribonuclease are illustrated in Fig. 19, and its amino acid sequence in Fig. 4.

As the temperature is increased, the region of the ribonuclease molecule that first becomes susceptible to proteolytic attack is the peptide bond between Tyr-25 and Cys-26, which is hydrolyzed by α-chymotrypsin at 45° and pH 6.8 and pH 8.0 (Rupley and Scheraga, 1963; Klee, 1967). Trypsin does not hydrolyze ribonuclease until higher temperatures than α-chymotrypsin, and its first sites of attack are the Lys-31–Ser-32 and Arg-33–Asn-34 bonds (Ooi *et al.*, 1963; Klee, 1967). Using data such as these, together with spectral and NMR data (Roberts and Benz, 1973), information about the importance of the C-terminal tetrapeptide for proper refolding of the protein during the formation of the disulfide bonds (Anfinsen, 1956), and the

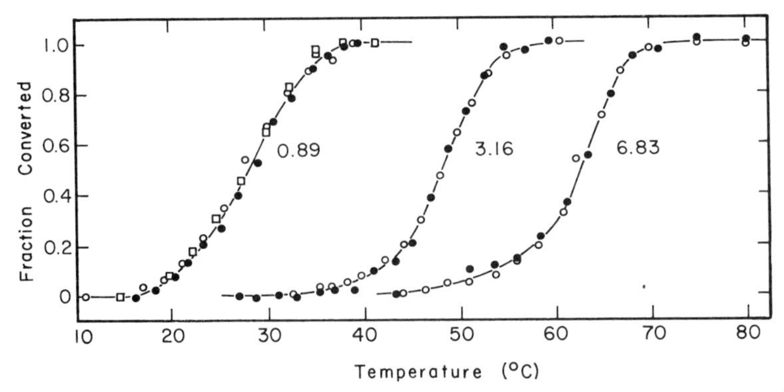

FIG. 19. Fraction coverted versus temperature at several pHs for reversible denaturation of ribonuclease. Open and filled symbols correspond to ultraviolet difference spectra and optical rotation measurements, respectively. From Hermans and Scheraga (1961).

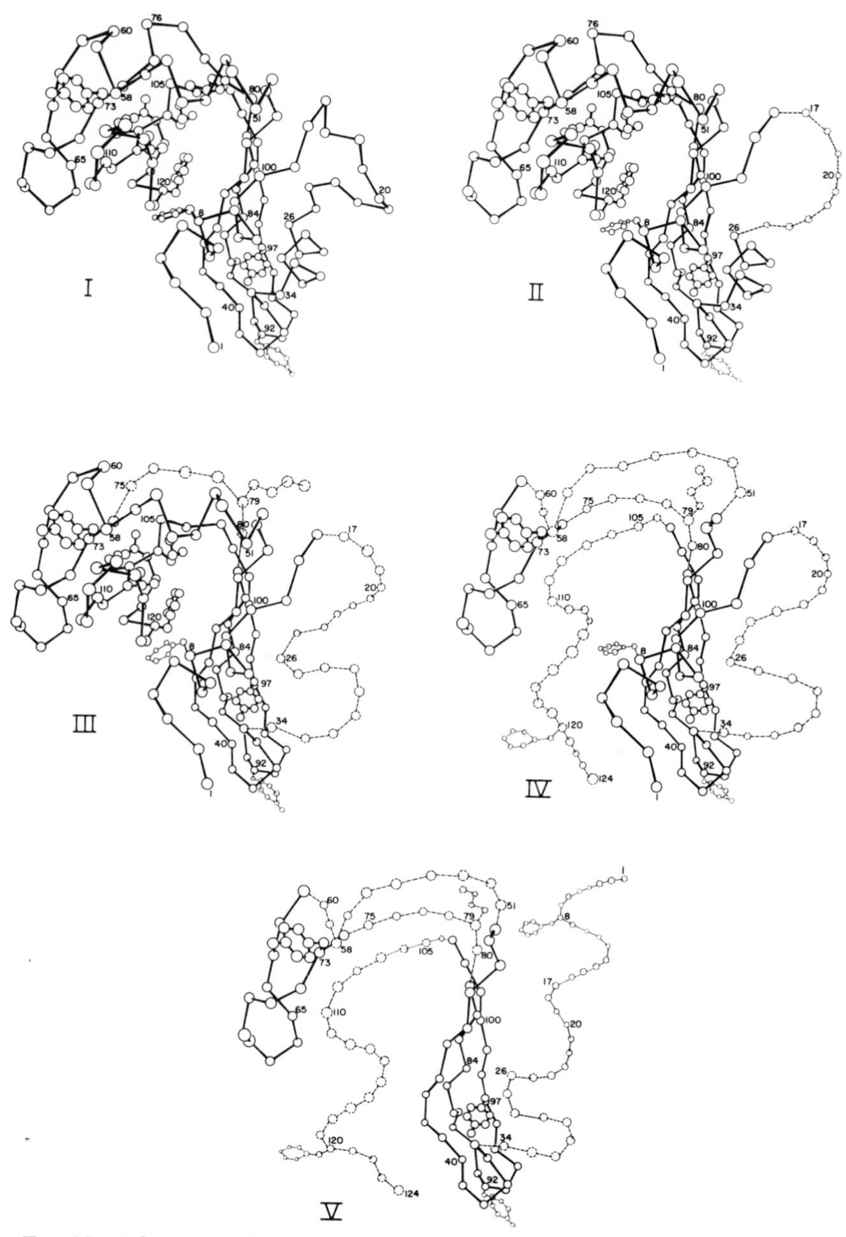

FIG. 20. Schematic pathway for the thermally induced unfolding of ribonuclease A, showing the first five of six identifiable stages (the sixth stage, not shown, being the completely unfolded chain). Disulfide bonds connect half-cystine residues 26–84, 40–95, 58–110, and 65–72. From Burgess and Scheraga (1975b).

X-ray structure of the native protein (Kartha *et al.*, 1967; Wyckoff *et al.*, 1970), Burgess and Scheraga (1975b) proposed the pathway of unfolding shown in Fig. 20. In this model, the C-terminus (being involved in internal interactions) unfolds relatively late in the transition. This aspect of the model for unfolding was confirmed by showing that the C-terminal valine residue becomes susceptible to attack by carboxypeptidase A relatively late in the thermal transition (Burgess *et al.*, 1975b). The diagram of Fig. 20 is not meant to imply that a *single* stable intermediate exists at the end of each stage of unfolding. On the contrary, the pathway of unfolding (which is assumed to be the same as that for folding of the disulfide-bonded molecules, in this reversible process) involves the existence of numerous species in equilibrium with each other at any given temperature. As the temperature changes, the distribution of species alters, and Fig. 20 is intended to represent the nature of the dominant species present at each temperature, on the pathway from the native to the denatured state.

Burgess and Scheraga (1975b) postulated that the scheme of Fig. 20 can be used to set up a model for the folding of the nascent chain *without disulfide bonds*. This does not necessarily imply that the steady-state concentration of these intermediates is high, but only that the folding proceeds through the sequence of events along this pathway; see the original paper for further discussion of this point. With this provision in mind, we see that part V of the diagram illustrates two initial nucleation sites for folding. In one site, there are two chain reversals at residues 36–39 and 91–94 which enable three extended chains (the anti-parallel β pleated sheet between residues 81–87 and 96–102, and the chain between residues 40 and 48) to participate in long-range interactions. The other nucleation site (62–74) involves a nonregular conformation with a chain reversal at residues 65–68. These nucleation sites (chain reversals and extended structures) arise from short- and medium-range interactions (Lewis *et al.*, 1970, 1971; Ponnuswamy *et al.*, 1973b; Burgess *et al.*, 1974). Part IV of Fig. 20 shows that the N-terminal helix (which has a *tendency* to form at residues 4–12 because of short-range interactions) then acquires stability from long-range interactions with one of the initial extended-chain nucleation sites. Part III of Fig. 20 illustrates a similar phenomenon in which the C-terminal region could condense between the two nucleation sites (and "tie" them together), accompanied by a stabilization of the helix at residues 51–60. Subsequently, the helix at residues 25–33 might be stabilized at one of the nucleation sites, and the extended chain at residues 75–80 at the other (see Part II of Fig. 20). Finally, the region of the chain between residues 17 and 25 would acquire its native conformation (Part I of Fig. 20), and the side chain of residue Tyr-92 would lose its mobility.

C. Immunological Approach to the Study of Conformation

The techniques described in Sections VII,A and B provide information about the pathway of folding. In Sections VII,C and D, we consider techniques designed primarily to study the conformation of fragments of native protein molecules, but which also can give information about the native protein.

The fraction of the molecular fragments that exist in the native format can be determined by an immunological method if there exists a convenient method for measuring the binding constant for the interaction between an antibody with the native protein and with an excised fragment thereof, or of directly measuring the competition between the native protein and the fragment for the antibody. Such a method is available for staphylococcal nuclease (Sachs *et al.*, 1972a,b,c) and for enzymes in general (Arnon, 1973) since many antibodies form inactive complexes with them. Radioactive labeling of antigens can also be used for quantitative studies of antigen–antibody interactions (Furie *et al.*, 1974a,b), because they allow the estimation of amounts of antigen–antibody complexes by precipitation with heterologous antiglobulins.

When goats are immunized with native nuclease, antibodies are produced in good yield and with high affinity. The portion of the immunoglobulin fraction that is specific for nuclease may be isolated by absorption on columns of agarose to which nuclease has been attached (Sachs *et al.*, 1972a,b,c). Many of the antibodies in this fraction are potent inhibitors of nuclease activity. Subfractions of the antibody population possessing specificity for limited regions of the native molecule may be isolated by further affinity chromatography of the antinuclease on columns to which peptide fragments of the enzyme have been attached. These subfractions are identified by the subscript N [e.g., anti-$(99–149)_N$], indicating a fraction of the total anti-native nuclease antibody which is bound by an agarose column bearing a portion of the sequence containing the indicated residues. If anti-$(99–149)_N$ is further fractionated by passage through a column of agarose-(127–149), two subfractions anti-$(127–149)_N$ and anti-$(99–126)_N$ are obtained. These form catalytically inactive, but soluble complexes with nuclease. The complexes of anti-$(99–126)_N$ with nuclease contain one molecule each of antibody and enzyme, as judged by ultracentrifugal analysis (Sachs *et al.*, 1972a). Scatchard analysis of the interaction of the anti-$(99\text{–}126)_N$ antibody population with nuclease shows that 50% of the antibody binds with an association constant of about $10^9\ M^{-1}$ and the other 50% binds with an association constant of $10^7\ M^{-1}$, at 25°, pH 7.2, and ionic strength 0.17 (Sachs *et al.*, 1972b). The experiments described below were carried out with excess antibody, so that only the population with the higher association constant was involved. The rate of association of this antibody with its antigenic determinant is very rapid (with a second-order rate constant of $4 \times 10^5\ M^{-1}\ \text{sec}^{-1}$) as determined by analysis of the kinetics of inactivation, while the rate of dissociation is very slow (with a rate constant of $5 \times 10^{-4}\ \text{sec}^{-1}$) (Sachs *et al.*, 1972b). These rates provide a basis for the method (described below) used to estimate the concentration of the complex between antibody and nuclease fragment.

Inactivating antibody fractions such as these may be used to estimate the percent of peptide fragments which exist in the native format. The equilibrium involving the sequence (99–149), which contains the determinant for the inactivating, but nonprecipitating, antibody anti-$(99\text{–}126)_N$, may be visualized, schematically, as shown in Fig. 21. This equilibrium

$$(99\text{–}149)_R \rightleftharpoons (99\text{–}149)_N \tag{1}$$

is characterized by the equilibrium constant

$$K_{\text{conf}} = \frac{[(99\text{–}149)_N]}{[(99\text{–}149)_R]} \tag{2}$$

where the brackets designate molar concentrations and K_{conf} is $K_{\text{conformation}}$. The term $(99\text{–}149)_N$ refers to the peptide (99–149) when it exists in a conformation close to, or identical with, that of this section in the native protein. The term $(99\text{–}149)_R$ refers to all the disordered, nonnative conformations assumed by this particular stretch of amino acids in solution.

The interaction of fragment (99–149) with anti-$(99\text{–}126)_N$ may be described in terms of the equilibrium:

$$\text{anti-}(99\text{–}126)_N + (99\text{–}149)_N \rightleftharpoons \text{anti-}(99\text{–}126)_N{\cdot}(99\text{–}149)_N \tag{3}$$

with an equilibrium constant ($K_{\text{ass}} = K_{\text{association}}$)

$$K_{\text{ass}} = \frac{[\text{anti-}(99\text{–}126)_N{\cdot}(99\text{–}149)_N]}{[\text{anti-}(99\text{–}126)_N]\ [(99\text{–}149)_N]} \tag{4}$$

It is assumed that the interaction of anti-$(99\text{–}126)_N$ with $(99\text{–}149)_R$ is

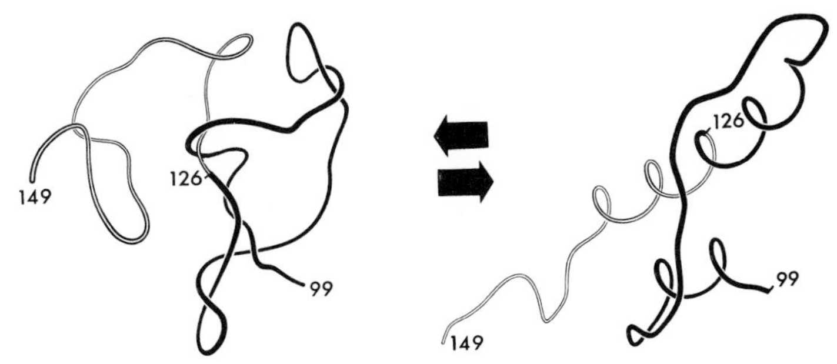

FIG. 21. Schematic representation of the postulated equilibrium between fragment (99–149) of nuclease in its predominantly random form (left) and that fraction of the population in the "native format" (right). From Sachs *et al.* (1972c).

insignificant (Sachs *et al.*, 1972a). It is also assumed that the association constant of the antibody in question is the same for the sequence (99–126) as a free fragment in its *native* format and as a part of the native structure of nuclease. Hence we may combine Eqs. (2) and (4) to obtain

$$K_{\text{conf}} = \frac{[\text{anti-}(99-126)_{\text{N}} \cdot (99-149)_{\text{N}}]}{K_{\text{ass}}[(99-149)_{\text{R}}][\text{anti-}(99-126)_{\text{N}}]} \tag{5}$$

When the conformational equilibrium (Eq. 1) between the random and native-format forms of (99–149) lies far in the direction of the random form as indicated by physical chemical studies (Taniuchi and Anfinsen, 1969) showing that (99–149) is disordered when free in solution, under conditions that the whole protein is native, i.e., $K_{\text{conf}} << 1$, the value of $[(99\text{–}149)_{\text{R}}]$ is essentially the same as [total(99–149)], the total concentration of peptide. Therefore, Eq. (5) may be rewritten as

$$K_{\text{conf}} = \frac{[\text{anti-}(99-126)_{\text{N}} \cdot (99-149)_{\text{N}}]}{K_{\text{ass}}[\text{total}(99-149)][\text{anti-}(99-126)_{\text{N}}]} \tag{6}$$

The value of K_{ass} has been found to be $10^9\ M^{-1}$, as described above. The value of [total(99–149)] is, of course, known from the amount of peptide used in the experiment. Thus, if $[\text{anti-}(99\text{–}126)_{\text{N}} \cdot (99\text{–}149)_{\text{N}}]$ and $[\text{anti-}(99\text{–}126)_{\text{N}}]$, the equilibrium concentrations of bound and free antibody, respectively, can be determined experimentally, we can compute K_{conf}. The total amount of antibody $[\text{total anti-}(99\text{–}126)_{\text{N}}]$ used in the experiment is known, and

$$[\text{total anti-}(99\text{–}126)_{\text{N}}] = [\text{anti-}(99\text{–}126)_{\text{N}}] + [\text{anti-}(99\text{–}126)_{\text{N}} \cdot (99\text{–}149)_{\text{N}}] \tag{7}$$

Hence, if we can determine, say, [anti-$(99\text{–}126)_N$], we will also know [anti-$(99\text{–}126)_N \cdot (99\text{–}149)_N$].

Sachs *et al.* (1972a) determined [anti-$(99\text{–}126)_N$] by a kinetic experiment, in which the inhibitory action of anti-$(99\text{–}126)_N$ on the nuclease digestion of denatured DNA is measured, i.e., the enzymic reaction is

$$\text{nuclease} + \text{denatured DNA} \xrightarrow{k_1} \text{products} \tag{8}$$

where k_1 is the observed rate constant; this reaction can be followed by observing the change in OD_{260} as denatured DNA is digested. This reaction was run under conditions that are first order in nuclease and zero order in DNA (i.e., the DNA is present in large excess). The reaction was studied in the presence of various amounts of anti-$(99\text{–}126)_N$ and (99–149). Some of the total anti-$(99\text{–}126)_N$ combines with $(99\text{–}149)_N$, according to Eq. (3), and the rest remains free. The free antibody, anti-$(99\text{–}126)_N$, is available, then, to inhibit the Reaction, Eq. (8), because an inactive complex is formed as follows:

$$\text{anti-}(99\text{–}126)_N + \text{nuclease} \rightleftharpoons \underset{\text{inactive}}{\text{anti-}(99\text{–}126)_N \cdot \text{nuclease}} \tag{9}$$

The experiment was carried out by incubating aliquots of the antibody with various amounts of the fragment in the assay mixture, containing the DNA, and then adding aliquots of nuclease after Reaction (3) had reached equilibrium. From the observed rate of enzymic attack of nuclease on DNA in the presence and in the absence of various levels of fragment, it was possible to determine the *forward* rate constant of Reaction (9) [expressed as a half-time, which is a pseudo-first-order one because anti-$(99\text{–}126)_N$ is present at much higher concentration than nuclease], and hence the concentration of free antibody [anti-$(99\text{–}126)_N$], as follows.

Since Reaction (8) is first order in nuclease and zero order in DNA, the concentration of hydrolytic products (expressed as ΔOD_{260}, which is the difference in OD_{260} at time t and that at the time that nuclease is added) is

$$\Delta OD_{260} = k_1 t [\text{nuclease}] \tag{10}$$

The value of [nuclease] depends on that of [anti-$(99\text{–}126)_N$], which in turn depends on that of $(99\text{–}149)_N$ according to Eq. (3). Since the backward rates in Eq. (3) and Eq. (9) are small ($k = 5 \times 10^{-4}\ \text{sec}^{-1}$), the concentration of anti-$(99\text{–}126)_N$ available for the (forward) association of Reaction (9) can be assumed to be the equilibrium concentration in the preincubation reaction of Eq. (3). Hence, the forward rate of Eq. (9) is

$$-\frac{d[\text{nuclease}]}{dt} = k_2[\text{anti-}(99\text{–}126)_{\text{N}}]\cdot[\text{nuclease}] \qquad (11)$$

where k_2 is the *forward* rate constant. Integration of Eq. (11) leads to

$$\ln[\text{nuclease}] = \ln[\text{nuclease}]_0 \quad k_2[\text{anti-}(99\text{–}126)_{\text{N}}]t \qquad (12)$$

Substituting the value of [nuclease] from Eq. (10) into Eq. (12), we obtain

$$\ln(\triangle \text{OD}_{260}/t) = \ln(\triangle \text{OD}_{260}/t)_0 - k_2[\text{anti-}(99\text{–}126)_{\text{N}}]t \qquad (13)$$

Thus, if data for OD_{260} vs t are replotted according to Eq. (13), then the slope gives the desired quantity, viz., $k_2[\text{anti-}(99\text{–}126)_{\text{N}}]$. Experiments were carried out at constant values of [total anti-$(99\text{–}126)_{\text{N}}$], but with increasing amounts of fragment, total(99–149).

Figure 22 illustrates the results of such experiments, with the data

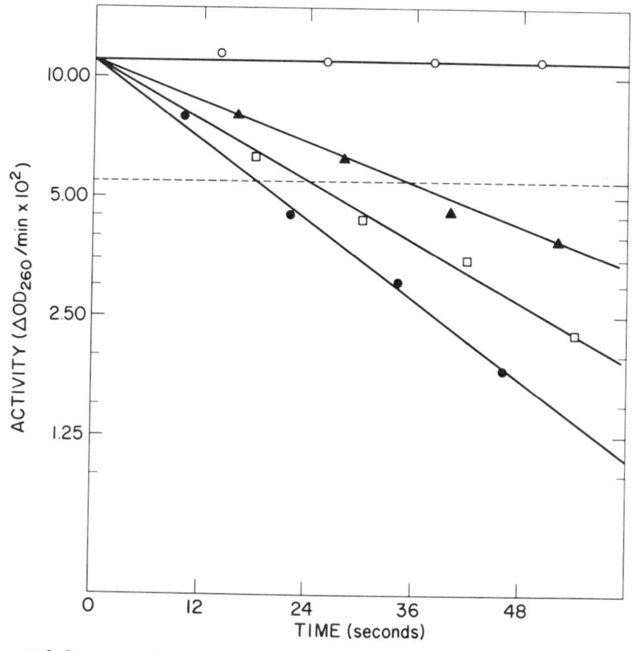

FIG. 22. Inhibition of antibody-induced inactivation. Semilogarithmic plot of activity vs time for assays of 0.05 μg of nuclease in the presence of: ○—○, no antibody or fragment; ●—●, 6 μg of anti-$(99\text{–}126)_{\text{N}}$ and no fragment; □—□, 6 μg of anti-$(99\text{–}126)_{\text{N}}$ plus 12 μg of (99–149); ▲—▲, 6 μg of anti-$(99\text{–}126)_{\text{N}}$ plus 48 μg of (99–149). The dashed line represents one-half of the initial activity. From Sachs *et al.* (1972c).

plotted semilogarithmically according to Eq. (13). A summary of the rate constants k_2, expressed in terms of a pseudo-first-order half-time defined as

$$t_{1/2} = (\ln 2/k_2)(1/[\text{anti-}(99\text{–}126)_\text{N}]) \tag{14}$$

is given in Table II. As [(99–149)] increases, the concentration of anti-$(99\text{–}126)_\text{N}$ decreases and the half-time increases. The value of k_2 can be ascertained when there is no peptide fragment present; thus, the concentration of free antibody, [anti-$(99\text{–}126)_\text{N}$], can be determined.

TABLE II

Inhibition by Fragments of Inactivation of Nuclease by Antibody[a,b]

Fragment(s) added	Concentration of fragments (μM)	$t_{1/2}$ (sec)	Concentration of free antibody (μM)	Concentration of bound antibody (μM)	$K_{conf} \times 10^4$	% Free fragment in native conformation
0	0	18[c]	0.076	0	—	0
(99–149)	0.65	20	0.068	0.0080	2.20	0.022
	2.0	24	0.057	0.019	2.02	0.020
	2.6	27	0.051	0.025	2.29	0.023
	6.5	33	0.042	0.034	1.51	0.015
	7.8	35	0.039	0.037	1.47	0.015
						Av: 0.019 ± 0.003
	0	20[c]	0.076	0	—	—
(50–149)	2.4	27	0.055	0.021	2.0	0.020
	4.7	40	0.038	0.038	2.6	0.026
						Av: 0.023 ± 0.003
(50–149) + syn(6–43)	2.4 / 0.47	44	0.034	0.042	6.5	0.065
(50–149) + syn(6–43)	2.4 / 0.95	86	0.017	0.059	18	0.18
(50–149) + syn(6–43)	2.4 / 1.9	226	0.0065	0.069	57	0.57

[a] From Sachs *et al.* (1972a).

[b] In all cases the *total* concentration of antibody was 0.076 μM and the *total* concentration of nuclease was 0.001 μM. The *total* concentration of DNA was 120 μM (as nucleotide residues).

[c] These (control) values differ because they pertain to different preparations of anti-$(99\text{–}126)_\text{N}$. Thus, small differences in antibody concentration or in degree of antibody denaturation may have been present.

From the data of Table II, it can be seen that, for fragment (99–149), the native format (or at least those formats that are sufficiently similar to that of the native structure in this portion of the total amino acid sequence to be recognized by the antibody) is present to the extent of approximately 0.02% of the total population. Such a fraction would be difficult to detect by existing physicochemical methods (but see Section VII,D).

The immunological approach almost certainly does not measure the presence of *fully* folded peptide fragments because the antibody might recognize an almost-native format. However, it seems likely that many intramolecular interactions would be required to form and stabilize a conformation that would be recognized by the highly specific antibody reagent and that would have a significant time of survival in solution. The small amount of folded form indicated by the experiments summarized above emphasizes that a tendency to form the native structure exists (because of short- and medium-range interactions) but that the longer-range interactions are necessary for folding to the stable form.

It had been expected that similar studies on the equilibria between random and "native" forms of fragments longer than (99–149) in which some of the longer-range interactions are present, might show greater proportions of the native conformation. However, the peptide (50–149), which is one of the two complementing fragments (see Section III on complementing fragments) that are produced upon trypsin digestion of nuclease in the presence of the ligands, calcium ions and pdTp, exhibits approximately the same degree of folding into "native format," as judged by the immunological test described above, i.e., the value of K_{conf} for the equilibrium between the random and "native" forms of this peptide was also about 2×10^{-4}.

A synthetic fragment containing residues 6 through 43 was available (Sanchez *et al.*, 1973) which could substitute for the tryptic peptide (6–48) in the complementation with (50–149) to yield a semisynthetic nuclease-T analog. This complex was enzymically inactive, thus permitting the accurate estimation of the inhibition of activity of native nuclease by that portion of anti-$(99–126)_N$ that was not bound by the complex, now presumably containing a higher percentage of molecules with the residues in the (99–126) sequence in their "native format." The data shown in Table II illustrate the markedly increased folding of the (50–149) fragment that is induced and stabilized by the added fragment, because more, and possibly most, of the longer-range interactions are now available.

More recently, the immunological approach was used to determine values of k_{conf} of the *native* staphylococcal nuclease (Furie *et al.*, 1974b). Antibodies were prepared against the excised fragment of the sequence (residues 99–149), that contains the above-mentioned univalent antigenic determinant located between residues 99 and 126. In this population of antibodies, those that recognized the (99–126) portion of the total sequence as a free peptide in solution were isolated by affinity chromatography on a column of Sepharose-nuclease[6] and then on Sepharose-(1–126). This immunoglobulin fraction, which was *not* able to inactivate nuclease except at very high concentrations, was termed anti-$(99–126)_R$ (Furie *et al.*, 1974a). Its specificity was presumably against one or more of the *random* forms of the sequence (99–126).

The anti-$(99–126)_R$ subpopulation produced a soluble complex with $(99–149)_R$. The value of K_{ass} for the equilibrium

$$\text{anti-}(99\text{–}126)_R + (99\text{–}149)_R \rightleftharpoons \text{anti-}(99\text{–}126)_R\cdot(99\text{–}149)_R \quad (15)$$

was determined by labeling $(99–149)_R$ with ^{14}C-labeled carbamyl groups at the NH_2-terminus, and then precipitating anti-(99–126)$_R\cdot(99–149)_R$ complexes with rabbit antibodies prepared against goat γ-globulins. The average values of K_{ass} at 25°, pH 8.1, ionic strength 0.17 were found to be $10^7\ M^{-1}$ for a high affinity population, and $10^5\ M^{-1}$ for a lower-affinity population. Further, it was found that nuclease at high concentrations could inhibit the reaction of anti-$(99–126)_R$ with labeled $(99–149)_R$. These observations were interpreted by assuming the existence of the following equilibria:

$$\text{nuclease}_R \overset{K_{conf}}{\rightleftharpoons} \text{nuclease}_N \quad (16)$$

$$\text{anti-}(99\text{–}126)_R + \text{nuclease}_R \rightleftharpoons \text{anti-}(99\text{–}126)_R\cdot\text{nuclease}_R \quad (17)$$

By an analysis analogous to that presented above, involving a study of the inhibition of Reaction (15) by nuclease, K_{conf} for Reaction (16) at 25°, pH 8.1, ionic strength 0.17 was estimated to be 2900. This is on the basis of the concentration of native nuclease necessary to produce 50% inhibition of Reaction (15). This means that only 0.03%

[6] Immunoabsorption of antibodies to columns of polypeptide antigens covalently bound to Sepharose can occur with antibodies to either the native or disordered antigen (Ab_N or Ab_R). It is presumed that the large excess of antigen bound to Sepharose is in a conformational equilibrium so that both the native and disordered antigenic determinants are available for binding to antibody. Since all antibodies to fragment (99–149) might not be capable of cross-reacting with the unfolded form of nuclease (nuclease_R) because some portions of the fragment occupy internal positions in the native protein, anti-$(99–126)_R$ was first isolated by immunoadsorption to columns of Sepharose-nuclease (Furie *et al.*, 1974b).

of the staphylococcal nuclease molecules are sufficiently unfolded so that the segment from residues 99 to 126 can be recognized by antibodies having a high specificity for the *unfolded* form of this portion of the protein sequence. Addition of the ligands, calcium ions and pdTp, increased K_{conf} of Reaction (16) by more than 10-fold, while an increase in temperature decreased the value of K_{conf}.

It is clear that the value of K_{conf}, measured for nuclease, or for a fragment thereof, pertains to that portion of the amino acid sequence which is the antigenic determinant, not to the entire polypeptide. Further, the conformational specificity of the antibodies, and the simple equilibria [Reactions (1) and (16)] assumed for the antigens are obviously approximations. However, because of the dominance of short-range interactions and also [in the case of Reaction (16)] the cooperativity of most protein folding processes, these are probably good approximations. More importantly, the immunological method provides a set of parameters that can be used for a quantitative description of protein folding in a range of concentrations of native format where it is difficult to apply other techniques.

Finally, it is of interest to consider the significance of a value of $K_{conf} = 2 \times 10^{-4}$ for the 99–126 portion of fragment 99–149, i.e., for a 28-residue fragment, to see whether this is to be regarded as a small or large number (assuming that 0.02% of the *whole* segment is native, and that the antibody interacts with the whole segment or, alternatively, with two or three residues far apart in the amino acid sequence but requiring a fixed conformation of 28 residues to place these two or three residues properly). As a very rough calculation, assume that each amino acid residue of the fragment can exist in five discrete backbone conformational states, each of which is accessible with equal probability. Then there are 5^{28} (or 10^{20}) possible conformations of this fragment, of which the native conformation is only one. Thus, if the native conformation were attained as a random event, K_{conf} would be 10^{-20}. (If the antibody interacts only with, say, 10 residues rather than 28, then K_{conf} would be 10^{-7}, on the same basis.) When viewed in this light, it can be seen that 2×10^{-4} is a very large number, arising from intramolecular interactions that make the native conformation a very probable (nonrandom) one for the excised fragment. A value of $K_{conf} = 2 \times 10^{-4}$ for Reaction (1) corresponds to a standard free-energy change of ~5 kcal at 300°K. In order to increase K_{conf} to 10^4 [i.e., to make the fragment become essentially completely native, as observed in Reaction (16)], an additional 10 kcal would be required from long-range interactions, or 0.37 kcal per residue. It is reasonable to expect that an additional

0.37 kcal per residue can be obtained from long-range interactions (when this fragment interacts with the rest of the protein chain) to form the native structure. In fact, energies of this magnitude were computed to account for the stabilization of two anti-parallel right-handed α-helices (with an intervening hairpin bend) of poly(L-alnine) in water (Silverman and Scheraga, 1972). Also, energies of the same order of magnitude can raise the melting point of a protein considerably; this was observed recently for D-glyceraldehyde-3-phosphate dehydrogenase from a thermophilic bacterium, where the extra stability (i.e., an increase of ~40° in the melting point due to a stabilization energy of ~10 kcal/mole) arises from subtle features of the molecular structure (Fujita and Imahori, 1974).

D. Examination of Small *Fragments*

It is well known (Schellman, 1955; Zimm and Bragg, 1959; Lifson and Roig, 1961) that short oligopeptides (4 or 5 residues in length) cannot form stable α-helices in aqueous solution. However, it is of interest to determine whether the four-residue segments that contribute β-bends or chain reversals (see Fig. 23) in proteins can exist to any significant extent in the native format as an excised fragment. The technique of Section VII,C may provide an answer to this question, and such experiments are in progress with a chain-reversal segment of staphylococcal nuclease (S. J. Leach, H. A. Scheraga, and C. B. Anfinsen, unpublished). In addition, a physicochemical ap-

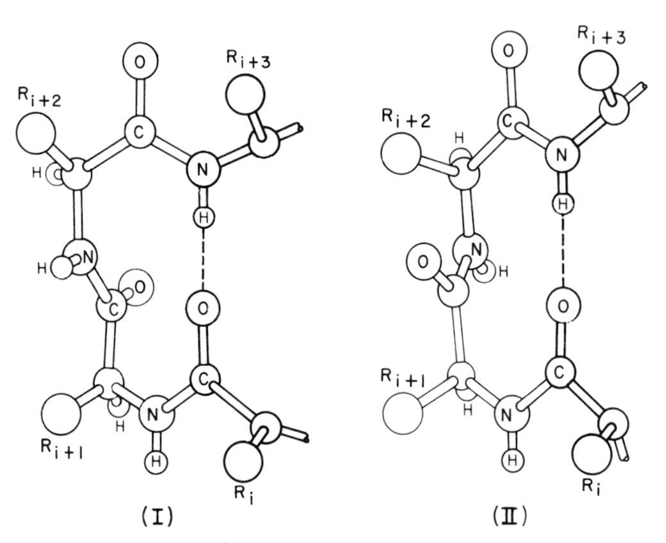

FIG. 23. Types I and II β-bends. From Venkatachalam (1968).

proach is being used to determine whether the properties of short oligopeptides can show that a detectable amount of native format is present in the total population of conformations (Scheraga *et al.*, 1973).

For this purpose, the sequence (1) Asp-Lys-Thr-Gly (residues 35–38 in α-chymotrypsin), which forms a chain reversal in this enzyme, has been considered along with the following variants of the sequence: (2) Gly-Thr-Asp-Lys, (3) Asp-Lys-Gly-Thr, and (4) Lys-Thr-Gly-Asp. To avoid end effects, these tetrapeptides were blocked with CH_3 CO—and—$NHCH_3$ groups at the N—and C— termini, respectively (Howard *et al.*, 1975). Using probabilities for the occurrence of the chain-reversal conformation (Lewis *et al.*, 1971), the relative probabilities of occurrence of bends in these four tetrapeptides are 20, 1, 40, and 40, respectively. Thus, peptides 3 and 4 have the highest probability of occurrence as a bend, the native

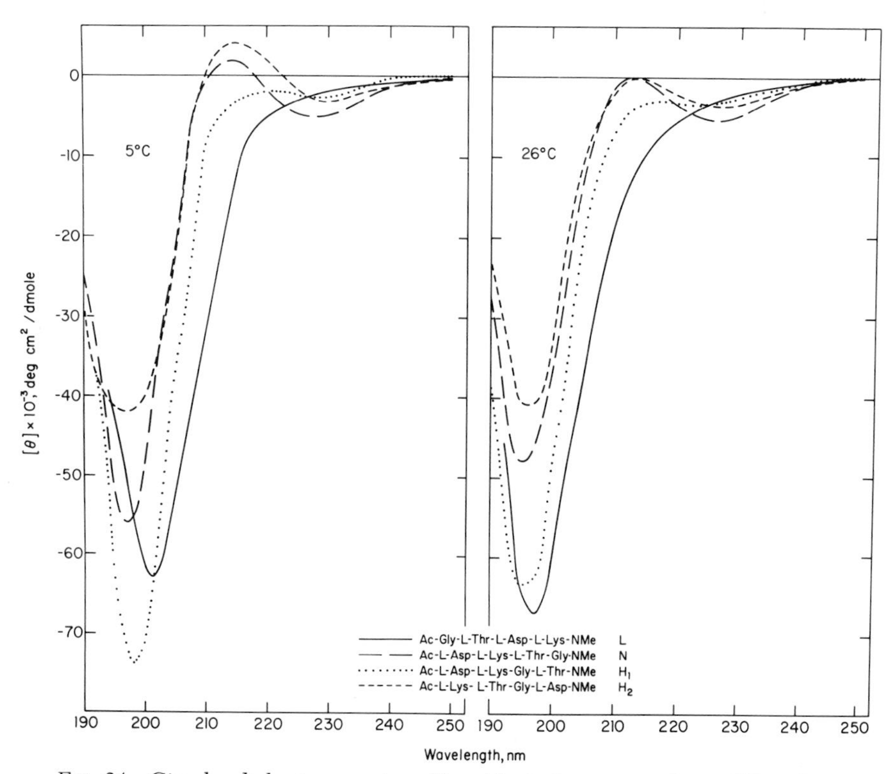

FIG. 24. Circular dichroism spectra of four blocked tetrapeptides at 5°C and 26°C. —, Ac-Gly-L-Thr-L-Asp-L-Lys-NMe (L); – –, Ac-L-Asp-L-Lys-L-Thr-Gly-NMe (N); ········, Ac-L-Asp-L-Lys-Gly-L-Thr-NMe (H_1); ------, Ac-L-Lys-L-Thr-Gly-L-Asp-NMe (H_2). From Howard *et al.* (1975).

one (No. 1) is next, and peptide 2 has the lowest probability. In other words, the probability of occurrence of a bend in a given tetrapeptide depends not only on its composition, but also on its amino acid sequence.

At 5°C in aqueous solution, peptides 1, 3, and 4 exhibit a detectable nuclear Overhauser effect (NOE) indicating that, in a small fraction of each of these peptides, the methyl protons of the CH_3CO- and -$NHCH_3$ end groups are near each other (which implies that a bend exists in some fraction of these molecules) (Howard *et al.*, 1975). No such effect is detected at 26°C. Further, peptides 1, 3, and 4 exhibit a different circular dichroism (CD) behavior from peptide 2 at 5°C (see Fig. 24). Making use of theoretical calculations of rotational strengths of such bend conformations (Woody, 1974), it has been concluded from these CD and NOE data that a small (but significant) fraction of the total population of conformations of peptides 1, 3, and 4 (but not peptide 2) exists in a chain-reversal conformation. This conclusion is supported by conformational energy calculations (Howard *et al.*, 1975) in which the energies of approximately 80 starting conformations per tetrapeptide (selected from high-probability conformations for the individual residues) were minimized. For peptides 1, 3, and 4, the minimum-energy conformations were bends, and were several kilocalories per mole lower in energy than any other conformation considered. For peptide 2, there were five low-energy conformations (including a bend), all within 3 kcal/mole of each other, indicating that there was no strong preference for this sequence to adopt a bend conformation, but rather to distribute among at least five different conformations.

VIII. Energetic Factors Determining Protein Folding

The foregoing experimental evidence indicates that a native protein has a conformation that is a stable minimum in a multidimensional energy surface which depends on all interactions between the atoms of the protein molecule and between the protein molecule and the solvent. Whether or not this minimum is the one of lowest free energy has not yet been determined unambiguously. If it were not the one of lowest free energy, then it would follow that unique pathways exist to reach the *same* native structure (for example, the same conformation of native ribonuclease is formed from the denatured protein with its disulfide bonds intact and from the more unfolded form with its disulfide bonds reduced, as discussed in Section II) and that very high energy barriers prevent the polypeptide chain from attaining the conformation of lowest free energy. The attainment of the native conformation (of a single chain) from *several* different initial states (Hantgan *et al.*, 1974) suggests that the pathways are not unique, and that the various regions of the conformational space are accessible from one another without insurmountable intervening energy barriers (except possibly those required to prevent passage to an aggregated denatured form). Thus, in our view, it appears likely, though by no means proved, that the native conformation of a protein is indeed the one of lowest free energy.

This is not to say that *all* of conformational space must be traversed in the folding of a polypeptide chain. If the chain explored all possible conformations at random by rotations about the various single bonds of the structure, it would take too long to reach the native conformation. For example, if the individual residues of an unfolded polypeptide chain can exist in only two states, which is a gross underestimate, then the number of possible randomly generated conformations is 10^{45} for a chain of 150 amino acid residues (although, of course, most of these would probably be sterically impossible ones). If each conformation could be explored with a frequency of a molecular rotation (10^{12} sec^{-1}), which is an overestimate, it would take approximately 10^{26} years to examine all possible conformations. Since the synthesis and folding of a protein chain such as that of ribonuclease or lysozyme can be accomplished in about 2 minutes (Lacroute and Stent, 1968; Wilhelm and Haselkorn, 1970; Hunt *et al.*, 1969; Goldberger *et al.*, 1963), it is clear that all conformations are not traversed in the folding process. Instead, it appears to us that, in response to local interactions, the polypeptide chain is directed along a variety of possible low-energy pathways (relatively small in number), possibly passing through unique intermediate states, toward the conformation of lowest free energy. *Evolution (with thermodynamics dictating the folding) has selected the amino acid sequence to form a biologically active molecule, with presumably a limited number of pathways from the unfolded form to a unique native structure of lowest free energy.*

A. *Thermodynamic Criterion for the Native State*

Accepting the hypothesis that the native conformation is the one of lowest free energy (Liquori, 1969; Ramachandran, 1969; Ramachandran and Sasisekharen, 1968; Scheraga, 1968, 1969, 1971, 1973a,b), it is important to realize that entropic as well as energetic factors contribute to the free energy. Since a macromolecule in solution is not a static entity, there are two major sources of entropy: (1) from the various possible configurations of the solvent molecules for a *given* conformation of the macromolecule, and (2) from the translational and overall rotational motion of the macromolecule and from its internal vibrations that arise from the stretching of bonds, the bending of bond angles, and the variations of dihedral angles for rotation around single bonds. The sum of the entropy of solvation and the potential energy of all intrapolypeptide interactions constitutes the "conformational energy" (the term "conformational energy," as defined and used in this context, is a free energy), which may be represented on a multidimensional surface as a function of the conformation of the macromolecule. This surface is a very complicated one. Stable macroscopic conformations of the macromolecule in solution correspond to local minima of the "conformational energy," and the macromolecule undergoes stable oscillations around each of these minima. These oscillations contribute a conformational entropy; hence every macroscopic conformation corresponds to a collection of neighboring microscopic states (those located around each minimum of the "conformational energy"), the free energy of each macroscopic conformation depending on both the magnitude of the "conformational energy" at the minimum and on the contribution from all the vibrational states (plus the overall rotational entropy) in the neighborhood of the minimum.

We assume that the native conformation of a protein is that which corresponds to the local minimum of the "conformational energy" which has the lowest free energy. As shown schematically in Fig. 25, this need not be the global minimum of the "conformational energy"; that is, a local minimum *near* the global one may have a lower free energy because of contributions from the librational motion. This effect has been observed in computations on poly(L-alanine) (Gibson and Scheraga, 1969a,b) and on cyclo(Gly_3Pro_2) (Go *et al.*, 1970).

The calculations of stable conformations of polypeptides and proteins then involves three different problems. First, the geometry (i.e., bond lengths and bond angles) of the polypeptide chain must be known, and the functional form (and the values of the parameters) of the "conformational energy" must be determined. Second, the "conformational energy" must be minimized to locate the various local minima. Third, the free energy (which includes the conformational entropy) of each minimum within a reasonable range of the global one, i.e., within about 1 kcal per residue, must be calculated. A discussion of this computational approach (including the nature and magnitude of the interaction energies involved), and its application to a variety of model polypeptides as well as proteins, has been

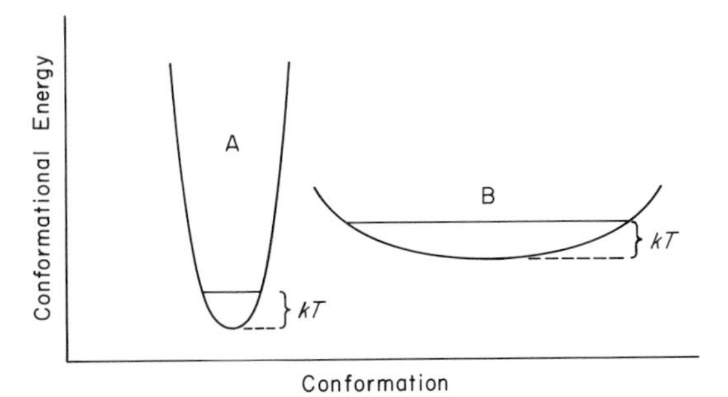

FIG. 25. Schematic representation (in two dimensions) of two local minima of a multidimensional surface. If the "conformational energy" can fluctuate by kT, the corresponding fluctuation in the conformation (and hence the conformational entropy) is greater in B than in A.

presented elsewhere (Ramachandran and Sasisekharan, 1968; Scheraga, 1968, 1969, 1971, 1974a). While a number of problems remain to be solved before one can compute the most stable conformation of a protein in water, those involving the nature of the "conformational energy" function and the calculation of the free energy have been resolved in principle, and in practice in some cases. However, the problem of the existence of many local minima in the multidimensional energy surface remains a formidable one. Since (energy) minimization algorithms lead to the nearest local minimum, depending on the starting point, resort has been had to the search for alternative methods that would lead to initial starting points which might have a reasonable chance of being in the correct potential energy well; then, energy minimization would lead to the desired local minimum. The search for alternative methods for obtaining an initial conformation of a protein (to be followed by energy minimization) has revealed some important energetic and stereochemical features of polypeptide chains, which provide insight into the folding process (Scheraga, 1974a,b).

For reasons outlined in Section VIII,B,1, one of the approaches for finding alternative methods was a consideration of the possible dominance of short-range interactions (Kotelchuck and Scheraga, 1968, 1969). This investigation led to the concept that the conformation of an amino acid residue in a polypeptide or protein is determined in very large measure (though not exclusively) by the short-range interactions between a side chain and the atoms of the backbone of the *same* amino acid residue, and is (again in first approximation) essentially independent of interactions with neighboring side chains or backbone portions of the chain. In a second approximation, medium-range interactions (those within four residues on either side of a given one) contribute additional stabilization (Ponnuswamy *et al.*, 1973b). The view that short-range interactions dominate receives further support from statistical analyses of the conformations of amino acid residues in globular proteins, and empirical conformation prediction schemes based thereon, as discussed in Section VIII,C. Therefore, in Section VIII,B, we shall trace the development of this concept (Scheraga, 1973b), in order to see how it may help overcome the multiple-minimum problem.

B. Role of Short-, Medium-, and Long-Range Interactions

In considering the energetic and entropic factors that determine the conformation of a polypeptide in a given solvent, it is useful to distinguish between short-, medium-, and long-range interactions since these contribute to the stability to differing extents, as stated above. In order to discriminate among different amino acid residues, the term "short-range," as used here, refers to an interaction between the side chain of an amino acid residue with the atoms of its own backbone, and does not include side chain–side chain interactions. The interactions between the atoms of a given residue with those of nearby residues (within four residues of the given one) are designated as "medium-range," and the interactions with more remote residues along the chain (even though, possibly, nearby in space), are termed "long-range."

1. The θ-Point

The basis for looking for a possible dominance of short-range interactions in proteins comes from the behavior of model systems, viz., randomly coiled polymers. Under appropriate conditions, the conformation of a randomly coiled homopolymer chain is determined exclusively by short-range interactions. Therefore, we consider first the conditions under which medium- and long-range interactions make no contribution to the conformation. The treatment of an ideal homopolymer chain by random-flight statistics leads to the conclusion that some average linear dimension of the chain, e.g., the root-mean-square end-to-end distance $<\overline{r^2}>^{1/2}$, varies with the square root of the molecular weight (Flory, 1953). While medium-, long-range, and excluded volume effects (not included in the random-flight calculation) tend to increase $<\overline{r^2}>^{1/2}$ beyond its *ideal* value, the choice of an appropriate (poor) solvent (in which polymer–polymer contacts are favored over polymer–solvent contacts) can reduce $<r^2>^{1/2}$ to its *ideal* value (Flory, 1953). Under these conditions (i.e., at the so-called θ-point), the polymer–polymer and polymer–solvent interactions compensate the medium-, long-range, and excluded volume effects, and the ideal value of $<\overline{r^2}>^{1/2}$ which results is determined exclusively by short-range interactions (Flory, 1953). Although a protein in aqueous salt solution may not be at the θ-point, the possibility existed that its conformation, while not determined exclusively by short-range interactions, might nevertheless be dominated by them. As will be shown here, the dominance of short-range interactions has been demonstrated for the formation of α-helical and nonhelical portions of proteins (Kotelchuck and Scheraga, 1968, 1969; Finkelstein and Ptitsyn, 1971), and for the formation of β-turns (Lewis *et al.*, 1971) and extended structures (Burgess *et al.*, 1974a).

2. *Empirical Procedures for Predicting Helical and Other Regions in Proteins*

The foregoing argument, that short-range interactions may be dominant, is the implicit basis for the success, to differing extents, of various empirical procedures for predicting the location of helical, extended, β-turns, and other regions in proteins; these prediction schemes are considered in Section VIII,C. In essentially all these procedures, known X-ray structures were examined, and attempts were made to assign specific conformational preferences (e.g., right-handed α-helix, extended structure, or β-turn) to each of the amino acid residues. Using these preferences, and various rules that differed from author to author, predictions were made about the existence or nonexistence of various conformational features in other proteins. Such empirical predictive schemes (based solely on the behavior of single residues, or on short sequences of residues) would not lead to meaningful conclusions if short-range interactions were not dominant.

3. *Conformational Preferences within a Single Peptide Unit*

To examine the energetic basis for the validity of the hypothesis that short-range interactions are dominant, a study was made (Kotelchuck and Scheraga, 1968, 1969) of the role of these interactions in helix formation for proteins of known structure. In particular, calculations were carried out to obtain the energy of interaction of individual side chains in lysozyme with side chains that are nearest neighbors along the backbone, as well as with the backbone groups themselves. It was found that, for various initial backbone conformations (viz., the right- and left-handed α-helices, α_R and α_L, respectively, and the antiparallel pleated sheet structure, β), the conformation of lowest energy after minimization was the same in most cases for a given amino acid residue and was independent of the nature of the next amino acid in the chain. Furthermore, the backbone structures corresponding to the lowest energy (i.e., α_R, β, or α_L) showed a high degree of correlation with the so-called helix-making or helix-breaking character of a residue, as determined by the earlier *empirical* studies (cited in Sections VIII,B,2 and VIII,C) on the identification of α-helical regions in proteins. In other words, it appears that the short-range interactions within a given peptide unit may be the physical origin of the so-called helical potential of a residue. In addition, since the side chain-side chain interactions do not play a major role in determining backbone conformation in most cases, the cooperativity among residues, which is necessary for the formation of a helical segment, may simply be the additive effect of placing some sequence of helix-making residues in a particular region. This suggested a model for helix formation in which each type of peptide unit in proteins of known amino acid sequence was assigned a designation *h* or *c* (helix-making or helix breaking, respectively), based on a study of the energy surface of the peptide unit. Then, from an examination of the *h* or *c* assignments for lysozyme, myoglobin, α-chymotrypsin, and ribonuclease, empirical rules were formulated to distinguish between helical and nonhelical regions. These rules are: (a) an α-helical segment will be nucleated when at least four *h* residues in a row appear in the amino acid sequence, and (b) this helical segment will continue growing toward the C-terminus of the protein until two *c* residues in a row occur, a condition that terminates the helical segment. These rules were consistent with the helical or nonhelical state of 78% of the residues of the four proteins mentioned above (Kotelchuck and Scheraga, 1968, 1969).

With the later availability of the X-ray structures of seven proteins, the validity of these rules was examined further (Kotelchuck *et al.*, 1969). It was observed that, if a nonhelical dipeptide ever occurred at the C-terminus of a helical region, it had a low probability of occurring elsewhere in a helical region and as high as a 90% probability of occurring elsewhere in nonhelical regions; i.e., two *c* residues in a row prevent further growth of a helical segment. It was also found that those residues designated as *c*'s tended to predominate at the C-termini of helical segments. These results constitute an experimental demonstration of the validity of rule b above. Finkelstein and Ptitsyn (1971) also made a statistical analysis of the conformations of amino acid residues in proteins of known structure, and came to similar conclusions, viz., that short-range interactions are dominant, in that single residues can be classified as helix-making or helix-breaking and that side chain–side chain interactions play a minor role in determining the conformational preference of a given amino acid residue. Similarly, short-range interactions have been shown to dominate in determining the preferences for certain sequences to form extended structures and β-turns (see Sections VIII,B,7 and B,8). The success of the more recent empirical rules for predicting conformational features of proteins, discussed in Section VIII,C, confirms the validity of the concept of the dominance of short-range interactions.

At this point, it is of interest to consider the factors that determine the conformational preference of a given amino acid residue. The conformational entropy of a residue in the random coil state[7] must be overcome by favorable energetic factors in order for the residue to be helix-making; otherwise, it will be helix-breaking. Glycyl residues, with no side chains, have no favorable energetic factors to enhance helix formation; thus, the entropy of the coil makes glycyl residues helix-breaking (M. Go *et al.*, 1971). When a β-CH_2 group is added, the resulting nonbonded interactions tend to favor the α_R conformation (Kotelchuck and Scheraga, 1968, 1969; M. Go *et al.*, 1971). Thus, alanine is a helix-making residue (M. Go *et al.*, 1971). While all amino acids besides glycine have a β-CH_2 group, they are not all helix-making because of interactions involving groups beyond the β-carbon; for example, in Asn which is helix-breaking, electrostatic interaction between a polar side-chain group and the polar backbone amide group destabilizes the α_R conformation relative to other conformations. In Gln and Glu, the electrostatic effect is weaker because of the greater distance between the backbone amide group and the polar side-chain group (resulting from entropically favored extended side-chain conformations); hence the preferred conformation for Gln and Glu is α_R. Recently, an extensive series of conformational energy calculations (including the computation of free energies) was carried out for the *N*-acetyl-*N'*-methyl amides of all twenty naturally occurring amino acids (Lewis *et al.*, 1973b). From these calculations, it is possible to assess how the various energetic factors con-

[7] When the protein is in its unfolded form, in principle every residue may take on an *ensemble* of conformations (the random coil state), which contributes a conformational entropy. When the protein folds to its native form, all residues (even the nonhelical or, in general, nonregular ones) adopt a specific conformation (not an ensemble), and this entropy is lost (except for the small conformational fluctuations of the native structure); thus, it is incorrect to refer to residues that are in nonregular conformations in the *native* protein as being in the random coil state.

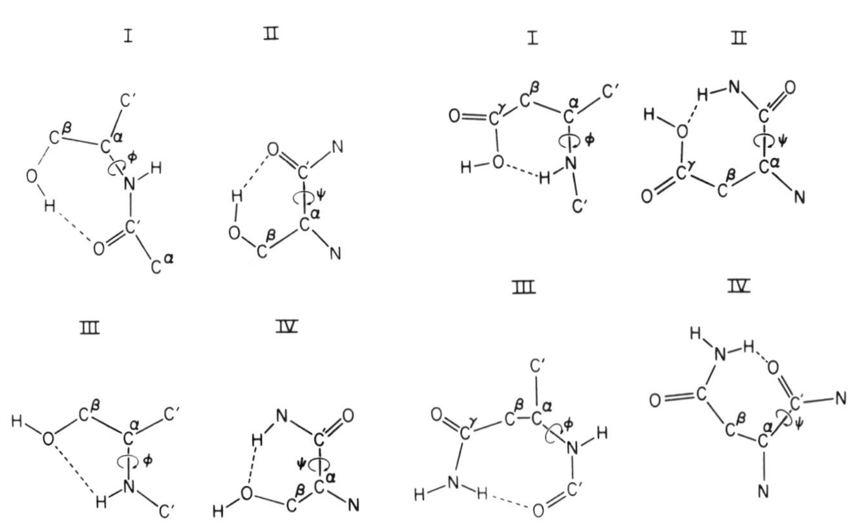

FIG. 26. Illustration of the types of hydrogen bonds between serine side chains and the backbone. From Lewis *et al.* (1973b).

FIG. 27. Illustration of the types of hydrogen bonds between aspartic acid and asparagine side chains and the backbone. From Lewis *et al.* (1973b).

tribute to the conformational preferences of each residue. For example, the side chains of both Ser and Asp can form hydrogen bonds with the nearby backbone amide groups when these residues are in nonhelical conformations, as illustrated in Figs. 26 and 27; thus, Ser and Asp are helix-breaking. The fact that various residues have different preferences for selected conformations is also indicated by ϕ, ψ diagrams for the 20 amino acids in proteins of known X-ray structure (Burgess *et al.*, 1974). From Figs. 28–31, it can be seen that the distribution of ϕ, ψ conformations varies for each amino acid and, in many cases, appears to reflect the particular characteristics of each amino acid; e.g., the restricted extended-structure region for Val (Fig. 30) is caused by the bulky C^{γ} methyl groups. The differences between the various amino acids tend to be lost when the ϕ, ψ distributions for all the amino acids are pooled (Pohl, 1971). The data of Figs. 28–31 have been used to compute the mean values of ϕ and ψ (and their relative probabilities of occurrence) for each of the 20 amino acids in several conformational states (Burgess *et al.*, 1974); similar data have been used by Chou and Fasman (1974a,b) to compute relative probabilities of occurrence. Such data are used in the empirical predictive schemes discussed in Section VIII,C.

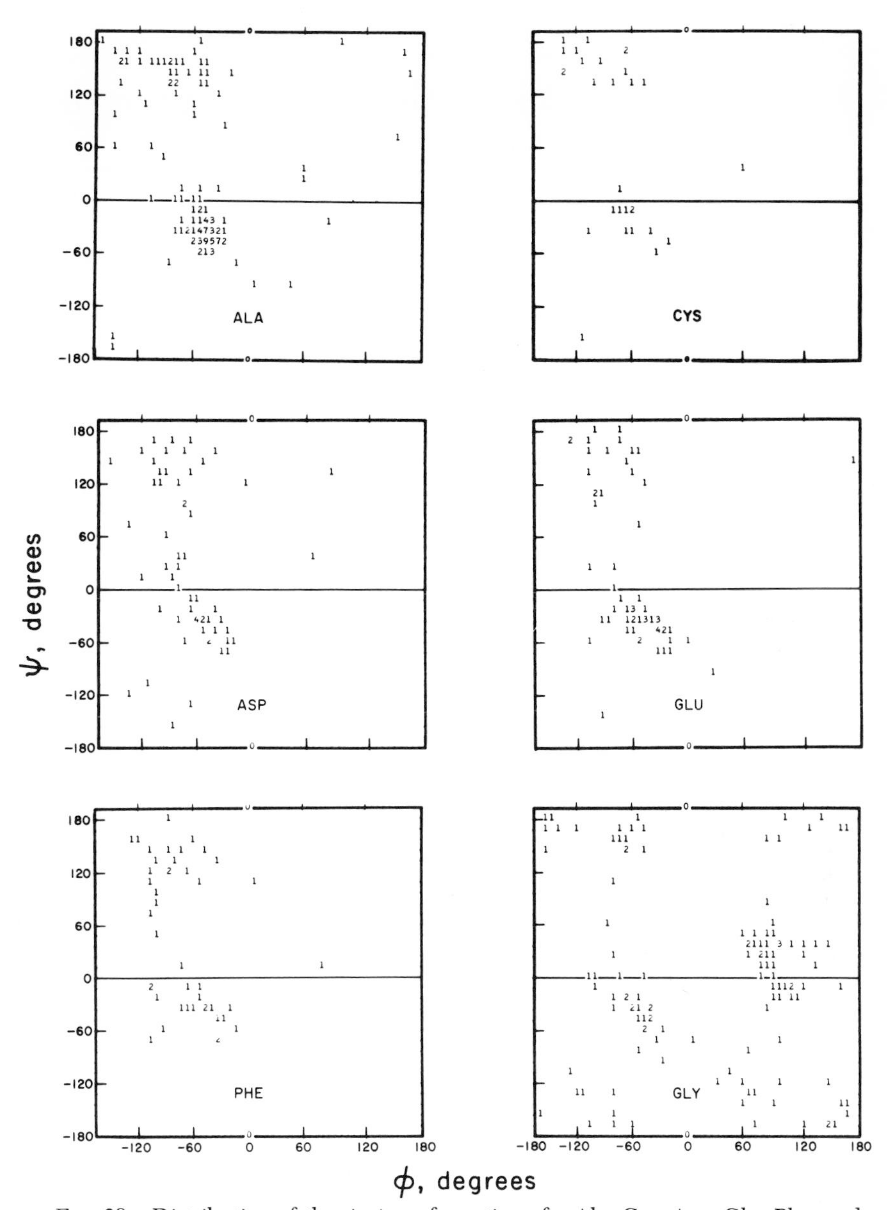

FIG. 28. Distribution of the ϕ, ψ conformations for Ala, Cys, Asp, Glu, Phe, and Gly in eight proteins. The numbers indicate the number of occurrences of each conformation corresponding to the ϕ, ψ dihedral angles at that position of the map. The ϕ, ψ domains for the most frequently occurring conformational states are as follows: right-handed α-helix ($-130° \leq \phi \leq -10°$ and $-90° \leq \psi \leq -10°$); left-handed α-helix ($10° \leq \phi \leq 130°$ and $10° \leq \psi \leq 90°$); extended structure ($-180° \leq \phi \leq -45°$ and $100° \leq \psi \leq 180°$ or $-180° \leq \psi \leq -140°$; and $140° \leq \phi \leq 180°$ and $100° \leq \psi \leq 180°$ or $-180° \leq \psi \leq -140°$). From Burgess *et al.* (1974).

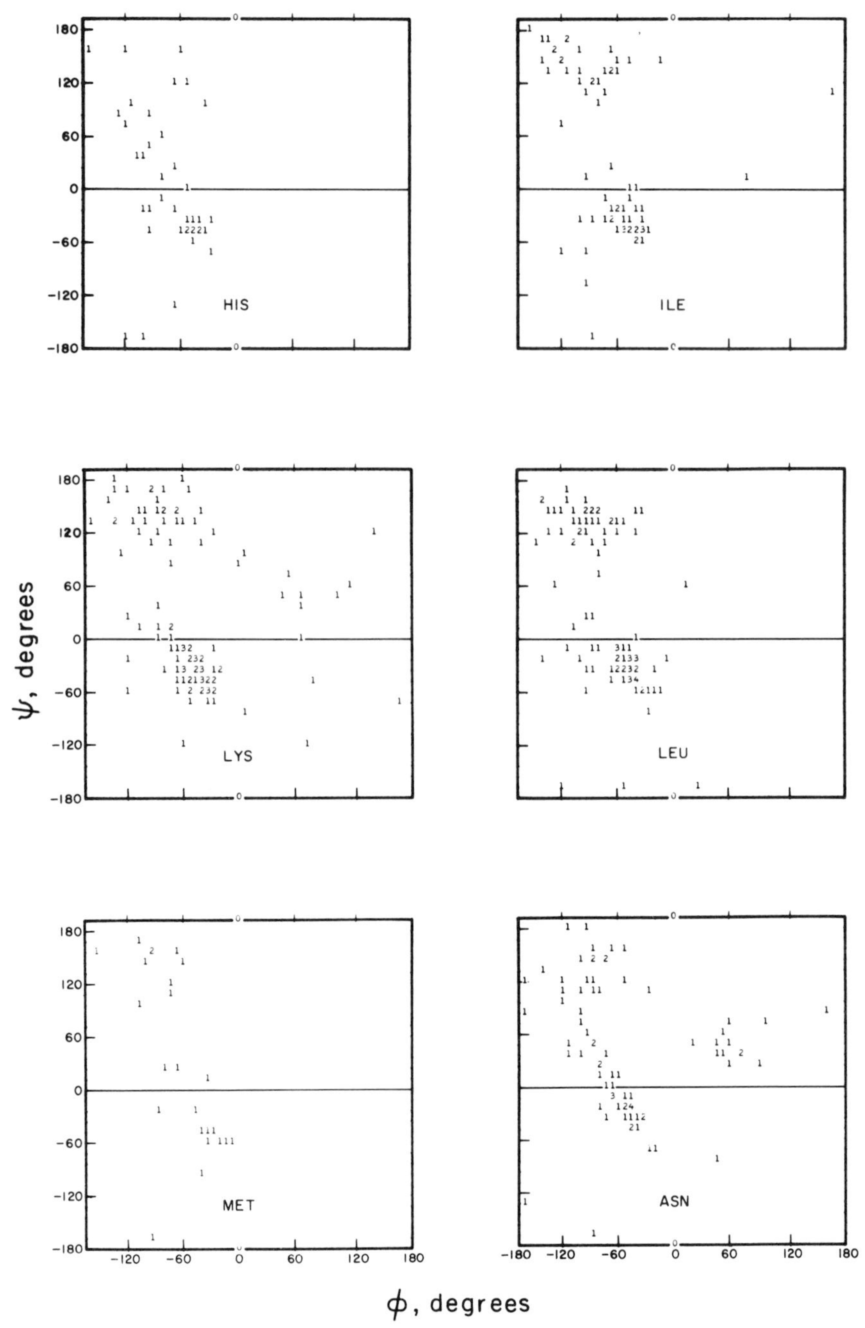

FIG. 29. Same as Fig. 28, for His, Ile, Lys, Leu, Met, and Asn. From Burgess *et al.* (1974).

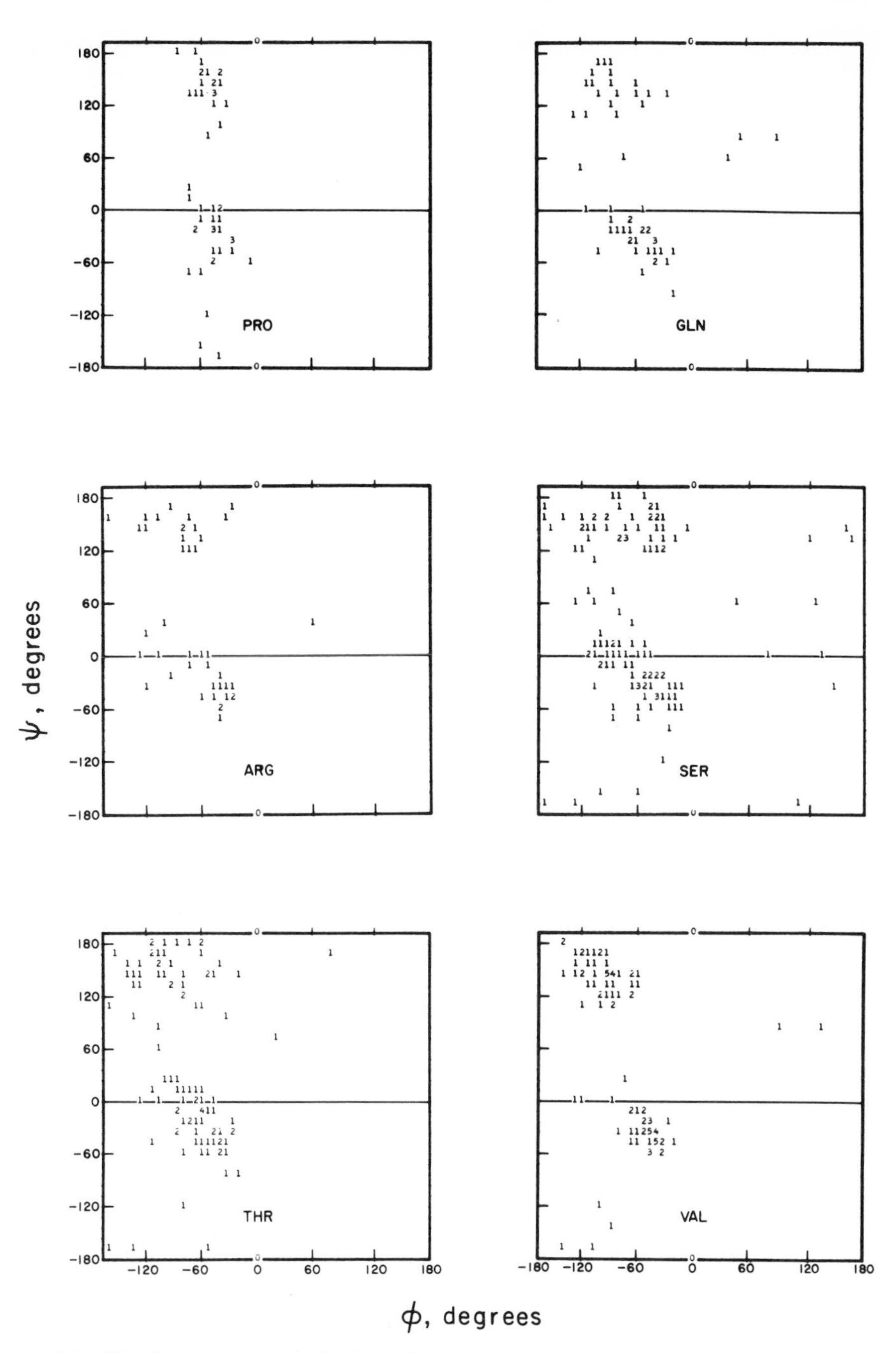

FIG. 30. Same as Fig. 28, for Pro, Gln, Arg, Ser, Thr, and Val. From Burgess *et al.* (1974).

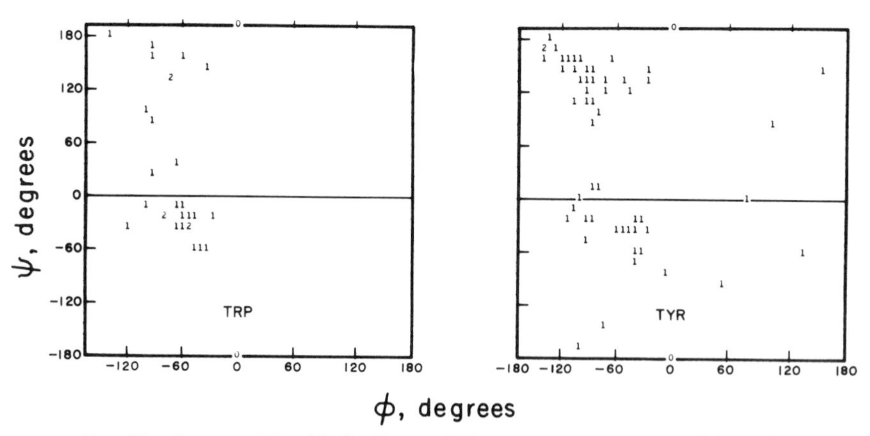

FIG. 31. Same as Fig. 28, for Trp and Tyr. From Burgess *et al.* (1974).

4. *Quantitative Specification of Helix-Making and Helix-Breaking Character*

Having demonstrated that the conformation of an amino acid residue in a protein is determined largely by short-range interactions, and, thus, in first approximation is essentially independent of the chemical nature of its neighbors, it becomes desirable to have a quantitative scale to specify, say, the helix-making and helix-breaking character of the twenty naturally occurring amino acids, instead of the earlier (Kotelchuck and Scheraga, 1968, 1969) assignment of all amino acids to two categories, h or c. We consider here a specific model for helix and coil, and in Section VIII,C we examine empirical rules for predicting various conformations (including helical ones). A model that suggests itself for specifying helix-making and helix-breaking character is the thermally induced helix-coil transition in homopolymers; that is, the Zimm–Bragg (1959) parameters σ and s, which characterize the transition curve, would appear to provide a quantitative basis for specifying the helix-making and -breaking tendency of any amino acid in its corresponding homopolymer (and, therefore, in a protein, since short-range interactions dominate in both cases). If the values of σ and s for a particular homopolymer are such that it melts at low temperature, then the corresponding residues would be helix-breaking; similarly, helix-making residues would have values of σ and s indicating that the corresponding homopolymer would melt at high temperatures.

The quantities σ and s are equilibrium constants for the elementary processes involved in the nucleation and growth, respectively, of helical sequences (Zimm and Bragg, 1959; Poland and Scheraga, 1970). If a long helical sequence exists in a homopolymer, the helical portion can be lengthened by one residue (the growth process) by restricting the dihedral angles of the residue to those characteristic of the α_R helix and, thereby, forming one additional hydrogen bond; the equilibrium constant for this process is s.

$$\cdots \mathrm{h\,h\,h\,h\,h\,h}\; c \cdots \overset{s}{\rightleftharpoons} \cdots \mathrm{h\,h\,h\,h\,h\,h}\; h \cdots \tag{18}$$

On the other hand, in order to initiate the formation of a helix in a nonhelical region (the nucleation process), the dihedral angles of *three* consecutive residues must be restricted to those of the α-helix, and only one hydrogen bond is formed; the equilibrium constant for this process is σs^3.

$$\cdots \mathrm{c\,c\,c} \cdots \overset{\sigma s^3}{\rightleftharpoons} \cdots \mathrm{h\,h\,h} \cdots \tag{19}$$

Since $s \sim 1$ in the transition region, the small value of σ ($\sim 10^{-4}$) expresses the fact that it is much more difficult to start a helix (Reaction 19) than to increase the size of an already existing one (Reaction 18). The parameter s depends on temperature, while the theoretically expected (Poland and Scheraga, 1965) temperature-dependence of σ is too small to detect experimentally. The curve for the melting of a homopolyamino acid α-helix to the random coil form (the helix-coil transition curve) can be computed from a knowledge of σ and the values of s at various temperatures. If the experimental transition curve is available, the values of σ and s can be obtained by fitting the experimental data by a theoretical curve.

In principle, then, σ and s could be determined for each naturally occurring amino acid by studying the thermally induced helix-coil transition of the corresponding homopolymer in water. For such a procedure to be feasible (Scheraga, 1971; von Dreele *et al.*, 1971a,b), the homopolymer must satisfy three requirements, viz., it must (1) be water soluble, (2) be helical in water at low temperature, and (3) melt to the random coil in water as the temperature is increased. In practice, none of the homopolymers of the naturally occurring amino acids satisfies these three requirements. Therefore, homopolymers cannot be used for this purpose, and resort is had instead to the use of random copolymers of two components—a helical host, e.g., poly(hydroxypropyl- or hydroxybutylglutamine) (which does satisfy these three requirements and for which σ and s are determinable) and the guest residues; from the effect of increasing amounts of the guest residues on the helix-coil transition curve of the homopolymer of the host residues, it is possible to determine σ and s for the guest

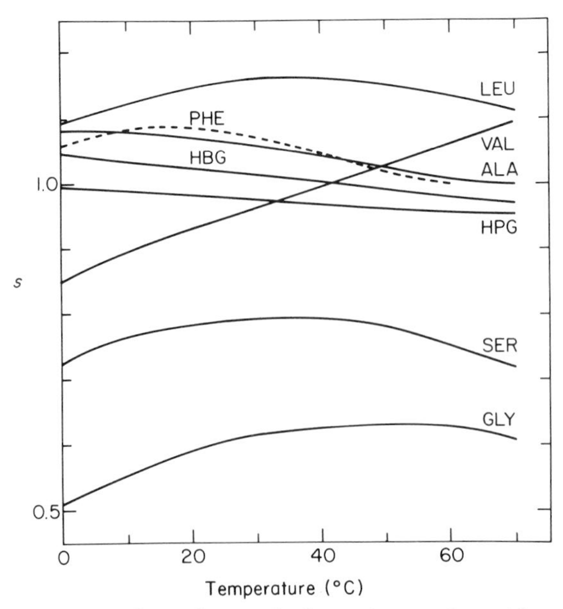

FIG. 32. Temperature dependence of s for various amino acid residues in water (Alter *et al.*, 1973). HBG and HPG refer to hydroxybutylglutamine and hydroxypropylglutamine, respectively, two different host residues.

residues. Thus far, these experiments have been carried out for the following guest residues: Gly (Ananthanarayanan *et al.*, 1971), Ala (Platzer *et al.*, 1972), Ser (Hughes *et al.*, 1972), Leu (Alter *et al.*, 1972), Phe (Van Wart *et al.*, 1973), and Val (Alter *et al.*, 1973), and the results are shown in Fig. 32. Further experiments on the remaining naturally occurring amino acids are in progress (H. A. Scheraga and co-workers, unpublished work). It can be seen that Gly and Ser are helix breakers, Gly more so than Ser because $s < 1$, and Ala and Leu are helix makers, Leu more so than Ala. Val is a helix breaker at low temperature and helix maker above 40°C. Since the experiments were carried out in aqueous solution, the resulting experimental values of σ and s contain all energetic and entropic contributions (including solvation) which determine the conformational preference. Also, the assumption that short-range interactions determine the behavior of the guest residue implies that the values determined for σ and s are independent of the nature of the host; the validity of this assumption is being tested experimentally (H. A. Scheraga and co-workers, unpublished work).

5. *Helix Probability Profiles*

The experimental values of σ and s can be used to obtain information about the conformation of any specific sequence of amino acids, e.g., that of a protein. However, since the values of σ and s were obtained from the Zimm–Bragg theory, which is based on the one-dimensional Ising model, we cannot treat the *native* protein molecule since its conformation is, in some measure, influenced by long-range interactions which are not taken into account in the Zimm–Bragg theory. Since the *denatured* protein is devoid of tertiary structure and hence, presumably, of long-range interactions other than excluded volume effects, the polypeptide conforms to the one-dimensional Ising model. Thus, above the denaturation temperature, we may apply the Zimm–Bragg formulation to this copolymer of ~20 amino acids to determine the probability that any given residue of the chain will be in the α_R or in the random coil conformation, respectively (Lewis *et al.*, 1970). It will then be shown that there is a correlation between the calculated α_R probability profile of the *denatured* protein and the experimentally observed α_R regions in the corresponding *native* structures; i.e., in many cases, those regions in the denatured protein which exhibit a propensity for being in the α_R conformation correspond to the α_R regions observed in the native protein.

The partition function Z, and the probability, $P_H(i)$, that the ith amino acid (of type A) in a chain of N residues is in the α_R conformation are given by

$$Z = (0,1)\left[\prod_{j=1}^{N} \mathbf{W}_A(j)\right]\begin{pmatrix}1\\1\end{pmatrix} \tag{20}$$

and

$$P_H(i) = (0,1)\left[\prod_{j=1}^{i-1} \mathbf{W}_A(j)\right]\frac{\partial \mathbf{W}_A(i)}{\partial \ln s_A(i)}\left[\prod_{j=i+1}^{N} \mathbf{W}_A(j)\right]\begin{pmatrix}1\\1\end{pmatrix}\Big/ Z \tag{21}$$

where $\mathbf{W}_A(j)$ is the matrix of statistical weights (i.e., measures of free energy) for the jth residue which is of amino acid type A, viz.,

$$\mathbf{W}_A(j) = \begin{pmatrix}s_A(j) & 1\\ \sigma_A(j)s_A(j) & 1\end{pmatrix} \tag{22}$$

$s_A(j)$ is the statistical weight assigned to this residue when it is in an α_R conformation and preceded by a residue in the α_R conformation, and $\sigma_A(j)s_A(j)$ is the statistical weight assigned to this residue when it is in an α_R conformation and preceded by a residue in the random coil conformation; Eqs. (20)–(22) pertain to the model of the helix-coil transition that is implied by Eqs. (18) and (19). The use of Eq. (21) to compute $P_H(i)$ automatically includes the cooperativity which is characteristic of the nearest-neighbor one-dimensional Ising model.

The values of σ and s, determined in aqueous solution (as indicated in Section VIII,B,4), apply to the *initial* folding of a polypeptide chain; that is, the groups of a denatured protein are exposed to water and the values of σ and s (obtained from random copolymers in water) determine their *tendency* to form helices in the *denatured* protein, *before* the onset of globularity buries the helix in the nonaqueous interior of the protein.

Pending the acquisition of data, such as those of Fig. 32, for the remainder of the twenty naturally occurring amino acids, the set of amino acids has been grouped into three categories (all with σ taken as 5×10^{-4}), viz., helix breakers (with $s = 0.385$), helix formers (with $s = 1.05$), and helix indifferent (with $s = 1.00$). Taking into account the limited data of Fig. 32, the earlier h and c assignments of Kotelchuck and Scheraga (1968, 1969), and the results of an information-theory analysis by Pain and Robson (1970), the amino acids are assigned as in Table III. It should be emphasized that these *tentative* values of σ and s are used here only pending completion of experiments that will complete Fig. 32 for the remaining amino acids.

Helix probability profiles for 11 proteins have been calculated from Eq. (21), using the values of σ and s discussed above and the as-

TABLE III
Assignment of Amino Acid Residues to Three Categories according to Helix-Forming Power[a]

Helix breaker	Helix indifferent	Helix former
Gly	Lys	Val
Ser	Tyr	Gln
Pro	Asp	Ile
Asn	Thr	His
	Arg	Ala
	Cys	Trp
	Phe	Met
		Leu
		Glu

[a] From Lewis and Scheraga (1971a).

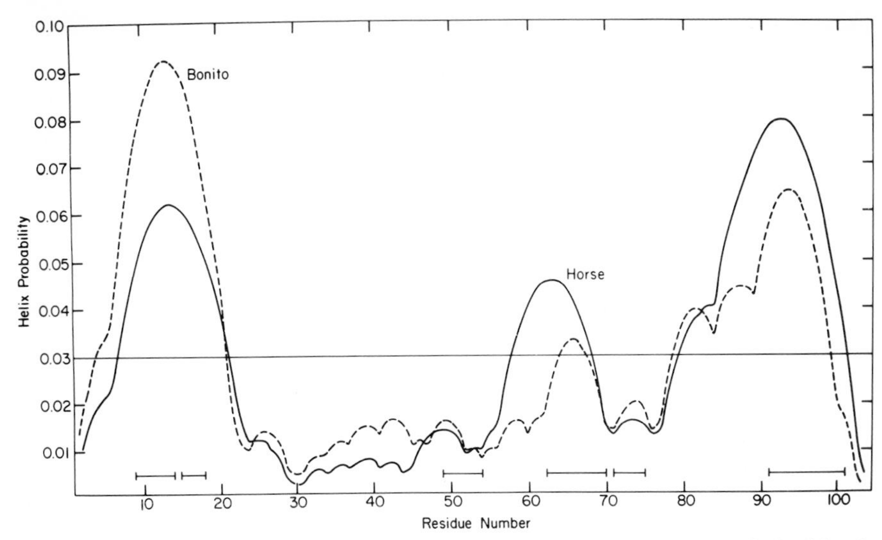

FIG. 33. Helix probability profiles for horse (solid line) and bonito (dashed line) ferricytochrome *c* proteins. The ordinate corresponds to helix probability, $P_H(i)$, and the abscissa to chain site. The horizontal bars (|—|) denote those regions of the protein found to be in the α_R conformation by X-ray diffraction analysis. The horizontal line is the calculated value of the overall mean helix content θ_H. From Lewis and Scheraga (1971a).

signments of Table III (Lewis and Scheraga, 1971a). [More-detailed information about the conformational state of each residue is provided by a recently developed eight-state model for the helix-coil transition in homopolymers and specific-sequence copolymers (M. Go *et al.*, 1971; N. Go *et al.*, 1971).] An example of some of these curves is shown in Fig. 33 for the horse and bonito ferricytochrome *c* proteins; the occurrence of helix in the *native* protein is indicated by the short horizontal lines in each diagram. From these and similar curves for other proteins, it appears that there is a close correlation between the propensity of a particular amino acid residue to be in the α_R conformation in the denatured protein and its occurrence in a helical region in the globular structure of the corresponding native protein. On this basis, it was suggested (Lewis *et al.*, 1970) that, during renaturation, the protein chain acquires *specific* long-range interactions which stabilize the helical regions that tend to form in certain portions of the chain; that is, folding of the polypeptide chain into the native conformation of a protein is thought to occur by incipient formation of nucleation sites (which may be helical if the constituent residues have a propensity for this conformation) stabilized

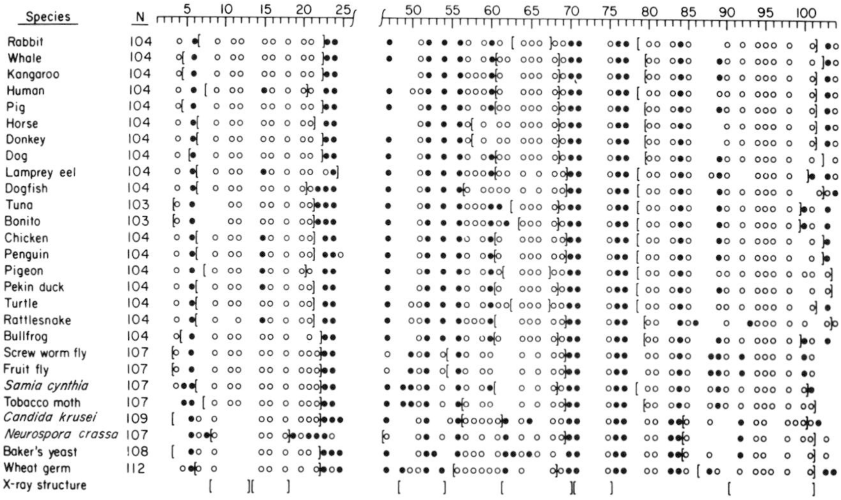

Species	N
Rabbit	104
Whale	104
Kangaroo	104
Human	104
Pig	104
Horse	104
Donkey	104
Dog	104
Lamprey eel	104
Dogfish	104
Tuna	103
Bonito	103
Chicken	104
Penguin	104
Pigeon	104
Pekin duck	104
Turtle	104
Rattlesnake	104
Bullfrog	104
Screw worm fly	107
Fruit fly	107
Samia cynthia	107
Tobacco moth	107
Candida krusei	109
Neurospora crassa	107
Baker's yeast	108
Wheat germ	112
X-ray structure	

FIG. 34. Predicted α_R helical segments for 27 cytochrome c proteins in the regions between residues 4–25 and 47–104. N is the total number of amino acid residues in each (single-chain) protein. Referring to Table III, the helix breakers are designated by ●, the helix formers by ○, and the helix indifferent residues by blank spaces. The predicted helical segments are indicated by square brackets; the observed ones are given at the bottom of the figure. From Lewis and Scheraga (1971a).

by specific long-range interactions, with the remainder of the protein molecule then folding around these stabilized nucleation sites.

Consistent with this view, it is found (Lewis and Scheraga, 1971a) that, despite amino acid substitutions in a series of 27 species of cytochrome c proteins, there is a striking similarity in their helix probability profiles, and a good correlation with the location of the helical regions in the X-ray-determined structure of the horse and bonito proteins (see Figs. 33 and 34). Among other things, the preponderance of *two* helix-breaking residues at the C-termini of helical sections (except at the C-terminus of the molecule where helix-breaking residues are not needed to break the helix), shown in Fig. 34, confirms the earlier, more primitive rule (b) mentioned in Section VIII,B,3). It appears that amino acid substitutions may be tolerated in evolution, provided that the helix-making or helix-breaking tendency (i.e., values of σ and s) of each amino acid residue is preserved, thereby enabling the altered protein to maintain the same three-dimensional conformation and, hence, the same biological function.

Application of this approach to lysozyme and α-lactalbumin (Lewis and Scheraga, 1971b), two different proteins with striking homo-

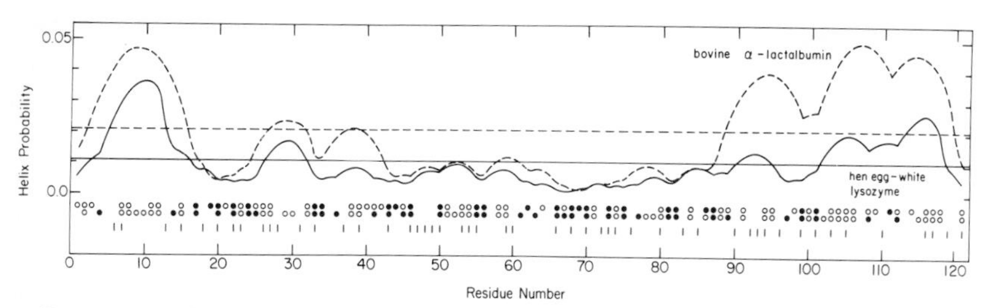

FIG. 35. Helix probability profiles for hen egg white lysozyme (solid line) and bovine α-lactalbumin (dashed line). The ordinate corresponds to helix probability, $P_H(i)$, and the abscissa to chain site. The horizontal solid and dashed lines are the calculated values of the overall mean helix content, θ_H, for the *complete* chains of lysozyme and α-lactalbumin, respectively. The helix-breakers are designated by ●, the helix-formers by ○, and the helix-indifferent residues by blank spaces (the upper line is for α-lactalbumin, and the lower one for lysozyme). The short vertical lines indicate those residues which are identical in both proteins. From Lewis and Scheraga (1971b).

logies in their amino acid sequence, led to very similar helix probability profiles (see Fig. 35). This result supports earlier suggestions (Brew *et al.*, 1967; Hill *et al.*, 1968; Browne *et al.*, 1969) that the two proteins might have similar three-dimensional structures, and again demonstrates the conservative nature of amino acid replacements (as far as helix-forming power in homologous proteins is concerned) which was found for the cytochrome *c* proteins. Interestingly, it has been possible to obtain a low-energy structure of α-lactalbumin by energy minimization, starting with a conformation which resembles that of lysozyme (Warme *et al.*, 1974).

6. *Implication for Nucleation and Folding of Protein Chains*

The results presented in Section VIII,B,5 are of importance for an understanding of how the three-dimensional structure of a protein is nucleated in specific regions so that the whole chain may then fold up into the native structure. As pointed out in the Introduction, the nucleation sites may be regular or nonregular.

In discussing the situation in which the nucleation site in the folding of a protein chain is a helical one, consider first the formation of an α-helical section of j residues in a *homopolymer* (Poland and Scheraga, 1970). In the nearest-neighbor Ising model, the statistical weight of such a section is σs^j [see Eqs. (18) and (19)]. In addition, there is a combinatorial entropy which stabilizes such a helical section; this entropy arises from the fact that this helical sequence can appear in many locations in the chain. However, in a copolymer, helical sequences tend to be more localized in specific parts of the chain, with a higher probability the greater the values of σ_A and s_A for the specific residues in the given local amino acid sequence (Poland and Scheraga, 1969a,b, 1970). In the absence of long-range interactions, the formation of a helical region in a *given part* of a *protein* chain depends primarily on the factor σs^j rather than on the combinatorial entropy. Since $\sigma \sim 10^{-4}$, σs^j can approach unity only if s^j approaches 10^4. But s varies only from ~ 1.02 to ~ 0.98 throughout the transition range of a homopolymer. Thus, even with the value of 1.02 for s, j would have to be ~ 500 in order that s^j approach 10^4. This is simply a statement of the well-known fact that helical sequences have to be very large to be stable—*in a system behaving according to the one-dimensional Ising model.* In an infinite chain conforming to this model, the most probable length of a helical sequence in the middle of the transition range (i.e., at $s = 1$) is $\sim \sigma^{-1/2}$ (or ~ 100 residues), and shorter helical sequences have a lower probability of occurrence. If one wanted to assign a value much greater than 1.02 to s, in order to achieve stability for *short* helical sections, one would encounter the dilemma that a chain with such helical sequences would not be predicted to denature in the accessible temperature range of 0–100°, whereas most proteins are observed to do so. However, short helical sequences ($\lesssim 10$ residues long) are known to occur in native proteins. How can we account for this and, at the same time, for the longer helical regions in myoglobin, etc.? Clearly, a short helical sequence cannot be stable in a polypeptide conforming to the one-dimensional Ising model. However, our view is that, because a protein is not a homopolymer, small regions in the amino acid sequence of the protein can be densely populated with helix-making residues, which make it more likely that an α-helix will form in that region compared to other regions, but that the stabilization of these incipient α-helical structures into actual short helical sequences in the native protein is accomplished by means of specific medium- and long-range interactions (Lewis *et al.*, 1970). It is just these medium- and long-range interactions that lead to globularity and thus the inapplicability of the one-dimensional Ising model in a native protein. On the other hand, this model is still applicable to a copolymer of, say, A and B units, in those cases in which the presence of a second component does not disrupt the one-dimensional character of the α-helical form of the first component, and indeed the data of Fig. 32 were obtained from such copolymers.

The correlations between the regions of higher helical propensity in the denatured state and the helical regions in the native structure (shown in Fig. 33) suggest that, during folding, the protein chain acquires *specific* medium- and long-range interactions that tend to stabilize these short regions, which have a tendency to be helical (Lewis *et al.*, 1970). The difficulty of forming this initial medium- and long-range interaction (nucleation) introduces the aspect of cooperativity with respect to the formation of the three-dimensional globular structure. The incipient formation of nucleation sites (among those residues with a propensity to adopt a particular local conformation, e.g., α-helices) stabilized by specific medium- and long-range interactions, may constitute the nucleation process for the refolding of the protein chain. The remainder of the protein molecule could then fold around these stabilized helical regions. Alternatively, in a protein with low helix content, other backbone conformations (regular, such as extended structures, or nonregular ones) may possibly serve as nucleation centers (Burgess *et al.*, 1974). Some experimental approaches which lead to this same point of view were discussed in Sections VII,C and D.

Again it should be noted that, although the one-dimensional Ising model can be applied to the denatured state (to assess the propensity toward helix formation), it is physically unrealistic for systems in which medium- and long-range interactions (beyond nearest or next-nearest neighbors) are operative. In this light, the correlations mentioned in Section VIII,B,2 lend further support to the substantial role of nearest-neighbor interactions in determining the overall three-dimensional structure of the native protein. However, if there were *no* long-range interactions, e.g., if a portion of a protein molecule were cleaved from the rest of the chain, the protein presumably could not assume its native conformation; this was a possible explanation provided earlier (Epand and Scheraga, 1968) for the failure of three isolated portions of the myoglobin chain to assume the conformation which they have in the intact native protein. One encounters numerous examples for this requirement, that most of the protein chain be present in order that the molecule fold into its native conformation. For example, the S peptide and S protein of ribonuclease do not separately have the conformations which they do in the native ribonuclease molecule; however, when added together, the two fragments interact to stabilize a three-dimensional structure which resembles that of native ribonuclease (Wyckoff *et al.*, 1970). Such a bimolecular reaction has a reasonable chance of leading to a properly stabilized structure. The extra translational and rotational entropy loss required in a trimolecular reaction makes it less likely that three fragments can associate to form a structure resembling the native protein. However, as discussed earlier (Section III), three large peptides derived from the sequences of both pancreatic ribonuclease and staphylococcal nuclease *do* interact to form enzymically active complexes (Lin *et al.*, 1970; Andria *et al.*, 1971).

In summary, folding of a given *intact* protein chain appears to require the information of nucleation sites in specific parts of the amino acid sequence. These nucleation sites would be stabilized cooperatively only in the folding process involving long-range interactions between various parts of the chain. A schematic representation of such a mechanism is illustrated for the folding of ribonuclease in Fig. 20. The removal of a significant part of a protein chain would prevent such long-range interactions from stabilizing the folded structure, and the remaining part of the protein chain could not acquire the stable conformation which it has in the intact native protein.

7. *Extended Structures*

While the early efforts in predicting the preferences for local conformations were confined to the α-helix, the extended structure has also received considerable attention recently (Ptitsyn and Finkelstein, 1970; Burgess *et al.*, 1974; Chou and Fasman, 1974a,b). It should be noted that "extended structure" is *not* β-structure, since the latter term implies information about the association of at least two extended structures in parallel or anti-parallel fashion. While conformational energy calculations on single residues (Lewis *et al.*, 1973b) provide information about the relative probabilities of occurrence of various conformations for each residue, and have been used in this manner (Burgess *et al.*, 1974), most reliance has thus far been placed on empirical probabilities of occurrence of these conformations based on observations of proteins of known X-ray structure. Thus, the probability of occurrence of extended structures, as well as helical structures, and chain reversals are treated empirically (see Section VIII,C). The success of these prediction schemes attests to the dominance of short-range interactions in determining the conformational preferences in extended structures, as well as in helical regions.

8. *Chain Reversals*

While it appears that short-range interactions are responsible for the incipient formation of α-helical and extended-structure regions, which acquire stability only after the introduction of medium- and long-range interactions, it has already been pointed out at the beginning of Section VIII that folding cannot occur by a random sampling of conformational space (Lewis *et al.*, 1971). Thus, the incipient local-ordered regions must be *directed* toward each other to acquire stability; that is, medium- and long-range interactions are not brought into play by a *chance* encounter of the ordered regions. Instead, there is a tendency for chain reversals or β-turns to occur among certain amino acid residues, thereby "directing" the encounter of the ordered regions. Idealized types of β-turns are illustrated in Fig. 23 (Venkatachalam, 1968). However, the actual turns occurring in proteins do not have these idealized shapes (see Figs. 36 and 37) and, indeed, conformational energy calculations indicate that the hydrogen bond is not required for stability (Lewis *et al.*, 1973a). The formation of a chain reversal should be regarded as an occurrence which depends on the proper juxtaposition of several residues, each of which has its own conformational preference (again, on the basis of short-range interactions) that enables it to participate in the chain reversal. From a statistical analysis of the amino acid compositions of the bends in three proteins, it has been possible to formulate rules for the prediction of the positions of the β-turns in protein sequences (Lewis *et al.*, 1971). Application of these rules to several other proteins led to a high degree of correlation between the predicted regions where the β-turns should appear, and their existence in the (X-ray determined) structure (Lewis *et al.*, 1971). A discussion of β-turns in proteins has also been presented by Popov and Lipkind (1971), Kuntz (1972), Esipova and Tumanyan (1972), Burgess *et al.* (1974), and Chou and Fasman (1974a,b). It is of interest that residues like Gly, Ser, and Asp, which have a low tendency toward helix formation, have a high propensity to form β-turns; that is, the preferred conformations, being ones which favor β-turns, make it less likely that these residues will accommodate to helix formation.

A study has been reported (Crawford *et al.*, 1973; Lewis *et al.*, 1973a) of the bends found in the native structures of eight proteins. The 135 bends which were located (Lewis *et al.*, 1973a) could be grouped among ten types (shown in Figs. 36 and 37), and over 40% of the bends did not possess a hydrogen bond between the C=O of residue i and the NH of residue $i + 3$; that is, many of the bends observed in proteins do not have the idealized shapes indicated in Fig. 23. In addition, conformational energy calculations were carried out (Lewis *et al.*, 1973a) on three tetrapeptides with amino acid sequences found as bends in the native structure of α-chymotrypsin. The results indicate that the bends occur not only in the whole mole-

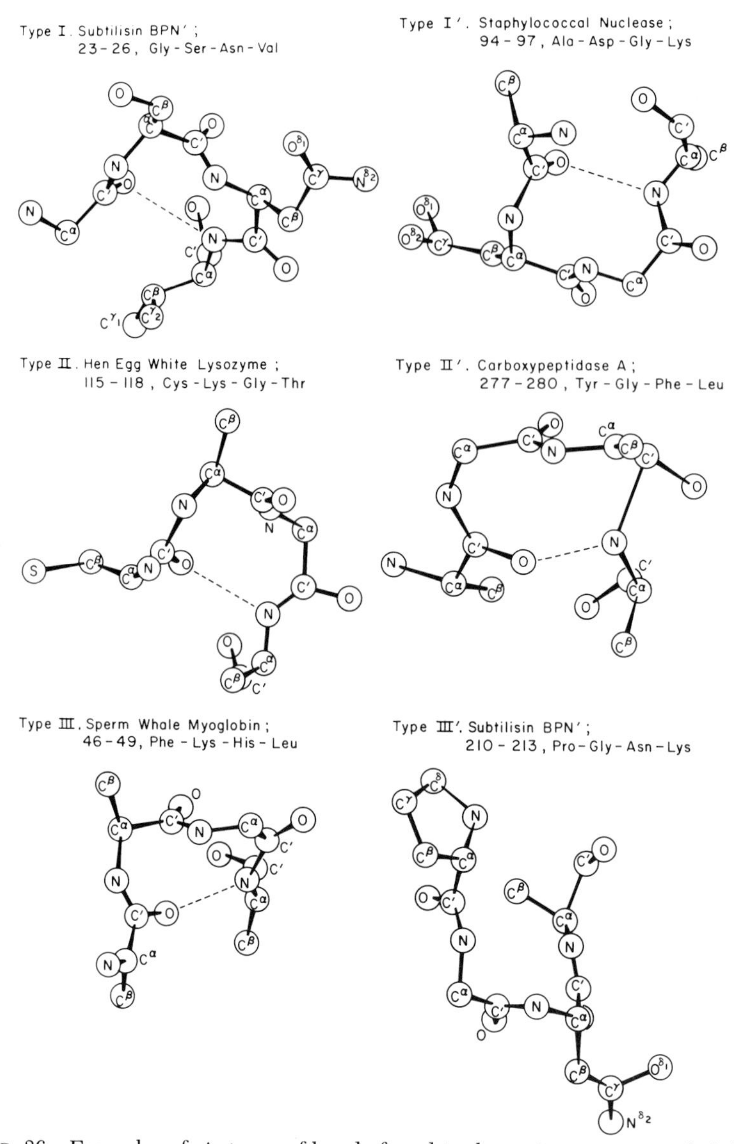

FIG. 36. Examples of six types of bends found in the native structures of globular proteins. When they occur, backbone–backbone hydrogen bonds are indicated by dashed lines. Only those portions of the side chains that do not interfere with the view of the backbone are shown (Lewis *et al.*, 1973a). These bends were taken from the X-ray structures of subtilisin BPN′ (Alden *et al.*, 1971), staphylococcal nuclease (E. E. Hazen, Jr., personal communication), hen egg white lysozyme (D. C. Phillips, personal communication), bovine carboxypeptidase A (Quiocho and Lipscomb, 1971), and sperm whale myoglobin (Watson, 1969).

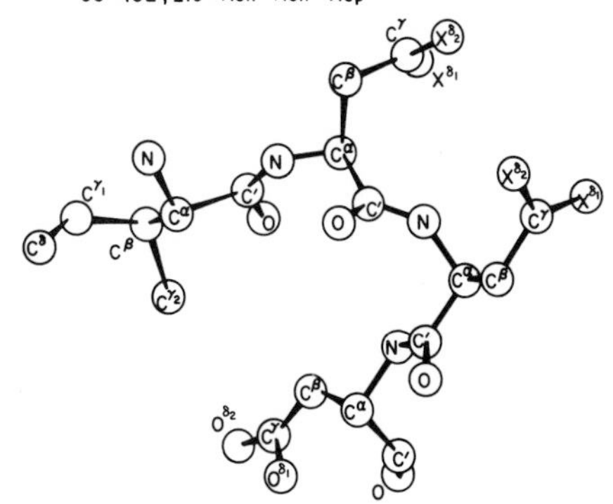

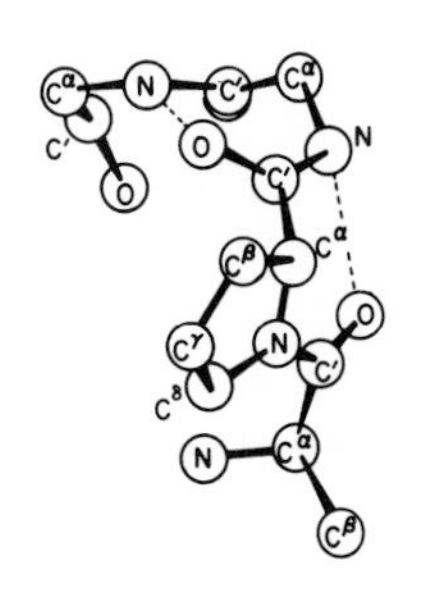

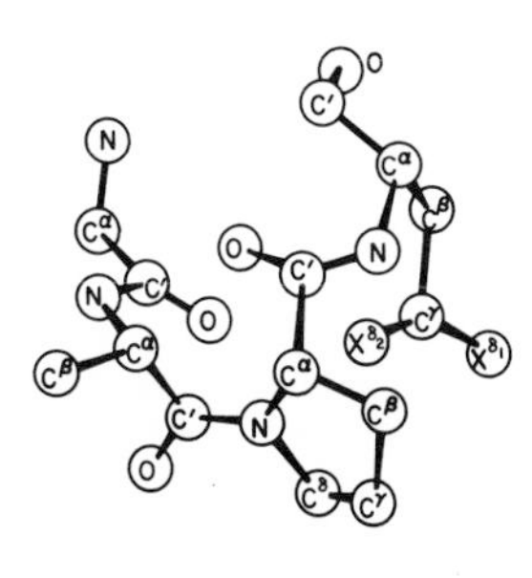

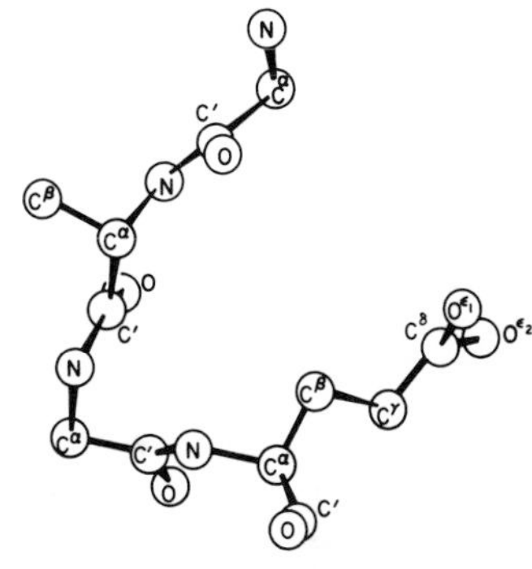

FIG. 37. Same as Fig. 36, but for four other types of bends (Lewis *et al.*, 1973a). These bends were taken from the X-ray structures of tosyl α-chymotrypsin (Birktoft *et al.*, 1969), horse ferricytochrome *c* (R. E. Dickerson, personal communication), and bovine ribonuclease S (Wyckoff *et al.*, 1970).

cule, but should also exist to some extent in the free tetrapeptide in solution; i.e., the observed bends were the conformations of lowest energy even in the tetrapeptides (see also the discussion of tetrapeptides in Section VII,D). From a study of the antibody-binding capacity of a loop portion of the lysozyme molecule (and several synthetic homologs of the loop), it appears that the biological activity of the excised loop can be correlated with the tendency toward preservation of the three β-bends in this peptide (Arnon *et al.*, 1974). The stability of the bends, compared to those of other structures, arises principally from side chain-backbone interactions (e.g., a hydrogen bond between the side-chain COO^- of Asp in position $i+3$ and the backbone NH of the residue in position i) rather than from i to $i+3$

backbone-backbone hydrogen bonds. This result is consistent with the observation (Lewis *et al.*, 1971) that residues with small polar side chains, such as Ser, Thr, Asp, and Asn, are found frequently in bends, presumably because these residues can interact most strongly with their immediate backbones (see also Figs. 26 and 27). Water may also play a role in stabilizing bends among these polar residues, especially since such bends usually occur on the surface of a protein.

From the above discussion, the following hypothesis for the folding of the polypeptide chain emerges: nucleation sites tend to form in certain regions of the amino acid sequence of the polypeptide chain, in response to short-range interactions. These are stabilized to yield a significant population of native structures, however, only when medium- and long-range interactions come into play (see also Section VII,C). This is brought about by the formation of chain reversals among *specific* amino acid residues, also on the basis of short-range interactions, thereby enabling the nucleation sites to approach each other. The remainder of the polypeptide chain then folds around these one or more regions of interacting nucleation sites. The dominance of short-range interactions in determining backbone comformation, demonstrated for α-helices (Kotelchuck and Scheraga, 1968, 1969), β-turns (Lewis *et al.*, 1971), and extended structures (Burgess *et al.*, 1974), is a manifestation of the simple way in which nature makes use of the strong interactions between a side chain and its own backbone.

9. *Medium-Range Interactions*

If short-range interactions play the dominant role in forming local ordered regions, with long-range interactions being required for their stability, the question immediately arises as to the necessary range of the long-range interactions—that is, do they encompass the

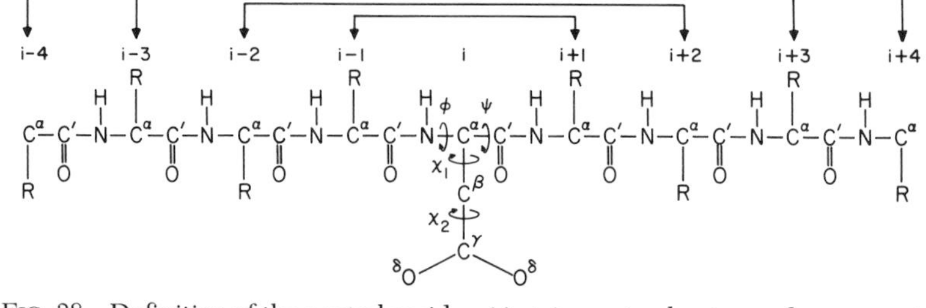

FIG. 38. Definition of the central residue i in tri-, penta-, hepta-, and nonapeptides. The dihedral angles ϕ, ψ, χ_1, and χ_2 of the central residue are indicated for Asp, as an example. From Ponnuswamy *et al.* (1973b).

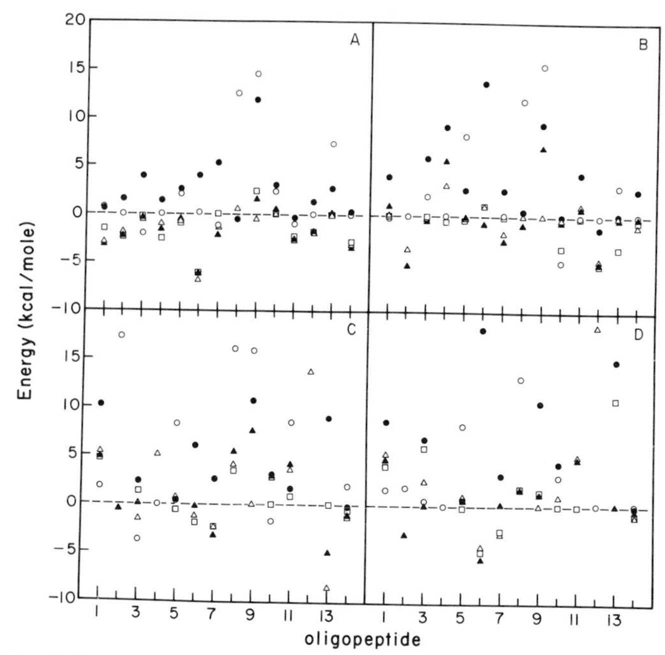

FIG. 39. Relative minimized energies of 14 oligopeptides of lysozyme. The dihedral angles of the central residue of each oligopeptide were varied during the energy minimization. The horizontal dashed line in each case is the energy obtained by minimizing from the observed (X-ray) conformation. The symbols ○, ●, △, ▲, □ pertain to minimization for the α_R, α_L, β_1, β_2, and C_7^{eq} conformations, respectively, of the central residue. Panels A, B, C, and D pertain to tri-, penta-, hepta-, and nonapeptides, respectively (Ponnuswamy *et al.*, 1973b). The 14 nonapeptides of lysozyme are nonoverlapping.

whole protein molecule or limited portions of it? To answer this question, conformational energy calculations were carried out on tri-, penta-, hepta-, and nonapeptide oligomers, as illustrated in Fig. 38 (Ponnuswamy *et al.*, 1973b). The amino acid sequences were selected from lysozyme, and all dihedral angles of the *central* residue of each oligopeptide were varied during energy minimization in which the conformations of the residues on both sides of the central one were held fixed at those obtained from a partial energetic refinement (Warme and Scheraga, 1973) of the X-ray structure. While several computed energy minima are found for the tripeptides, the observed one (whose energy is indicated by the dashed line of Fig. 39, for each of 14 oligopeptides) becomes the preferred one (i.e., the one of lowest energy), in most cases, in the calculations involving the longer sequences (hepta- and nonapeptides), as illustrated in Fig. 39

(Ponnuswamy *et al.*, 1973b). This demonstrates that medium-range interactions (those present in the hepta- and nonapeptides) provide the additional interactions required to determine the conformation of the central residue (as the one observed in the X-ray structure of the whole protein). The four discrepancies which are still evident in the nonapeptides (Fig. 39D) may indicate that, in these few cases, longer-range interactions (involving more remote parts of the protein chain) are additionally required to determine the conformation of the central residue. Essentially similar results were obtained with nonapeptide segments of bovine pancreatic trypsin inhibitor (Burgess and Scheraga, 1975a,c).

While short-range interactions play the dominant role in the formation of *regular* nucleation sites (e.g., α-helices and extended structures), medium-range interactions *may* contribute significantly in the formation of β-turns and of nonregular, but specifically folded structures. The latter, illustrated in Fig. 20, usually involve medium-range hydrophobic interactions between nonpolar side chains in water.

Besides demonstrating that short-range interactions are the most important ones, and medium-range interactions less so (but more important than longer-range ones) in determining protein backbone conformation, the computations described above may provide a way to circumvent the multiple-minimum problem in conformational energy calculation. Perhaps, by treating overlapping nonapeptides (whereby the energy is minimized by varying the conformation of the central residue), it might be possible to cycle several times through the whole protein to arrive at a rough conformation, which would then serve as a starting point for conventional energy minimization; such studies are in progress (see Section VIII,D). The selection of the initial conformation for energy minimization is based not only on these computations on nonapeptides, but also on the information gained about the regions where α_R helical, extended structures and chain reversals are likely to occur.

C. Empirical Predictive Schemes

As mentioned in Section VIII,B,2, many empirical schemes (based on short- and medium-range interactions) have been proposed for the prediction of the gross backbone conformation of each amino acid residue in a native protein. However, as indicated in Figs. 28–31, the actual backbone conformation of a given residue may differ considerably from the mean value of the distribution in any given conformational region. Thus, even if the mean values could be predicted with 100% accuracy, a protein built from such *mean* values for each of its residues would not resemble (even in gross overall dimensions) the native structure. Instead, these predictive schemes may be useful in obtaining *initial* conformations for energy minimization (in which interactions over the whole protein molecule are then included). In the minimization, the medium- and long-range interactions would induce the ϕ, ψ values of each residue to move away from the mean values, and thus enable the protein to attain the native structure. Thus, the predictive schemes would have contributed to the solution of the multiple-minimum problem.

It is in this light that empirical predictive schemes should be considered. Many such schemes have been published, and an extensive critique of these has been presented by Burgess *et al.* (1974). These authors have proposed the use of a quantitative prediction index P to evaluate the various predictive schemes, where

$$P = N_s Q - 1 \qquad \text{for} \qquad 0 \leqslant Q \leqslant 1/N_s \tag{23}$$

$$P = (N_s Q - 1)/(N_s - 1) \qquad \text{for} \qquad 1/N_s \leqslant Q \leqslant 1 \tag{24}$$

and

$$Q = \frac{\text{number of residue states assigned correctly}}{\text{number of residues in the chain}} \tag{25}$$

i.e., the fraction of correctly predicted states, and N_s is the number of states possible for each residue in the prediction model. The index P varies from -1 to $+1$ depending on the value of Q; if no states are predicted correctly, then $P = -1$, if the fraction of states predicted correctly (Q) is the same as expected for a random assignment of states (i.e., $1/N_s$ for a model of N_s states), then $P = 0$, and if all states are assigned correctly $P = 1$. The larger the value of N_s in the model, the more is the information about the topography of the chain that is provided by a high value of P. Several of the currently used prediction algorithms provide fairly high values of P (say, ~0.7) when averaged over many proteins. Since the primary use of these predictive schemes is to obtain a starting conformation for energy minimization, a fairly high value of P (but still less than 1) may possibly suffice to provide a suitable initial structure for subsequent energy minimization. However, this purpose will be better served if the prediction algorithms can be improved so that P approaches 1; using an optimization technique to obtain the probabilities of occurrence of various conformations (in a three-state model, viz., helix, extended structure, and coil), it has been possible to improve the

TABLE IV: *Evaluation of Models for*

	Values of prediction index *P*						
	α-Helix						
Protein	Low *et al.* (1968)	Guzzo (1965)	Prothero (1966)	Schiffer and Edmundson (1967)	Kotelchuck and Scheraga (1968, 1969)	Leberman (1971)	Lewis *et al.* (1970)
Lysozyme	0.05	−0.1	0.7	0.5	0.6	0.6	0.5
Ribonuclease	0.7	−0.2	0.5	0.5	0.6	0.8	0.2
Myoglobin	−0.5	−0.1	0.6	0.4	0.4	0.0	0.2
α-Chymotrypsin							
B chain	0.6	0.7	0.0	0.4	0.8	0.9	0.3
C chain	0.4	−0.5	0.1	0.0	0.3	0.6	0.7
Subtilisin	–	–	–	–	0.5	0.5	0.4
Carboxypeptidase	–	–	–	–	–	–	–
Staphylococcal nuclease	–	–	–	–	–	0.5	0.5
Cytochrome *c*	0.6	–	–	0.1	–	0.4	0.2
Carp myogen	–	–	–	–	–	–	–
Cytochrome b_5	–	–	–	–	–	–	–
Rubredoxin	–	–	–	–	–	–	–
Papain	–	–	–	–	–	–	–
Concanavalin A	–	–	–	–	–	–	–

[a] From Burgess *et al.* (1974).
[b] Calculated from both single- and double-residue information (column 4 of Table 3 of Robson and Pain).
[c] Calculated from single-residue information (columns 6 and 7 of Table 3 of Robson and Pain).

prediction scheme to the extent that the average value of *P* for 15 proteins is ~0.9 (Burgess *et al.*, 1975a). Making use of the prediction index *P*, the various predictive schemes in the literature have been compared. This comparison is shown in Table IV (Burgess *et al.*, 1974), from which it can be seen that no one scheme has achieved perfection.[8] On the one hand, one should not expect *P* to equal 1 unless the predictive schemes are modified to include longer-range interactions; on the other hand, the fact that most values of *P* are significantly greater than zero, and approach 1, indicates that short-range interactions play a dominant role in determining the conformational preference of each residue.

[8] After the preparation of Table IV, the paper of Nagano (1974) appeared. Using his earlier rules (Nagano, 1973) [which gave two-state average values of *P* of 0.7, 0.3, and 0.8 for α-helix, bends, and β structure, respectively (Burgess *et al.*, 1974), for proteins *within* the same set used to obtain the rules], the four-state values of *P* for several proteins outside the set used to obtain the rules are: 0.3 for carp myogen; 0.0 for rubredoxin; 0.3 for cytochrome c_2; and 0.0 for concanavalin A (A. W. Burgess, private communication). An example of a successful application of empirical rules is provided by the recent work of Carter *et al.* (1974).

Prediction of Protein Backbone Structure[a]

Values of prediction index *P*										
α-Helix				Bends			Extended or β-structure		Three state[d]	Four state[e]
Robson and Pain (1971)[b]	Robson and Pain (1971)[c]	Ptitsyn and Finkelstein (1970)	Burgess *et al.* (1974)	Lewis *et al.* (1971)	Kuntz (1972)	Burgess *et al.* (1974)	Ptitsyn and Finkelstein (1970)	Burgess *et al.* (1974)	Chou and Fasman (1974a,b)	Burgess *et al.* (1974)
0.7	0.5	0.5	0.6	0.5	0.1	0.6	0.4	0.6	0.8	0.3
0.9	0.6	0.9	0.6	0.3	0.4	0.4	–	0.5	0.8	0.2
0.7	0.3	0.6	0.2	–	0.2	0.8	1.0	0.9	0.8	0.4
1.0	1.0	0.7	0.8	0.5	0.4	0.6	0.4	0.4	0.7[f]	0.3
0.7	0.7	0.3	0.5	0.5	0.2	0.5	0.4	0.4	–	0.2
0.8	0.1	0.6	0.7	0.0	–	0.5	0.5	0.5	0.8	0.2
–	–	–	0.4	0.2	–	0.4	–	0.4	0.7	0.1
–	–	–	0.5	0.4	–	0.6	–	0.5	0.5	0.3
–	–	–	0.5	0.4	–	0.5	–	1.0	0.5	0.4
–	–	–	0.3	–	–	0.6	–	0.9	0.5	0.2
–	–	–	0.4	–	–	0.5	–	0.8	0.7	0.4
–	–	–	0.8	–	–	0.6	–	0.2	–	0.1
–	–	–	0.7	–	–	–	–	0.4	0.7	–
–	–	–	0.8	–	–	–	–	0.2	–	–

[d] The three states are α-helix, extended, and coil.
[e] The four states are α-helix, extended, bend, and coil.
[f] This applied to the combined B and C chains.

The empirical predictive procedure of Kabat and Wu (1973a,b) and Wu and Kabat (1973), based on tripeptide sequences, has not been included in Table IV because the nature of their predictions does not lend itself to this type of analysis. They presented two different procedures, for two different purposes. In the first procedure (Kabat and Wu, 1973a,b), they tabulated frequences of occurrence of helix, extended structure and coil in the $(n-1)$th and $(n+1)$th positions of tripeptides from proteins of known structure, irrespective of the conformation at the nth position. Applying these frequencies to tripeptide portions of proteins of unknown structure, Kabat and Wu predict which central residue of each tripeptide can disrupt either a helix or an extended structure when the $(n-1)$th and $(n+1)$th pair have low frequencies of occurrence for helix and extended structure. In their procedure, the portion of an amino acid sequence between their assigned helix-disrupting residues is permissively helix, if long enough, and likewise for permissive extended-structure regions. Thus, this procedure only *rules out* certain conformations. However, after helix-breaking and extended-structure-breaking residues are assigned by the procedure of Kabat and Wu (1973a,b), other methods (such as those summarized in Table IV) can be used for predicting the conformation of the intervening residues. The second procedure of Wu and Kabat (1973) is applicable only to families of homologous protein sequences, not to an individual protein that is not a member of a homologous family. All members of the family of a homologous set of sequences are assumed to have the same backbone structure. In this procedure, they average the dihedral angles of proteins of known structure over many tripeptide conformations, including those not containing the central residue under consideration, i.e., dipeptide components of tripeptides. After observing which conformations of the central residue occur most frequently for each of several sequence variants, Kabat and Wu predict the value of the dihedral angles of the central residue in tripeptide sequences of all the proteins of a homologous family, e.g., cytochrome *c*. In several cases, more than one choice is possible for a given residue, and Kabat and Wu decide between these choices by model building.

D. A Possible Algorithm to Fold a Protein

If the predictive schemes can be improved to predict the average conformational state of each residue from among, say, five frequently observed states [envisaged (Burgess *et al.*, 1974) as right- and left-handed α-helical conformations (α_R and α_L, respectively), extended conformations (ϵ), and conformations in the bridge regions between α_R and ϵ (ζ_R) and between α_L and ϵ (ζ_L), although, by far, the most frequently occurring states are α_R and ϵ], then use can be made of the information contained in the nonapeptide scheme of Fig. 38. Initially, one of five states (with an average value of ϕ and ψ for each type of residue) can be assigned to each residue in a nonapeptide in a protein. These assignments are based on conformational energy calculations, experimental determinations of σ and s and helix-probability profiles computed therefrom (Sections VIII,B,4 and 5), and empirical parameters expressing the tendencies toward formation of the above five conformations. The conformational energy of the whole nonapeptide can then be minimized with respect to the dihedral angles of only the central residue. Then, an overlapping nonapeptide can be selected (shifted by one residue toward the C-terminus of the chain), and the five-state model can be applied again to all residues except the one whose dihedral angles were varied in the previous step; for this residue, the minimum-energy *dihedral angles* would be assigned. This process is illustrated schematically in Fig. 40 (Scheraga, 1974a,b). Thus, by moving down the whole length of a protein chain (and possibly repeating the process several times) it may be possible to determine whether the native conformation is being approached by successive approximation. Once the

Step	1	2	3	4	5	6	7	8	9	10	11	12	13 ....
1	β	β	α	α	α	α	α	β	β				
2	β	β	α	α	ϕ=-52° ψ=-48°	α	α	β	β				
3		β	α	α	ϕ=-52° ψ=-48°	α	α	β	β	β			
4		β	α	α	ϕ=-52° ψ=-48°	ϕ=-51° ψ=-49°	α	β	β	β			
5			α	α	ϕ=-52° ψ=-48°	ϕ=-51° ψ=-49°	α	β	β	β	β		
6			α	α	ϕ=-52° ψ=-48°	ϕ=-51° ψ=-49°	ϕ=-52° ψ=-49°	β	β	β	β		

etc.

FIG. 40. Schematic illustration of successive steps in the computation of the dihedral angles of a protein. The amino acid sequence is specified by the numbers 1, 2, . . . , 13, . . . , and the backbone conformations of each residue are specified *initially* as α, β, etc., but in terms of the dihedral angles ϕ and ψ after each computational step. In step 1, one of the five conformational states is assigned to each residue in a nonapeptide. In step 2, the energy of the nonapeptide is minimized with respect to the dihedral angles of residue 5, etc. From Scheraga (1974a,b).

structure begins to resemble that of the native protein, it would be in the correct potential energy well; then the minimization of the conformational energy of the whole protein, not just that of nonapeptide segments, could be carried out. This procedure is now being tested, with the inclusion of an additional step between steps 1 and 2 of Fig. 40, viz., variation of the average values of the dihedral angles of each residue in response to the requirement that the disulfide bonds be formed between the proper half-cystine residues (Burgess and Scheraga, 1975a,c). The final minimization over *all* pair interactions in the *whole* protein molecule makes use of computational procedures similar to those already used for the energetic refinement of the X-ray structures of actinomycin D (Ponnuswamy *et al.*, 1973a), lysozyme (Warme and Scheraga, 1974), and rubredoxin (Rasse *et al.*, 1974); Levitt and Lifson (1969) and Nishikawa and Ooi (1972) have also carried out energetic refinements of X-ray structures of proteins.

IX. Concluding Remarks

It is now well documented experimentally that, in the proper environment, a polypeptide chain can fold spontaneously to the three-dimensional structure of a native protein. It seems that, despite the possible presence of barriers in conformational space, the chain can find its way around these barriers to reach the structure of lowest free energy. While the validity of this thermodynamic hypothesis is not yet established unambiguously, it nevertheless seems to have a sufficiently reasonable chance of holding true so that it provides a useful basis for treating the folding problem theoretically. Thus, the interatomic interactions within the chain, and between the chain and the solvent, dictate the folding. The range of the forces involved is such that near-neighbor interactions are dominant, and the backbone conformations of local segments of the chain are determined largely by such interactions. However, the native conformation becomes more stable in response, first, to medium- and second, to longer-range interactions. Folding is envisaged as taking place by the formation of nucleation sites in various parts of the chain, in response to near-neighbor (and similarly to medium-range) interactions; the various nucleation sites then become stabilized when they are brought into proximity (perhaps by formation of β-turns, which also depend on short-range interactions for their stability) so that long-range interactions can become operative. Empirical prediction algorithms, which incorporate not only short-range but also medium-range interactions, would seem to offer the hope of leading to an approximately correct three-dimensional structure (i.e., one within the correct potential energy well), so that subsequent energy minimization (which includes interactions between *all* pairs of atoms in the *whole* molecule) can lead to the native structure. The experiments described herein provide a reasonable basis for this point of view, and it remains to be seen whether this concept and the further development of conformational energy calculation procedures will enable the three-dimensional structure of a native protein to be predicted from a knowledge of its amino acid sequence and its interactions with the solvent in which it is dissolved.

Acknowledgments

We are indebted to our colleagues, Drs. A. W. Burgess, B. Furie, and A. Schechter for reading this manuscript and making many helpful suggestions, and to Mrs. Dorothy Stewart for her expert assistance in compiling and typing.

References[9]

Alden, R. A., Birktoft, J. J., Kraut, J., Robertus, J. D., and Wright, C. S. (1971). *Biochem. Biophys. Res. Commun.* **45**, 337.

Alter, J., Taylor, G. T., and Scheraga, H. A. (1972). *Macromolecules* **5**, 739.

Alter, J., Andreatta, R. H., Taylor, G. T., and Scheraga, H. A. (1973). *Macromolecules* **6**, 564.

Ananthanarayanan, V. S., Andreatta, R. H., Poland, D., and Scheraga, H. A. (1971). *Macromolecules* **4**, 417.

Andria, G., Taniuchi, H., and Cone, J. L. (1971). *J. Biol. Chem.* **246**, 7421.

[9] The reader is advised that Chinese authors may be referenced under the given, rather than family, name.

Anfinsen, C. B. (1956). *J. Biol. Chem.* **221**, 405.
Anfinsen, C. B. (1966). *Harvey Lect.* **61**, 95.
Anfinsen, C. B. (1973). *Science* **181**, 223.
Anfinsen, C. B., and Haber, E. (1961). *J. Biol. Chem.* **236**, 1361.
Anfinsen, C. B., Haber, E., Sela, M., and White, F. H., Jr. (1961). *Proc. Nat. Acad. Sci. U. S.* **47**, 1309.
Anson, M. L. (1945). *Advan. Protein Chem.* **2**, 361.
Anson, M. L., and Mirsky, A. E. (1934a). *J. Gen. Physiol.* **17**, 393.
Anson, M. L., and Mirsky, A. E. (1934b). *J. Gen. Physiol.* **17**, 399.
Arnon, R. (1973). *In* "The Antigens" (M. Sela, ed.), Vol. 1, p. 87. Academic Press, New York.
Arnon, R., Teicher, E., and Scheraga, H. A. (1974). *J. Mol. Biol.* **90**, 403.
Arnone, A., Bier, C. J., Cotton, F. A., Day, V. W., Hazen, E. E., Jr., Richardson, D. C., Richardson, J. S., and Yonath, A. (1971). *J. Biol. Chem.* **246**, 2302.
Bellamy, G., and Bornstein, P. (1971). *Proc. Nat. Acad. Sci. U. S.* **68**, 1138.
Birktoft, J. J., Matthews, B. W., and Blow, D. M. (1969). *Biochem. Biophys. Res. Commun.* **36**, 131.
Bishop, J., Leahy, J., and Schweet, R. (1960). *Proc. Nat. Acad. Sci. U. S.* **46**, 1030.
Bodanszky, M., and Williams, N. J. (1967). *J. Amer. Chem. Soc.* **89**, 685.
Bodanszky, M., Ondetti, M. A., Levine, S. D., and Williams, N. J. (1967). *J. Amer. Chem. Soc.* **89**, 6753.
Bohnert, J. L., and Taniuchi, H. (1972). *J. Biol. Chem.* **247**, 4557.
Bradbury, J. H., and Scheraga, H. A. (1966). *J. Amer. Chem. Soc.* **88**, 4240.
Bränden, C., Eklund, H., Nordström, B., Boiwe, T., Söderlund, G., Zeppezaur, E., Ohlsson, I., and Åkeson, Å. (1973). *Proc. Nat. Acad. Sci. U. S.* **70**, 2439.
Brew, K., and Campbell, P. N. (1967). *Biochem. J.* **102**, 258.
Brew, K., Vanaman, T. C., and Hill, R. L. (1967). *J. Biol. Chem.* **242**, 3747.
Brew, K., Castellino, F. J., Vanaman, T. C., and Hill, R. L. (1970). *J. Biol. Chem.* **245**, 4570.
Brown, J. E., and Klee, W. A. (1969). *Biochemistry* **8**, 2876.
Brown, J. E., and Klee, W. A. (1971). *Biochemistry* **10**, 470.
Browne, W. J., North, A. C. T., Phillips, D. C., Brew, K., Vanaman, T. C., and Hill, R. L. (1969). *J. Mol. Biol.* **42**, 65.
Burgess, A. W., and Scheraga, H. A. (1975a). To be published.
Burgess, A. W., and Scheraga, H. A. (1975b). *J. Theor. Biol.* (in press).
Burgess, A. W., and Scheraga, H. A. (1975c). *Proc. Nat. Acad. Sci. U. S.* (in press).
Burgess, A. W., Ponnuswamy, P. K., and Scheraga, H. A. (1974). *Isr. J. Chem.* **12**, 239.
Burgess, A. W., Skavlinsky, K., and Scheraga, H. A. (1975a). To be published.
Burgess, A. W., Weinstein, L. I., Gabel, D., and Scheraga, H. A. (1975b). *Biochemistry* **14**, 197.
Canfield, R. E., and Anfinsen, C. B. (1963). *Biochemistry* **2**, 1073.
Carter, C. W., Jr., Kraut, J., Freer, S. T., Xuong, N. H., Alden, R. A., and Bartsch, R. G. (1974). *J. Biol. Chem.* **249**, 4212.
Chaiken, I. M. (1971). *J. Biol. Chem.* **246**, 2948.
Chaiken, I. M., and Anfinsen, C. B. (1971). *J. Biol. Chem.* **246**, 2285.
Chance, R. E., Ellis, R. M., and Bromer, W. W. (1968). *Science* **161**, 165.
Chantrenne, H. (1961). *In* "Biosynthesis of Proteins," p. 122. Pergamon, Oxford.
Ching-I, Niu, Kung, Yueh-Ting, Huang, Wei-Teh, Ke, Lin-Tsung, Chen, Chan-Chin, Chen, Yuang-Chung, Du, Yu-Cang, Jiang, Rong-Qing, Tsou, Chen-Lu, Hu, Shih-Chuan, Chu, Shang-Quan, and Wang, Keh-Zhen. (1964). *Sci. Sinica (Peking)* **13**, 1343.

Chou, P. Y., and Fasman, G. D. (1974a). *Biochemistry* **13**, 211.
Chou, P. Y., and Fasman, G. D. (1974b). *Biochemistry* **13**, 222.
Cohen, J. S., and Jardetzky, O. (1968). *Proc. Nat. Acad. Sci. U. S.* **60**, 92.
Cone, J. L., Cusumano, C. L., Taniuchi, H., and Anfinsen, C. B. (1971). *J. Biol. Chem.* **246**, 3103.
Cooke, J. P., Anfinsen, C. B., and Sela, M. (1963). *J. Biol. Chem.* **238**, 2034.
Crawford, J. L., Lipscomb, W. N., and Schellman, C. G. (1973). *Proc. Nat. Acad. Sci. U. S.* **70**, 538.
Creighton, T. E. (1974). *In* "Peptides, Polypeptides and Proteins" (E. R. Blout *et al.*, eds.), p. 201. Wiley, New York.
De Lorenzo, F., Goldberger, R. F., Steers, E., Givol, D., and Anfinsen, C. B. (1966). *J. Biol. Chem.* **241**, 1562.
Dintzis, H. M. (1961). *Proc. Nat. Acad. Sci. U. S.* **47**, 247.
Doolittle, R. F. (1973). *Advan. Protein Chem.* **27**, 1.
Doty, P., and Katz, S. (1950). *Abstr., 118th Meet., Amer. Chem. Soc.* p. 14C.
Dreizen, P., Gershman, L. C., Trotta, P. P., and Stracher, A. (1967). *J. Gen. Physiol.* **50**, 85.
Dunnill, P. (1965). *Sci. Progr. (London)* **53**, 609.
Dunnill, P. (1967). *Nature (London)* **215**, 621.
du Vigneaud, V., Ressler, C., Swan, J. M., Roberts, C. W., and Katsoyannis, P. G. (1954). *J. Amer. Chem. Soc.* **76**, 3115.
Epand, R. M., and Scheraga, H. A. (1968). *Biochemistry* **7**, 2864.
Epstein, C. J., Goldberger, R. F., and Anfinsen, C. B. (1963). *Cold Spring Harbor Symp. Quant. Biol.* **28**, 439.
Epstein, H. F., Schechter, A. N., Chen, R. F., and Anfinsen, C. B. (1971a). *J. Mol. Biol.* **60**, 499.
Epstein, H. F., Schechter, A. N., and Cohen, J. S. (1971b). *Proc. Nat. Acad. Sci. U. S.* **68**, 2042.
Esipova, N. G., and Tumanyan, V. G. (1972). *Mol. Biol.* **6**, 840.
Ferry, J. D., and Morrison, P. R. (1947). *J. Amer. Chem. Soc.* **69**, 388.
Finkelstein, A. V., and Ptitsyn, O. B. (1971). *J. Mol. Biol.* **62**, 613.
Flory, P. J. (1953). "Principles of Polymer Chemistry," Chapter 14. Cornell Univ. Press, Ithaca, New York.
Flory, P. J. (1956). *J. Amer. Chem. Soc.* **78**, 5222.
Freedman, M. H., and Sela, M. (1966). *J. Biol. Chem.* **241**, 2383.
Fuchs, S., De Lorenzo, F., and Anfinsen, C. B. (1967). *J. Biol. Chem.* **242**, 398.
Fujita, S. C., and Imahori, K. (1974). *In* "Peptides, Polypeptides, and Proteins" (E. R. Blout *et al.*, eds.), p. 217. Wiley, New York.
Furie, B., Schechter, A. N., Sachs, D. H., and Anfinsen, C. B. (1974a). *Biochemistry* **13**, 1561.
Furie, B., Schechter, A. N., Sachs, D. H., and Anfinsen, C. B. (1974b). *J. Mol. Biol.* (in press).
Gibson, K. D., and Scheraga, H. A. (1969a). *Proc. Nat. Acad. Sci. U. S.* **63**, 9.
Gibson, K. D., and Scheraga, H. A. (1969b). *Proc. Nat. Acad. Sci. U. S.* **63**, 242.
Givol, D., De Lorenzo, F., Goldberger, R. F., and Anfinsen, C. B. (1965). *Proc. Nat. Acad. Sci. U. S.* **53**, 676.
Go, M., Go, N., and Scheraga, H. A. (1971). *J. Chem. Phys.* **54**, 4489.
Go, N., and Scheraga, H. A. (1969). *J. Chem. Phys.* **51**, 4751.
Go, N., Lewis, P. N., and Scheraga, H. A. (1970). *Macromolecules* **3**, 628.
Go, N., Lewis, P. N., Go, M., and Scheraga, H. A. (1971). *Macromolecules* **4**, 692.
Godfrey, J., and Harrington, W. F. (1970). *Biochemistry* **9**, 894.

Goldberger, R. F., Epstein, C. J., and Anfinsen, C. B. (1963). *J. Biol. Chem.* **238**, 628.
Gutte, B., and Merrifield, R. B. (1969). *J. Amer. Chem. Soc.* **91**, 501.
Gutte, B., Lin, M. C., Caldi, D. G., and Merrifield, R. B. (1972). *J. Biol. Chem.* **247**, 4763.
Guttmann, S., Pless, J., Huguenin, R. L., Sandrin, E., Bossert, H., and Zehnder, K. (1971). *In* "Peptides 1969" (E. Scoffone, ed.), p. 54. North-Holland Publ., Amsterdam.
Guzzo, A. V. (1965). *Biophys. J.* **5**, 809.
Haber, E., and Anfinsen, C. B. (1962). *J. Biol. Chem.* **237**, 1839.
Hall, C. E. (1949). *J. Biol. Chem.* **179**, 857.
Hantgan, R. R., Hammes, G. G., and Scheraga, H. A. (1974). *Biochemistry* **13**, 3421.
Hartley, B. S. (1970). *Phil. Trans. Roy. Soc. London, Ser. B* **257**, 77.
Hauschka, P. V., and Harrington, W. F. (1970a). *Biochemistry* **9**, 3734.
Hauschka, P. V., and Harrington, W. F. (1970b). *Biochemistry* **9**, 3745.
Hauschka, P. V., and Harrington, W. F. (1970c). *Biochemistry* **9**, 3754.
Helinski, D. R., and Yanofsky, C. (1963). *J. Biol. Chem.* **238**, 1043.
Hermans, J., and Scheraga, H. A. (1961). *J. Amer. Chem. Soc.* **83**, 3283.
Hill, E., Tsernoglu, D., Webb, L., and Banaszak, L. J. (1972). *J. Mol. Biol.* **72**, 577.
Hill, R. L., Brew, K., Vanaman, T. C., Trayer, I. P., and Mattock, P. (1968). *Brookhaven Symp. Biol.* **21**, 139.
Hirs, C. H. W., (1960). *Ann. N. Y. Acad. Sci.* **88**, 611.
Hirschmann, R. (1971). *In* "Peptides 1969" (E. Scoffone, ed.), p. 138. North Holland Publ., Amsterdam.
Hofmann, K., Finn, F. M., Limetti, M., Montibeller, J., and Zanetti, G. (1966). *J. Amer. Chem. Soc.* **88**, 3633.
Howard, J. C., Ali, A., Scheraga, H. A., and Momany, F. A. (1975). *Macromolecules* (to be submitted).
Hughes, L. J., Andreatta, R. H., and Scheraga, H. A. (1972). *Macromolecules* **5**, 187.
Hunt, T., Hunter, T., and Munro, A. (1969). *J. Mol. Biol.* **43**, 123.
Hvidt, A., and Linderstrøm-Lang, K. (1954). *Biochim. Biophys. Acta* **14**, 574.
Hvidt, A., and Nielsen, S. O. (1966). *Advan. Protein Chem.* **21**, 287.
Jardetzky, O., Thielmann, H., Arata, Y., Markley, J. L., and Williams, M. N. (1971). *Cold Spring Harbor Symp. Quant. Biol.* **36**, 257.
Jörnvall, H. (1973). *Proc. Nat. Acad. Sci. U. S.* **70**, 2295.
Kabat, E. A., and Wu, T. T. (1973a). *Biopolymers* **12**, 751.
Kabat, E. A., and Wu, T. T. (1973b). *Proc. Nat. Acad. Sci. U. S.* **70**, 1473.
Kartha, G., Bello, J., and Harker, D. (1967). *Nature (London)* **213**, 862.
Kato, I., and Anfinsen, C. B. (1969a). *J. Biol. Chem.* **244**, 1004.
Kato, I., and Anfinsen, C. B. (1969b). *J. Biol. Chem.* **244**, 5849.
Katsoyannis, P. G., Gish, D. T., and du Vigneaud, V. (1957). *J. Amer. Chem. Soc.* **79**, 4516.
Katsoyannis, P. G., Tometsko, A., and Fukuda, K. (1963). *J. Amer. Chem. Soc.* **85**, 2863.
Katsoyannis, P. G., Fukuda, K., Tometsko, A., Suzuki, K., and Tilak, M. (1964). *J. Amer. Chem. Soc.* **86**, 930.
Kauzmann, W. (1959a). *In* "Sulfur in Proteins" (R. Benesch *et al.*, eds.), p. 93. Academic Press, New York.
Kauzmann, W. (1959b). *Advan. Protein Chem.* **14**, 1.
Kerwar, S. S., Kohn, L. D., Lapiere, C. M., and Weissbach, H. (1972). *Proc. Nat. Acad. Sci. U. S.* **69**, 2727.

Klee, W. A. (1967). *Biochemistry* **6,** 3736.
Klee, W. A. (1968). *Biochemistry* **7,** 2731.
Kotelchuck, D., and Scheraga, H. A. (1968). *Proc. Nat. Acad. Sci. U. S.* **61,** 1163.
Kotelchuck, D., and Scheraga, H. A. (1969). *Proc. Nat. Acad. Sci. U. S.* **62,** 14.
Kotelchuck, D., Dygert, M., and Scheraga, H. A. (1969). *Proc. Nat. Acad. Sci. U. S.* **63,** 615.
Krakow, W., Endres, G. F., Siegel, B. M., and Scheraga, H. A. (1972). *J. Mol. Biol.* **71,** 95.
Kuntz, I. D. (1972). *J. Amer. Chem. Soc.* **94,** 4009.
Lacroute, F., and Stent, G. S. (1968). *J. Mol. Biol.* **35,** 165.
Layman, D. L., McGoodwin, E. B., and Martin, G. R. (1971). *Proc. Nat. Acad. Sci. U. S.* **68,** 454.
Leberman, R. (1971). *J. Mol. Biol.* **55,** 23.
Lee, B., and Richards, F. M. (1971). *J. Mol. Biol.* **55,** 379.
Levinthal, C. (1968). *J. Chim. Phys.* **65,** 44.
Levitt, M., and Lifson, S. (1969). *J. Mol. Biol.* **46,** 269.
Lewis, P. N., and Scheraga, H. A. (1971a). *Arch. Biochem. Biophys.* **144,** 576.
Lewis, P. N., and Scheraga, H. A. (1971b). *Arch. Biochem. Biophys.* **144,** 584.
Lewis, P. N., Go, N., Go, M., Kotelchuck, D., and Scheraga, H. A. (1970). *Proc. Nat. Acad. Sci. U. S.* **65,** 810.
Lewis, P. N., Momany, F. A., and Scheraga, H. A. (1971). *Proc. Nat. Acad. Sci. U. S.* **68,** 2293.
Lewis, P. N., Momany, F. A., and Scheraga, H. A. (1973a). *Biochim. Biophys. Acta* **303,** 211.
Lewis, P. N., Momany, F. A., and Scheraga, H. A. (1973b). *Isr. J. Chem.* **11,** 121.
Lifson, S., and Roig, A. (1961). *J. Chem. Phys.* **34,** 1963.
Light, A. (1974). "Proteins, Structure and Function." Prentice Hall, Englewood Cliffs, New Jersey.
Lin, M. C., Gutte, B., Moore, S., and Merrifield, R. B. (1970). *J. Biol. Chem.* **245,** 5169.
Liquori, A. M. (1969). *Symmetry Funct. Biol. Syst. Macromol. Level, Proc. Nobel Symp., 11th, 1968* p. 101.
Low, B. W., Lovell, F. M., and Rudko, A. D. (1968). *Proc. Nat. Acad. Sci. U. S.* **60,** 1519.
McDonald, C. C., and Phillips, W. D. (1967). *J. Amer. Chem. Soc.* **89,** 6332.
McGuire, R. F., Vanderkooi, G., Momany, F. A., Ingwall, R. T., Crippen, G. M., Lotan, N., Tuttle, R. W., Kashuba, K. L., and Scheraga, H. A. (1971). *Macromolecules* **4,** 112.
Markley, J. L., Putter, I., and Jardetzky, O. (1968). *Z. Anal. Chem.* **243,** 367.
Markley, J. L., Williams, M. N., and Jardetzky, O. (1970). *Proc. Nat. Acad. Sci. U. S.* **65,** 645.
Meadows, D. H., Jardetzky, O., Epand, R. M., Ruterjans, H. A., and Scheraga, H. A. (1968). *Proc. Nat. Acad. Sci. U. S.* **60,** 766.
Meienhofer, J., Schnabel, E., Bremer, H., Brinkhoff, O., Zabel, R., Sroka, W., Klostermeyer, H., Brandenburg, D., Okuda, T., and Zahn, H. (1963). *Z. Naturforsch. B* **18,** 1120.
Merrifield, R. B. (1965). *Science* **150,** 178.
Mihalyi, E., and Harrington, W. F. (1959). *Biochim. Biophys. Acta* **36,** 447.
Miller, M., and Scheraga, H. A. (1975). To be published.
Nagano, K. (1973). *J. Mol. Biol.* **75,** 401.
Nagano, K. (1974). *J. Mol. Biol.* **84,** 337.

Naughton, M. A., and Dintzis, H. M. (1962). *Proc. Nat. Acad. Sci. U. S.* **48**, 1822.
Nishikawa, K., and Ooi, T. (1972). *J. Phys. Soc. Jap.* **32**, 1338.
Northrop, J. H. (1932). *J. Gen Physiol.* **16**, 333.
Ontjes, D. A., and Anfinsen, C. B. (1969). *J. Biol. Chem.* **244**, 6316.
Ooi, T., and Scheraga, H. A. (1964). *Biochemistry* **3**, 648.
Ooi, T., Rupley, J. A., and Scheraga, H. A. (1963). *Biochemistry* **2**, 432.
Pain, R. H., and Robson, B. (1970). *Nature (London)* **227**, 62.
Parikh, I., Corley, L., and Anfinsen, C. B. (1971). *J. Biol. Chem.* **246**, 7392.
Perutz, M. F., and Lehmann, H. (1968). *Nature (London)* **219**, 902.
Perutz, M. F., Kendrew, J. C., and Watson, H. C. (1965). *J. Mol. Biol.* **13**, 669.
Phillips, D. C. (1967). *Proc. Nat. Acad. Sci. U. S.* **57**, 484.
Piez, K. A. (1972). *In* "Current Topics in Biochemistry" (C. B. Anfinsen, R. F. Goldberger, and A. N. Schechter, eds.), p. 101. Academic Press, New York.
Piez, K. A., and Sherman, M. (1970). *Biochemistry* **9**, 4132.
Platzer, K. E. B., Ananthanarayanan, V. S., Andreatta, R. H., and Scheraga, H. A. (1972). *Macromolecules* **5**, 177.
Pohl, F. M. (1971). *Nature (London), New Biol.* **234**, 277.
Poland, D., and Scheraga, H. A. (1965). *Biopolymers* **3**, 283, 305, 315, and 335.
Poland, D., and Scheraga, H. A. (1969a). *Biopolymers* **7**, 887.
Poland, D., and Scheraga, H. A. (1969b). *Physiol. Chem. Phys.* **1**, 389.
Poland, D., and Scheraga, H. A. (1970). "Theory of Helix-Coil Transitions in Biopolymers," Chapters 4, 8, and 9. Academic Press, New York.
Ponnuswamy, P. K., McGuire, R. F., and Scheraga, H. A. (1973a). *Int. J. Peptide Protein Res.* **5**, 73.
Ponnuswamy, P. K., Warme, P. K., and Scheraga, H. A. (1973b). *Proc. Nat. Acad. Sci. U. S.* **70**, 830.
Popov, E. M., and Lipkind, G. M. (1971). *Mol. Biol.* **5**, 624.
Potts, J. T., Young, D. M., and Anfinsen, C. B. (1963). *J. Biol. Chem.* **238**, 2593.
Prothero, J. W. (1966). *Biophys. J.* **6**, 367.
Ptitsyn, O. B., and Finkelstein, A. V. (1970). *Biofizika*, **15**, 757.
Putter, I., Barreto, A., Markley, J. L., and Jardetzky, O. (1969). *Proc. Nat. Acad. Sci. U. S.* **64**, 1396.
Quiocho, F. A., and Lipscomb, W. N. (1971). *Advan. Protein Chem.* **25**, 1.
Ramachandran, G. N. (1969). *Symmetry Funct. Biol. Syst. Macromol. Level, Proc. Nobel Symp., 11th, 1968* p. 79.
Ramachandran, G. N., and Sasisekharan, V. (1968). *Advan. Protein Chem.* **23**, 283.
Rasse, D., Warme, P. K., and Scheraga, H. A. (1974). *Proc. Nat. Acad. Sci. U. S.* **71**, 3736.
Richards, F. M. (1958). *Proc. Nat. Acad. Sci. U. S.* **44**, 162.
Ristow, S. S., and Wetlaufer, D. B. (1973). *Biochem. Biophys. Res. Commun.* **50**, 544.
Roberts, D. E., and Mandelkern, L. (1958). *J. Amer. Chem. Soc.* **80**, 1289.
Roberts, G. C. K., and Benz, F. W. (1973). *Ann. N. Y. Acad. Sci.* **222**, 130.
Robson, B., and Pain, R. H. (1971). *J. Mol. Biol.* **58**, 237.
Rong-Qing, Jiang, Du, Yu-Cang, and Tsou, Chen-Lu. (1963). *Sci. Sinica (Peking)* **12**, 452.
Rossmann, M. G., Adams, M. J., Buehner, M., Ford, G. C., Hackert, M. L., Lentz, P. J., Jr., McPherson, A., Jr., Schevitz, R. W., and Smiley, I. E. (1971). *Cold Spring Harbor Symp. Quant. Biol.* **36**, 179.
Rupley, J. A., and Scheraga, H. A. (1963). *Biochemistry* **2**, 421.
Sachs, D. H., Schechter, A. N., Eastlake, A., and Anfinsen, C. B. (1972a). *Proc. Nat. Acad. Sci. U. S.* **69**, 3790.

Sachs, D. H., Schechter, A. N., Eastlake, A., and Anfinsen, C. B. (1972b). *J. Immunol.* **109**, 1300.
Sachs, D. H., Schechter, A. N., Eastlake, A., and Anfinsen, C. B. (1972c). *Biochemistry* **11**, 4268.
Sachs, D. H., Schechter, A. N., Eastlake, A., and Anfinsen, C. B. (1974). *Nature (London)* **251**, 242.
Sanchez, G. R., Chaiken, I. M., and Anfinsen, C. B. (1973). *J. Biol. Chem.* **248**, 3653.
Sanger, F. (1956). *In* "Currents in Biochemical Research" (D. E. Green, ed.), p. 434. Wiley (Interscience), New York.
Schechter, A. N., Moravek, L., and Anfinsen, C. B. (1969). *J. Biol. Chem.* **244**, 4981.
Schellman, J. A. (1955). *C. R. T. Lab. Carlsberg, Ser. Chim.* **29**, 230.
Scheraga, H. A. (1963). *In* "The Proteins" (H. Neurath, ed.), 2nd ed., Vol. 1, pp. 477 and 542. Academic Press, New York.
Scheraga, H. A. (1968). *Advan. Phys. Org. Chem.* **6**, 103.
Scheraga, H. A. (1969). *Symmetry Funct. Biol. Syst. Macromol. Level, Proc. Nobel Symp., 11th, 1968* p. 43.
Scheraga, H. A. (1971). *Chem. Rev.* **71**, 195.
Scheraga, H. A. (1973a). *Jerusalem Symp. Quantum Chem. Biochem.* **5**, 51.
Scheraga, H. A. (1973b). *Pure Appl. Chem.* **36**, 1.
Scheraga, H. A. (1974a). *In* "Current Topics in Biochemistry, 1973" (C. B. Anfinsen and A. N. Schechter, eds.), p. 1. Academic Press, New York.
Scheraga, H. A. (1974b). *In* "Peptides, Polypeptides and Proteins" (E. R. Blout *et al.*, eds.), p. 49. Wiley, New York.
Scheraga, H. A., and Mandelkern, L. (1953). *J. Amer. Chem. Soc.* **75**, 179.
Scheraga, H. A., Lewis, P. N., Momany, F. A., von Dreele, P. H., Burgess, A. W., and Howard, J. C. (1973). *Fed. Proc., Fed. Amer. Soc. Exp. Biol.* **32**, 495.
Schiffer, M., and Edmundson, A. B. (1967). *Biophys. J.* **7**, 121.
Schwyzer, R., and Sieber, P. (1963). *Nature (London)* **199**, 172.
Scoffone, E., Rocchi, R., Marchiori, F., Moroder, L., Marzotto, A., and Tamburro, A. M. (1967). *J. Amer. Chem. Soc.* **89**, 5450.
Scott, R. A., and Scheraga, H. A. (1963). *J. Amer. Chem. Soc.* **85**, 3866.
Sela, M., and Lifson, S. (1959). *Biochim. Biophys. Acta* **36**, 471.
Sela, M., White, F. H., Jr., and Anfinsen, C. B. (1957). *Science* **125**, 691.
Silverman, D. N., and Scheraga, H. A. (1972). *Arch. Biochem. Biophys.* **153**, 449.
Silverman, D. N., Kotelchuck, D., Taylor, G. T., and Scheraga, H. A. (1972). *Arch. Biochem. Biophys.* **150**, 757.
Slayter, H. S., and Lowey, S. (1967). *Proc. Nat. Acad. Sci. U. S.* **58**, 1611.
Smith, B. D., Byers, P. H., and Martin, G. R. (1972). *Proc. Nat. Acad. Sci. U. S.* **69**, 3260.
Smyth, D. G., Stein, W. H., and Moore, S. (1962). *J. Biol. Chem.* **237**, 1845.
Smyth, D. G., Stein, W. H., and Moore, S. (1963). *J. Biol. Chem.* **238**, 227.
Speakman, P. T. (1971). *Nature (London)* **229**, 241.
Steiner, D. F. (1967). *Trans. N. Y. Acad. Sci.* [2] **30**, 60.
Sturtevant, J. M., Laskowski, M., Jr., Donnelly, T. H., and Scheraga, H. A. (1955). *J. Amer. Chem. Soc.* **77**, 6168.
Sykes, B. D., and Scott, M. D. (1972). *Annu. Rev. Biophys. Bioeng.* **1**, 27.
Taniuchi, H. (1970). *J. Biol. Chem.* **245**, 5459.
Taniuchi, H., and Anfinsen, C. B. (1968). *J. Biol. Chem.* **243**, 4778.
Taniuchi, H., and Anfinsen, C. B. (1969). *J. Biol. Chem.* **244**, 3864.
Taniuchi, H., and Anfinsen, C. B. (1971). *J. Biol. Chem.* **246**, 2291.

Taniuchi, H., and Bohnert, J. L. (1973). *Fed. Proc., Fed. Amer. Soc. Exp. Biol.* **32**, No. 3, 458.
Taniuchi, H., Anfinsen, C. B., and Sodja, A. (1967). *Proc. Nat. Acad. Sci. U. S.* **58**, 1235.
Taniuchi, H., Moravek, L., and Anfinsen, C. B. (1969). *J. Biol. Chem.* **244**, 4600.
Taniuchi, H., Davies, D. R., and Anfinsen, C. B. (1972). *J. Biol. Chem.* **247**, 3362.
Van Wart, H. E., Taylor, G. T., and Scheraga, H. A. (1973). *Macromolecules* **6**, 266.
Venetianer, P., and Straub, F. B. (1963). *Biochim. Biophys. Acta* **67**, 166.
Venkatachalam, C. M. (1968). *Biopolymers* **6**, 1425.
von Dreele, P. H., Poland, D., and Scheraga, H. A. (1971a). *Macromolecules* **4**, 396.
von Dreele, P. H., Lotan, N., Ananthanarayanan, V. S., Andreatta, R. H., Poland, D., and Scheraga, H. A. (1971b). *Macromolecules* **4**, 408.
Vuust, J., and Piez, K. A. (1972). *J. Biol. Chem.* **247**, 856.
Warme, P. K., and Scheraga, H. A. (1973). *J. Comput. Phys.* **12**, 49.
Warme, P. K., and Scheraga, H. A. (1974). *Biochemistry* **13**, 757.
Warme, P. K., Momany, F. A., Rumball, S. S., Tuttle, R. W., and Scheraga, H. A. (1974). *Biochemistry* **13**, 768.
Watson, H. C. (1969). *Progr. Stereochem.* **4**, 299.
Weeds, A. G. (1969). *Nature (London)* **223**, 1362.
Westmoreland, D. G., and Matthews, C. R. (1973). *Proc. Nat. Acad. Sci. U. S.* **70**, 914.
Wetlaufer, D. B., and Ristow, S. (1973). *Annu. Rev. Biochem.* **42**, 135.
White, F. H., Jr. (1961). *J. Biol. Chem.* **236**, 1353.
Wilhelm, J. M., and Haselkorn, R. (1970). *Proc. Nat. Acad. Sci. U. S.* **65**, 388.
Woody, R. W. (1974). *In* "Peptides, Polypeptides and Proteins" (E. R. Blout *et al.*, eds.), p. 338. Wiley, New York.
Wu, T. T., and Kabat, E. A. (1973). *J. Mol. Biol.* **75**, 13.
Wünsch, E., Jaeger, E., and Scharf, R. (1968a). *Chem. Ber.* **101**, 3664.
Wünsch, E., Wendlberger, G., Jaeger, E., and Scharf, R. (1968b). "Peptides 1968" (E. Bricas, ed.), p. 229. North-Holland Publ., Amsterdam.
Wyckoff, H. W., Tsernoglou, D., Hanson, A. W., Knox, J. R., Lee, B., and Richards, F. M. (1970). *J. Biol. Chem.* **245**, 305.
Yu-Cang, D., Yu-Shang, Z., Zi-Xian, L., and Chen-Lu, T. (1961). *Sci. Sinica (Peking)* **10**, 84.
Zahn, H., Meienhofer, J., and Klostermeyer, H. (1964). *Z. Naturforsch. B* **19**, 110.
Zimm, B. H., and Bragg, J. K. (1959). *J. Chem. Phys.* **31**, 526.

Note Added in Proof

Some preliminary tests of the algorithm mentioned in Section VIII,D (and variants thereof) have been made (Burgess and Scheraga, 1975c), using bovine pancreatic trypsin inhibitor as a model. A starting conformation [referred to as a topographical structure (Burgess *et al.*, 1974)] was obtained by assigning one of five states (α_R, α_L, ϵ, ζ_R, ζ_L) to each residue, according to the observed X-ray structure; thus, the topographical structure can be thought of as the result of a *perfect* prediction scheme. However, since the values of the backbone dihedral angles, ϕ and ψ, assigned to each residue of the topographical structure were *average* values for each of the five states for each residue (obtained from Figs. 28–31), the overall topographical structure did *not* resemble the native one. In other words, the topographical structure had the correct short-range conformation, but not the proper long-range arrangement. The energy of the *whole* topographical structure was then minimized by allowing successive non-overlapping nonapeptide segments (in contact with the rest of the molecule) to undergo changes in conformation. While the energy minimization was not carried to completion because of the large amount of computer time required, the preliminary results were encouraging in that, as the proper half-cystines were brought together to form disulfide bonds, one of the key features of the structure (an anti-parallel β conformation) began to form; also, the α-helix at the C-terminus was preserved, the N- and C-termini were near each other (as in the native structure), and no high-energy contacts occurred between distant parts of the chain. This indicates that energy minimization has the potential to lead from a topographical structure to the native one, and emphasizes the need to improve the reliability of the empirical prediction schemes discussed in Section VIII,C in order to obtain good starting conformations for energy minimization.

AUTHOR INDEX

Numbers in italics refer to the pages on which the complete references are listed.

I

J

K

M

N

S

T

Z

SUBJECT INDEX

CONTENTS OF PREVIOUS VOLUMES

Volume 4

Volume 5

Volume 6

Volume 7

Volume 8

Volume 9

Volume 10

Volume 15

Volume 16

Volume 17

Volume 18

Volume 19

Volume 20

Volume 21

Volume 22

Volume 23

Volume 24

Volume 25

Volume 26

A 5 B 6 C 7 D 8 E 9 F 0 G 1 H 2 I 3 J 4

Volume 27

Volume 28